Sustainable Construction Engineering

Sustainable Construction Engineering

Edited by **Alistair Doyle**

Willford Press
New York

Published by Willford Press,
118-35 Queens Blvd., Suite 400,
Forest Hills, NY 11375, USA
www.willfordpress.com

Sustainable Construction Engineering
Edited by Alistair Doyle

International Standard Book Number: 978-1-68285-274-3 (Hardback)

Printed in China.

Contents

Preface

This book gives a comprehensive overview of sustainable construction engineering and its applications which benefit human kind. The book presents an advanced understanding of sustainability in the field of construction. It integrates concepts of sustainability into structural engineering standards and practices. It covers researches on topics such as sustainable designs and constructions, sustainable strategies, including relevant case studies as well as applications for future. This book will be beneficial for architects, civil engineers, students and professionals alike.

Significant researches are present in this book. Intensive efforts have been employed by authors to make this book an outstanding discourse. This book contains the enlightening chapters which have been written on the basis of significant researches done by the experts.

Finally, I would also like to thank all the members involved in this book for being a team and meeting all the deadlines for the submission of their respective works. I would also like to thank my friends and family for being supportive in my efforts.

Editor

1

FABRICATION AND EFFECT OF SULFONATED POLY (ETHER ETHER KETONE) WITH CLOISITE15A® NANOCLAYS FOR MICROBIAL FUEL CELL APPLICATION

Abdul Halim Mohd Yusof[1,3*], Mohamed Afizal Mohamed Amin [1,2,4], Ahmad Fauzi Ismail[1], Mohd Noorul Anam[1] and Nik Azmi Nik Mahmud[2]

[1]Advanced Membrane Technology Research Centre (AMTEC), UniversitiTeknologi Malaysia, 81310 Skudai, Johor DarulTa'zim, Malaysia,
[2]Faculty of Chemical Engineering (FKK), UniversitiTeknologi Malaysia, 81310 Skudai, Johor DarulTa'zim, Malaysia,
[3]Department of Biology, Building 68-370, Massachusetts Institute of Technology, Cambridge, Massachussetts 02139, United States of America,
[4]Department of Chemical Engineering and Sustainability Energy, Faculty of Engineering, Universiti Malaysia Sarawak, 94300 Kota Samarahan, Sarawak, Malaysia

*Corresponding E-mail : halimy@cheme.utm.my

Abstract

Ion exchange membranes have been used in microbial fuel cell (MFC) because it can prevent high oxygen permeability and the migration of substrate instead of protons from one chamber to another. Despite all these advantages the usage of this type of membrane suffers several operational problems such as high dissolve oxygen permeability, adverse deformation and proton conductivity which can affect the MFC performance. In order to overcome this problem, new membrane for microbial fuel cell system was fabricated using sulfonated polyether ether ketone (SPEEK) and Cloisite 15A®. The membranes were characterized based on their electrochemical and physical properties. The characteristics of the SPEEK nanocomposite membranes were then compared with the unmodified SPEEK membrane and the commercially available Nafion117 membrane. Based on the result, SPEEK/Cloisite 15A® contributed high energy generated across the load compare to SPEEK without additive and Nafion 117. The incorporation of Cloisite 15A® improved the water uptake from 56 to 78.2 wt% and reduced the dissolve oxygen permeability of the membrane from 1.5 to 0.7 cm/s but the tensile strength behaviour was reduced from 11.2 to 7.5 MPa.

Keywords: *Cloisite 15A®, sulfonated poly (ether ether ketone) (SPEEK), ion exchange membrane, microbial fuel cell (MFC)*

1.0 Introduction

MFC is a system that utilises microorganism activity to generate power. Numerous studies have been conducted in order to optimise the MFC performance based on possible factors that contribute such as type of microorganism, electrode, electron acceptor, design and membrane [1-4]. In this system, membrane plays a role as a separator between anode and cathode chamber and also as an ion exchange medium.

Previous studies with different types of membranes have been conducted in order to figure out the most outstanding power generation in MFC system such as anion exchange membrane (AEM), cation exchange membrane (CEM), bipolar membrane (BPM) and charge mosaic membrane (CMM) [5-14]. Most of the studies focused on the performance in term of dissolves oxygen permeability and migration of protons because, ideally, the proton conductive membrane must be able to inhibit the oxygen to migrate and allow high proton conductivity [15].

The coulombic efficiencies (CEs) of MFC can be affected when oxygen replaces the cathode as a terminal electron acceptor. Besides, the exposure of microbe to oxygen as a terminal electron acceptor may lead to an adjustment of the enzymatic machinery of the microbial cells. The consequence of this adjustment can increase the redox potential within a single biological catalytic entity hence reduces the power generation. Previous application of MFC normally used the Nafion membrane as a separator. However, there are some drawbacks associates with Nafion such as high cost of material and high dissolve oxygen permeability.

In response to above mentioned problem, this studied was carried out using SPEEK as a membrane based that possesses not only good electrochemical performance but inexpensive material and lower oxygen permeability. SPEEK is a hydrophilic polymer produced from the sulfonation reaction. SPEEK exhibited many essential properties as an ideal proton exchange membrane in fuel cell application such as good thermal stability, chemical inertness, good mechanical properties, low cost and adequate conductivity depending on the degree of sulfonation (DS) [16-17].

Different kinds of modification in SPEEK polymer matrix had been made in order to improve the proton conductivity performance. Numerous modification of SPEEK polymer can be generally achieved by addition of organic-inorganic component and by cross linking different polymers [18]. Cloisite is one of the materials that have been used to modify perflourinated inomer membrane (PFI) like Nafion and non-flourinated membrane such as SPEEK and have been proved to give a significant effect on the properties of membrane such as mechanical strength, water uptake and proton conductivity [19-20]. Apart from that, the compatibilizer which is 2,4,6-triaminopyrimidine (TAP) was added to enhance the interaction and dispersion of cloisite with SPEEK polymer [22]. Therefore, the main purpose of this study is to prepare SPEEK membrane incorporated with Cloisite 15A® and TAP for MFC application.

2.0 Methodology

For membrane preparation, fine powder PEEK polymer was obtained from Victrex US Inc. Ltd. Sulfuric acid H_2SO_4 (95-97%) and N,N-dimethylacetamide (DMAc) were supplied by Merck KGaA. Cloisite 15A® and 2,4,6-triaminopyrimidine (TAP) were purchased from Southern Clay Products, Inc. and Sigma-Aldrich Inc. respectively. Reference membrane (Nafion 117) was purchased from DuPont. For MFC setup, *Saccharomyces cerevisiae* was obtained from AB Mauri Malaysia Sdn. Bhd. Methylene blue and potassium permanganate were obtained from Sigma–Aldrich Inc. in powder form. Both Sodiumhydrogenphosphate and D-glucose were were supplied by Merck KGaA.

2.1 Preparation and Characterization of SPEEK

Sulfonation process was started by mixing the concentrated sulphuric acid (95-97%) with polyether ether ketone (PEEK) polymer powder in a round bottom flask. The solution was stirred 1 hour to get the homogenous solution. The mixture was then heated up to 60 °C for 3 hours. During this process, the colour of solution changed from light brown to dark brown indicated the progress of sulfonation reaction. The changes of functional group were characterized by using the nuclear magnetic resonance (^{1}H NMR).

The degree of sulfonation (DS) of SPEEK was confirmed by using ^{1}H NMR spectroscopy method (Bruker Avance 300 NMR). The resonance frequency of this instrument was adjusted at 400 MHz at room temperature. Prior to analysis, the dried purified SPEEK sample was prepared 3 wt% and dissolved in deutrated dimethyl sulfoxide (DMSO-*d6*).

2.2 Membrane Preparation

Nanocomposite was prepared by mixing 0.25 g of Cloisite15A® clays and 0.5 g of 2,4,6-triaminopyrimidine (TAP) with 15 mL N,N-dimethylacetamide (DMAc) in different bottle respectively and vigorously stirred for 24h at room temperature. To prepare polymer solution, 10 wt% of SPEEK was dissolve in 60 mL DMAc and stirred for 12h. Then, the prepared Cloisite and TAP solution were added to this solution and once again vigorously stirred for 24 h.

A pneumatic casting machine was used to cast the polymer solution onto a glass plate. The membrane was kept in an oven at 60 °C for 6 hr and further dried for the next 12 hr at 100°C to evaporate the residual solvent. The membrane was removed from the oven and allowed to cool to room temperature before it was immersed in deionized water overnight so it would peel off from glass surface and remove residual solvent. For pre treatment, the prepared membrane was immersed into 1M sulfuric acid H_2SO_4 solution overnight at room temperature. Finally, the membrane was rinsed several time with deionized water to remove the excess acid then blotted with absorbent paper.

2.3 Membrane Characterization

2.3.1 Water Uptake

This analysis was carried out to measure the ability of membrane to absorb water. This measurement was conducted by following the procedure of Ismail [17]. First, the membrane sample was dried at 60°C for 48h and the weight was measured (W_d). The membrane was then kept in distilled water at room temperature overnight and the weight once again being measured (W_w). The calculation of water uptake was carried out using equation 1.

$$\text{Water uptake \%} = \frac{W_w - W_d}{W_d} \times 100\% \tag{1}$$

where W_w is weight of membrane in wet condition while W_d is weight of membrane in dried condition.

2.3.2 Dissolve Oxygen Permeability

All figures must carry numbers in the text (e.g. Fig. 1) and captions.

The dissolved oxygen (Do) measurement was done using un-inoculated MFC system constructed for the membrane test. The method of this analysis was followed the procedure of Kim [29]. First and foremost, the desired membrane was fitted between two chambers of un-inoculated MFC. Both of the chambers were then filled with distilled water and then dissolved oxygen probe was inserted inside cathode chamber. The condition in anode chamber was kept free from dissolve oxygen by purged the nitrogen gas until the reading of Do concentration is zero. Finally, the Do concentration of cathode chamber over time (C_2) was taken. The mass transfer coefficient was calculated from the equation 2.

$$K_O = -\frac{V}{At}\ln\left[\frac{C_{1,0} - C_2}{C_{1,0}}\right] \tag{2}$$

where V is volume of water, A is the surface area of membrane fitted between two chambers, $C_{1.0}$ is the Do in water at anode chamber and C_2 is the measured Do in the cathode chamber at time t. The rate of oxygen diffusion (Do cm^2 s^{-1}) for each prepared membrane and Nafion 117 was calculated using equation 3.

$$D_o = K_o L_t \tag{3}$$

2.3.3 Mechanical Strength

The tensile strength of each membrane was measured using mechanical testing instrument, MTS (LRX 5kN, Llyod instruments). Prior to test, 5 samples from each types of membrane were cut into dumbbell shape with gauge length and the width were 15 mm and 4 mm respectively. The membranes thicknesses were also being measured before the test. Then, the membrane was fitted between the grips of the instrument and run at rate of speed 5 mm min^{-1}.

2.4 Membrane Morphology Structure

The morphology of the prepared membrane was investigated using a field emission scanning electron microscope (FESEM). The FESEM (JSM-6701F) is a useful technique to determine the membrane structures. The samples were prepared by immersed it under the liquid nitrogen to freeze and dry the membrane so that the clean break can be done. The samples are taken out and mounted on sample stubs using double surface scotch tape. Then, cross-sectional fractures of the membranes were vacuum sputtered with a thin layer of gold. Various images of cross-sectional of the membranes were taken and analyzed.

2.5 Electrochemical Measurement

The MFC was of semi-circle construction cylinder divided into anode and cathode chamber. The anode solution consisted of methylene blue, D-glucose, yeast and phosphate buffer solution (PBS). Meanwhile, the cathode solution consisted of potassium permanganate and phosphate buffer solution (PBS). The PBS for each compartment was made by dissolving inorganic salts (NaH_2PO_4 - 8.16 g/L; $Na_2H_2PO_4$ – 6.58 g/L) in distilled water. The pH of PBS was adjusted to 7 by adding 1M of sodium hydroxide (NaOH). For the anode solution, 50 mM of methylene blue and 1 % (wt/v) of glucose-D were added into the buffer solution. On the other hand, the cathode buffer solution was added with 50 mM potassium permanganate ($KMnO_4$) solution. The inoculum was prepared by dissolving 2g *Saccharomyces cerevisiae* (baker yeast) into 100 mL of phosphate buffer solution.

For the measurement, MFC was first run under open circuit voltage (OCV) until the voltage was stabilized at room temperature. Then, the polarization curves were obtained by applying a various external resistance (2 MΩ - 1 kΩ) loaded acros s the MFC with each resistance being connected until the maximum sustainable voltage was achieved. The voltages were recorded using a digital multimeter (Fluke 259). Current was calculated from I=E/R, where I represent the current, E is the measured voltage across the resistor and R the external resistance while power was calculated using P=IE. Both current and power density was normalized by the projected cathode surface area.

3.0 Results and Discussions

3.1 Nuclear Magnetic Resonance

In ^1H NMR spectrum, the intensity of H_G confirmed the presence of sulfonic acid group (SO_3H) in SPEEK polymer. The nomenclature of the characteristic proton in the aromatic ring in the SPEEK repeat unit is illustrated in Fig. 1. Based on this spectrum, the presence of sulfonic acid group causes 0.25 ppm down-field chemical shift of the hydrogen H_G compared with H_E and H_F in the ring and created a distinct resonance signal for protons at the G position [17]. Peak at 7.25 ppm in this spectrum represents the H_E protons in the PEEK repeat unit. Normally, when the sulfonation reaction occurred, the introduction of sulfonic functional group to the aromatic ring will cause the proton (H^+) differentiate into 3 resonance signals which are H_G (peak at ~ 7.5 ppm),

H_E (peak at ~ 7.22 ppm) and the H_F (peak at ~ 7.12 ppm). The DS at reaction temperature 60 °C was found to be 76 %.

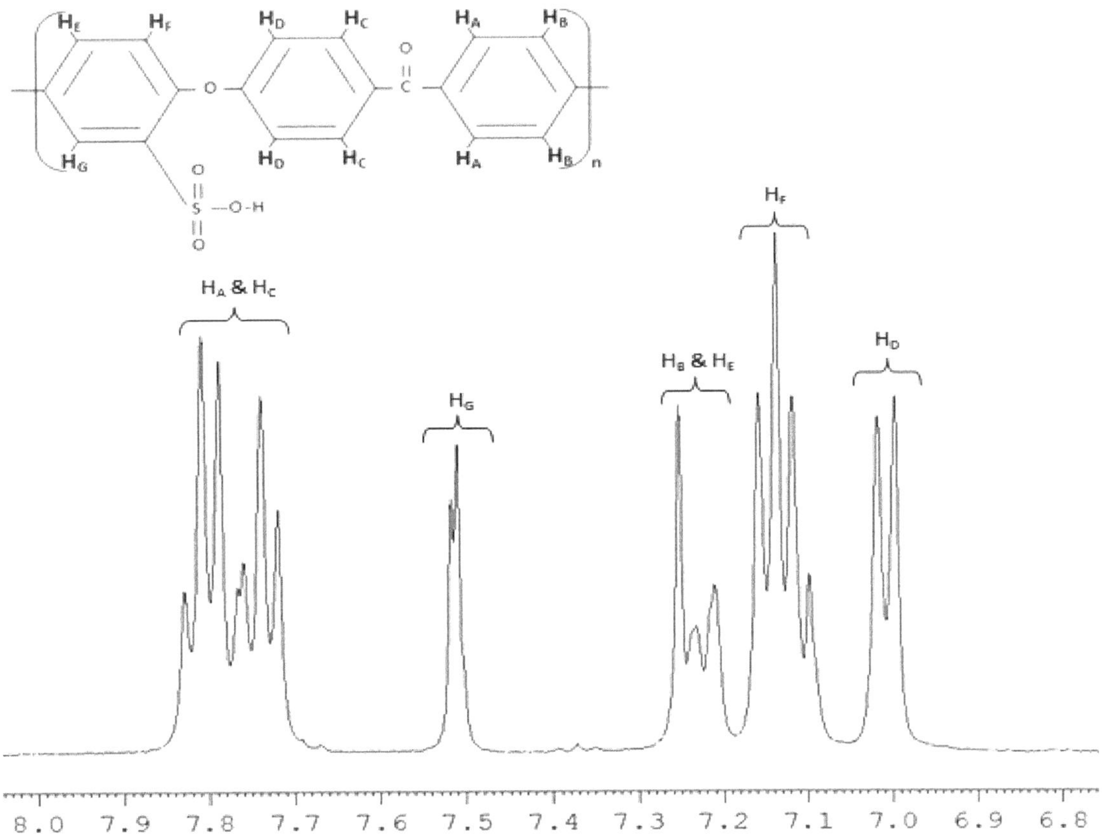

Figure 1: ^{1}H NMR spectrum of SPEEK

3.2 Water Uptake

Water plays an important role for proton conductivity membrane. Adequate water adsorption will exhibit good proton conductivity for membrane [17]. Water uptake rate also plays an essential role in the membrane characteristic and related to the basic membrane properties. The amount of water absorbed by the membrane can provide the pathway for proton conductivity across it. Water involve in modifying the mechanical properties by influence the inomer microstructure, creating clusters and altering channel sizes. However, an excess of water uptake would affect the mechanical behaviour of the membrane.

Table 1: Water uptake of the membranes

Membrane	Water uptake (wt%)
Nafion 117	16
SPEEK	56
SPEEK/Cl/TAP	78.2

The water uptake measurement for prepared membrane were analysed at room temperature. Based on Table 2, it is shown that when the Cloisite 15A® and TAP was incorporated into SPEEK membrane, the water uptake increased from 56 to 78.2 wt%. Previous research reported that the incorporation of Cloisite 15A® can successfully improve the water uptake of the SPEEK membrane because of the hydrophilicity properties in that material as well as TAP. Both of these

Cloisite 15A® and TAP material have this characteristic due to the existing stronger hydrogen bond created by hydrophilic groups OH and NH_2 in the chemical structure [19]. Between these materials, the main contributor for high water uptake is Cloisite 15A® [20]. Juhana claimed that SPEEK incorporated with Cloisite15A® show better wettability (34 wt%) compare with TAP (29 wt%). This might be due to the oxygen molecule in the Cloisite 15A® (Figure 2) that has more and higher affinities for electron or polarity and expect to form more hydrogen bond compare to the nitrogen in the TAP [21].

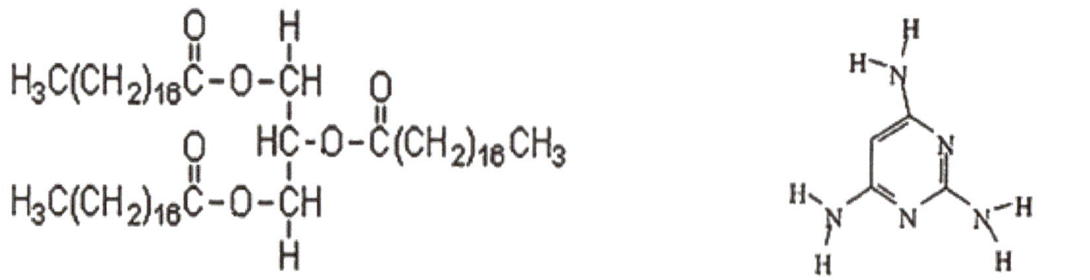

Cloisite15A **2,4,6-triaminopyrimidine (TAP)**

Figure 2: The chemical structure of Cloisite 15A and TAP

3.3 Dissolve Oxygen Permeability

The details of oxygen transfer rate (K_O) and diffusivities of oxygen (D_O) are listed in Table 2. From the experimental data, the highest K_O (2.4×10^{-3} cm/s) was provided by Nafion 117 and the lowest K_O (0.7×10^{-3} cm/s) was recorded by SPEEK/Cloisite 15A® membrane. The K_O of SPEEK was reduced by 53.3 % (from 0.7×10^{-3} to 1.5×10^{-3} cm/s) when Cloisite 15A® was introduced in the matrix membrane. Generally, the Cloisite 15A® plays a role as a filler. When it was introduced within the SPEEK matrix (Figure 3), the size of voids started to decrease and tend to increase the resistance toward oxygen that can freely pass through. When the stabilizer was added, it can finely dispersed the Cloisite 15A® in the SPEEK causing the arrangement of solid Cloisite 15A® to become closer and less aggregate, thus limiting the oxygen transfer. The high value of K_O explains that the membrane will have higher potential of allowing oxygen to pass through it. Nafion was known to exhibit high oxygen transfer and the results had been reported previously [8].

Table 2: Oxygen-transfer properties of the membranes

Membrane	Permeability rate of dissolve oxygen, K_O ($\times 10^{-3}$ cm s^{-1})	Oxygen diffusion coefficient, D_O ($\times 10^{-5}$ cm^2 s^{-1})
Nafion 117	2.4	1.9
SPEEK	1.5	1.0
SPEEK/Cl/TAP	0.7	0.6

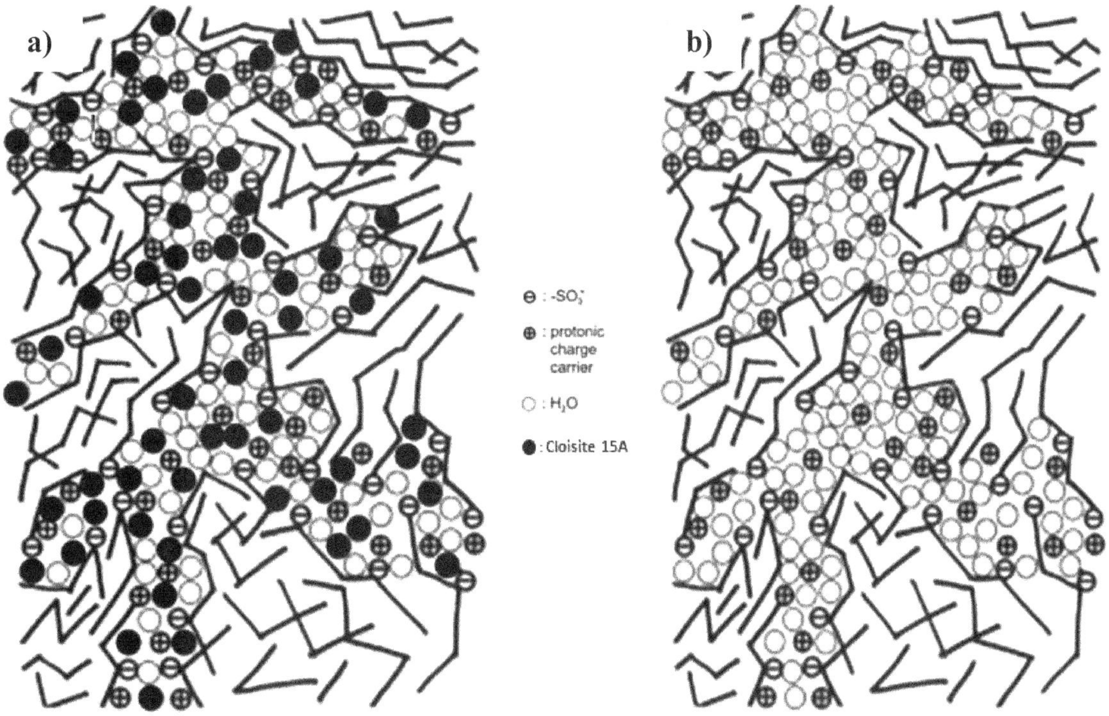

Figure 3: The comparison between (a) SPEEK/Cl/TAP and (b) SPEEK structure and the mechanism of proton transport (modified from Kreuer, 2001) [24].

3.4 Mechanical Strength

Table 3 shows the result of tensile strength tested on each membrane. From the result, pure SPEEK has the highest (11.2 Mpa) mechanical strength compare to the other prepared membrane and Nafion 117. This might be due to the SPEEK structure that has more branches compares to the Nafion 117 (Figure 4) [28]. The increasing of branch might cause the bonding of atom within the membrane become stronger and decrease the flexibility of it. Ironically, there is no significant improvement in the tensile strength when the Cloisite 15A® was load in the SPEEK matrix. The mechanical strength of the SPEEK membrane reduces from 11.2 to 7.5 MPa but nevertheless it is still higher 71 % than that of Nafion 117. The resulting lower tensile strength of SPEEK membrane incorporated with an additive might due to the reduction in the polymer rigidity. The additional of solid material inside the membrane disturb the periodic arrangement of SPEEK atom. Furthermore, it was observed during the experiments that SPEEK with additive membrane shown increase deformity of membrane compared to original SPEEK.

Table 3: Mechanical strength of the membranes

Membrane	Tensile strength (MPa)
Nafion 117	2.2
SPEEK	11.2
SPEEK/Cl/TAP	7.5

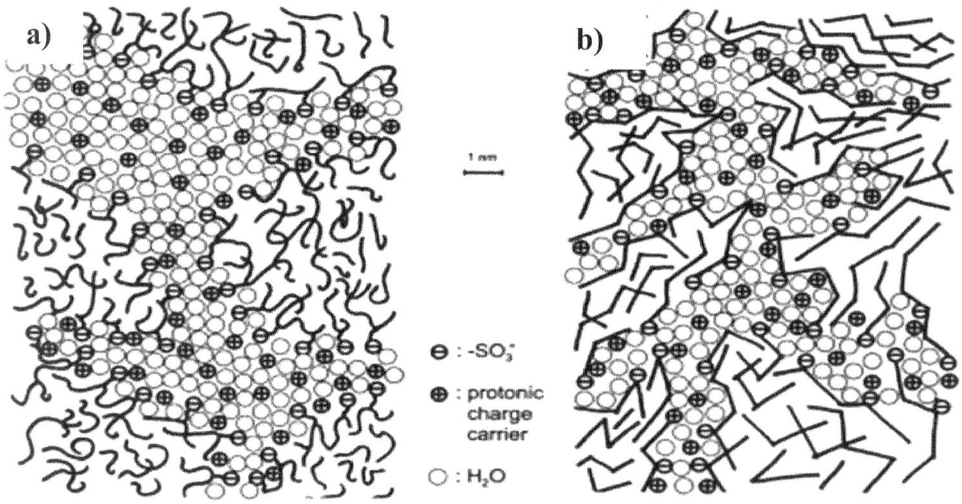

Figure 4: The comparison between (a) Nafion and (b) SPEEK structure and the mechanism of proton transport [28].

3.5 Membrane Morphology

Figure 5a, 5b and 5c shows surface image of SPEEK, SPEEK/Cl/TAP and Nafion117 membranes respectively at magnification 25 KX. Based on these images, the short crack on the surface of SPEEK/Cl/TAP and Nafion117 membranes is higher compare to the SPEEK. This crack indicated the number pores within the membrane. Compare to SPEEK membrane, the crack for SPEEK/Cl/TAP is short and uniformly distribute. This might be due to the role of Cloisite load in the membrane that can create a lot of small pores and uniformly filled by adding the TAP compatibilizer. TAP also might be a reasoned for the smooth surface of SPEEK/Cl/TAP membrane compare to the SPEEK.

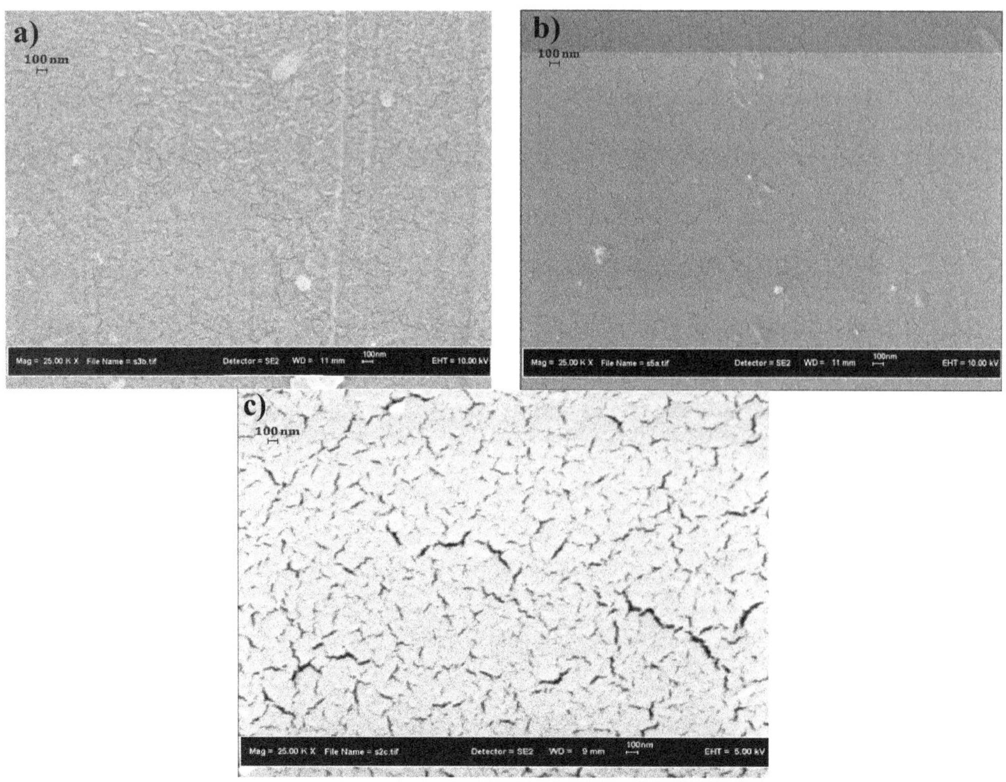

Figure 5 : The surface image of membrane at magnification 25 KX where (a) SPEEK, (b) SPEEK/Cloisite 15A® and (c) Nafion117

The cross-sectional and surface of membrane images of prepared membranes were investigated by using field electric scanning electron microscopy (FESEM). Figure 6a, 6b and 6c shows micrograph of cross-section for SPEEK, SPEEK/Cl/TAP and Nafion117 membranes respectively at low magnification (2.5 KX) and high magnification (25 KX). Based on these images, SPEEK and SPEEK/Cl/TAP membrane proved to have a dense and homogenous structure similar to the Nafion membrane. The images also show that, the number of pores for Nafion is quit high compared to the prepared membranes. At high magnification, the cross section of SPEEK/Cl/TAP membrane quite rough compare to the SPEEK. This roughness was signified the finely dispersed of the Cloisite 15A® in the SPEEK matrix [23-24]. This proved that TAP successfully plays a role as a compatibilizer in the membrane.

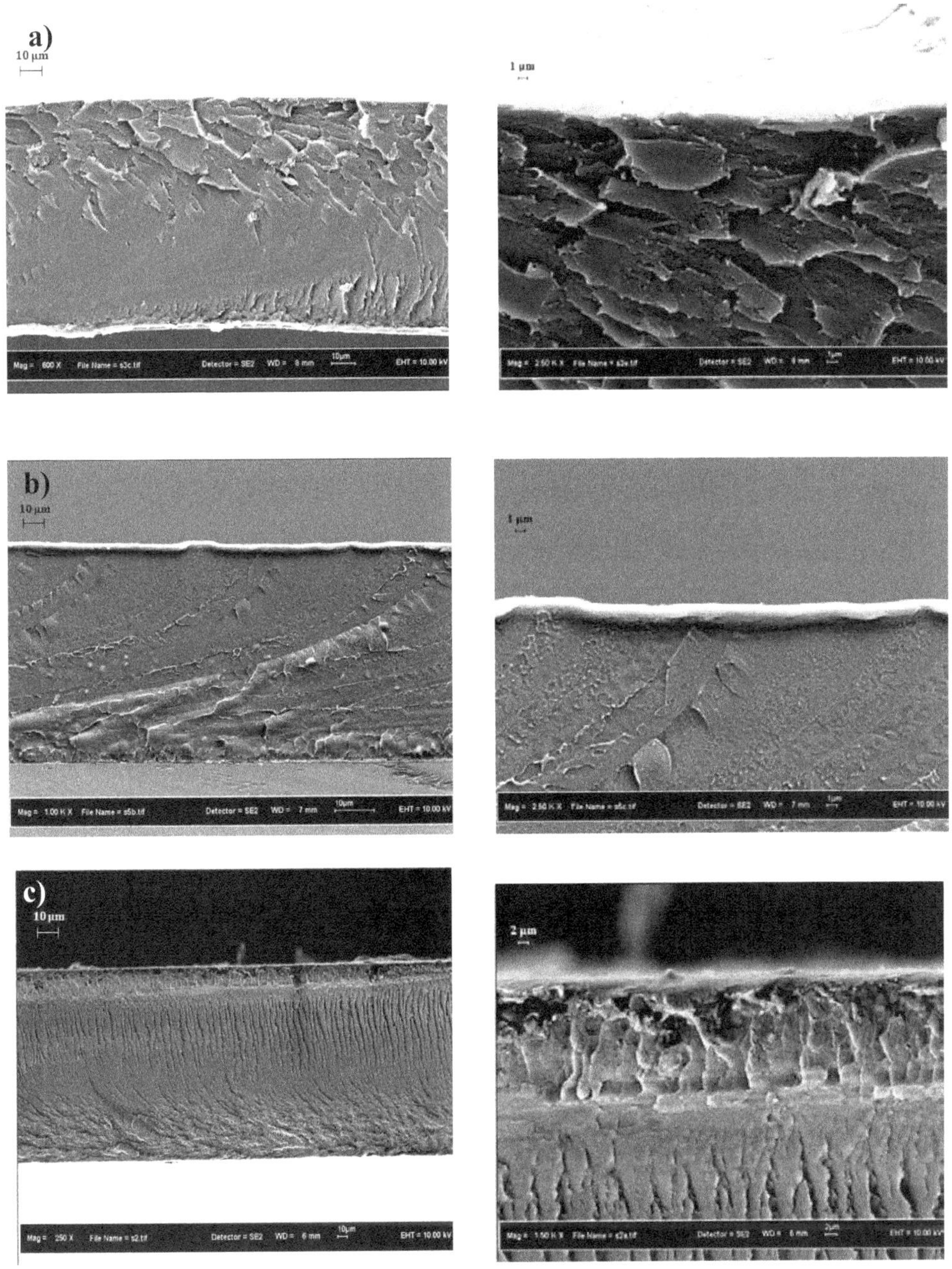

Figure 6 : The cross-section of membrane (a) SPEEK,(b) SPEEK/Cloisite 15A® and (c) Nafion117

3.6 Electrochemical Measurement

The optimum performance can be achieved when appropriate amount of Cloisite is added into the matrix of membrane. However, inappropriate amount of Cloisite 15A® incorporated within SPEEK membrane can give adverse effect on proton conductivity and water uptake [25]. The maximum amount of Cloisite for the optimum performance in MFC for this study was prepared by following the previous research which is 0.25 wt% [20].

Figure 7 shows the cell voltages and power densities as a function of current density for the MFC by using SPEEK/Cloisite 15A® membrane followed by pure SPEEK and Nafion117 as a comparison. Based on the result, it show that when the cloisite was load in the SPEEK membrane, the power density increase from 17.5 to 19.3 mW/m^2. Higher electrochemical performance of this membrane might be due to the conductivity of Cloisite 15A® [21].

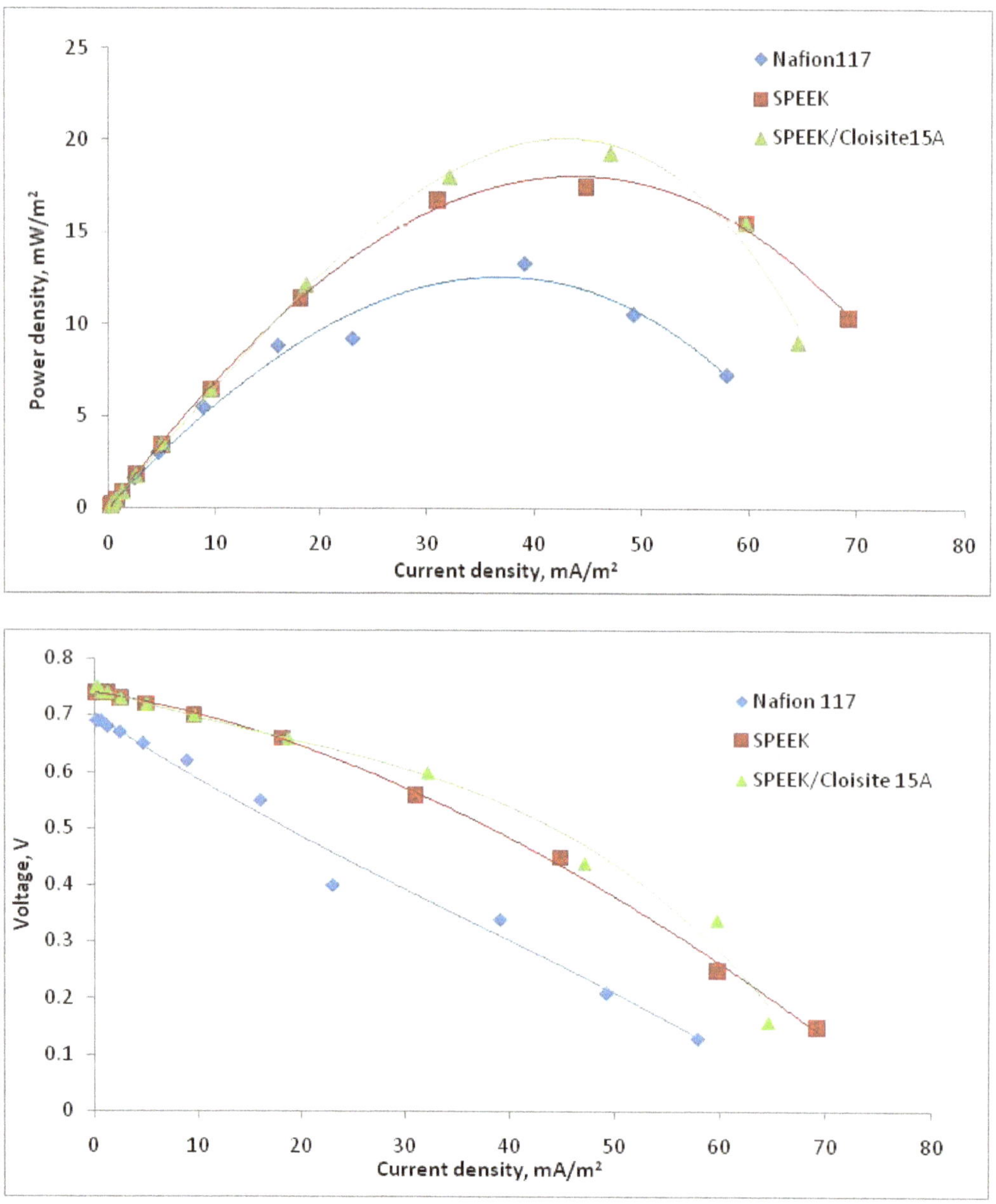

Figure 7 : Polarization curves of membranes

For further explanation, the mechanism of proton transfer is discussed. Generally, there are two main factors that can give a significant effect on proton conduction. First is vehicular mechanism and second, Grotthus mechanism. For vehicular mechanism, water plays a role as a vehicle by carrying the protons from anode to cathode. As for Grotthus mechanism, the proton was transfer from anode to cathode by hopping from stationary water molecule to another and transport along hydrogen-bonded ionic channels [26]. For this membrane, the mechanism that might be a main contributor for high electrochemical is through proton hopping along hydrogen-bonded. As mentioned before, the Cloisite and TAP can enhance the hydrogen bond in the SPEEK membrane. Besides, another factor that contribute to the performance of proton conductivity is the uniformly dispersion or less agglomeration of Cloisite 15A® particles within the membrane causing a lot of ions freely passed through the membrane with less resistance [27].

4.0 Conclusion

SPEEK membrane modified with Cloisite 15A® and TAP were successfully prepared and used in MFC system. Based on the experimental results and analyses, the following conclusions were derived.

- The SPEEK incorporated with Cloisite 15A® and TAP exhibited excellent power generation performance compared to the SPEEK membrane without additive and Nafion117.
- Permeability of dissolve oxygen from cathode chamber to anode was decreased using the membrane with both additives.
- The mechanical strength for SPEEK is not improved through the additional of Cloisite 15A® and TAP.

Based on the result and conclusions, the performance of the SPEEK membrane in MFC can be improved if the membrane with high proton conductivity, low oxygen permeability and good mechanical strength is figured out. For that, SPEEK with medium degree of sulfonation might be a solution because when the DS is decrease the mechanical behaviour will increase. As for high proton conductivity and low oxygen permeability, these properties might be improved by adding sufficient amount of Cloisite 15A® and TAP.

Acknowledgement

The authors gratefully acknowledge the financial support given for this work by the UniversitiTeknologi Malaysia under the GUP Research Grant number of Q.J130000.7125.02H21, Advanced Membrane Science Institute (AMTEC), Universiti Malaysia Sarawak (UNIMAS) and Minister of Higher Education for Scholarship.

References

[1] B. H. Kim, I. S. Chang, and G. M. Gadd, "Challenges in microbial fuel cell development and operation," Application of Microbiology and Biotechnology, vol. 76, pp. 485–494, 2007.

[2] B. E. Logan, B. Hemelers, R. Rozandel, U. Schroder, J. Keller, S. Freguia, P. Aelterman, W. Verstraete, and K. Rabaey, "Microbial fuel cells methodology and technology," Environmantal science and technology, vol. 40 (17), 2006

[3] D. R. Lovely, "Microbial fuel cell: novel microbial physiologies and engineering approaches," Current opinion in biotechnology, vol. 17, pp. 327–332, 2006

[4] B. E. Logan, and J. M. Regan, "Microbial fuel cell – challenges and application," Environmental science & technology, vol. 40, pp. 5172-5180, 2006

[5] B. Min, S. Cheng, and B. E. Logan, "Electricity generation using membrane and salt bridge microbial fuel cells," Water research, vol. 39, pp. 1675–1686, 2005

[6] R. A. Rozendal, H. V. M. Hamelers, and C. J. N. Buisman, "Effects of membrane cation transport on pH and microbial fuel cell performance," Environment science and technology, vol. 40, pp. 5206-5211, 2006

[7] R. A. Rozendal, T. H. J. A. Sleutels, H. V. M. Hamelers, and C. J. N. Buisman, "Effect of the type of ion exchange membrane on performance, ion transport, and pH in biocatalyzed electrolysis of wastewater," Water science & technology, vol. 57, pp. 1757-1762, 2008
[8] O, Lefebvre, Y. Shen, Z. Tan, A. Uzabiaga, I. S. Chang, and H. Y. Ng, "A comparison of membranes and enrichment strategies for microbial fuel cells," Bioresource technology, vol. 102, pp. 6291–6294, 2011
[9] W. W. Li, G. P. Sheng, X. W. Liu, and H. Q. Yu, "Recent advances in the separators for microbial fuel cells," Bioresource technology, vol. 102, pp. 244–252, 2011
[10] E. Ji, H. Moon, J. Piao, P. T. Ha, J. An, D. Kim, J. J. Woo, Y. Lee, S. H. Moon, B. E. Rittmann, and S. I, Chang, "Interface resistance of anion exchange membranes in microbial fuel cells with low ionic strength," Biosensors and Bioelectronics, vol. 26, pp. 3266-3271, 2011
[11] T. Xu, "Ion exchange membranes: state of their development and perspective," Journal of Membrane Science, vol. 263(1–2), pp. 1–29, 2005
[12] T. Saito, M. D. Merrill, V. J. Watson, B. E. Logan, and M. A. Hickner, "Investigation of ionic polymer cathode binders for microbial fuel cells," Electrochimica Acta, vol. 55, pp. 3398–3403, 2010
[13] A. T. Heijne, H. V. M. Hamelers, V. D. Wilde, R. A. Rozendal, and C. J. N. Buisman, "A bipolar membrane combined with ferric iron reduction as an efficient cathode system in microbial fuel cells," Environmental Science and Technology, vol. 40, pp. 5200-5205, 2006
[14] X. Zhang, S. Cheng, X. Wang, X. Huang, and B. E. Logan, "Separator characteristics for increasing performance of microbial fuel cells," Environmental Science and Technology, vol. 43, pp. 8456–8461, 2009
[15] F. Harnisch, and U. Schrçder, "Selectivity versus mobility: separation of anode and cathode in microbial bioelectrochemical systems," ChemSusChem, vol. 2, pp. 921 – 926, 2009
[16] M. H. D. Othman, A. F. Ismail, and A. Mustafa, "Proton conducting membrane from sulfonated poly (ether ether ketone) and boron orthophosphate for direct methanol fuel cell application," Journal of membrane science, vol. 299, pp. 156-165, 2007
[17] A. F. Ismail, N. H. Othman, and A. Mustafa, "Sulfonated polyether ether ketone composite membrane using tungstosilicic acid supported on silica–aluminium oxide for direct methanol fuel cell (DMFC)," Journal of membrane science, vol. 329, pp. 18–29, 2009
[18] M. L. Vona, D. Marani, C. D'ottavi, M. Trombetta, E. Traversa, and I. Beurroies, "A simple new route to covalent organic/inorganic hybrid proton exchange polymeric membranes," Chemistry of materials, pp. 69-75, 2006
[19] R. Gosalawit, S. Chirachanchai, S. Shishatskiy, and S. P. Nunes, "Krytox–montmorillonite–Nafion® nanocomposite membrane for effective methanol crossover reduction in DMFCs", Solid state ionics, vol. 178, pp. 1627–1635, 2007
[20] J. Jaafar, A. F. Ismail, T. Matsuura, K. Nagai, "Performance of SPEEK based polymer–nanoclay inorganic membrane for DMFC", Journal of membrane science, vol. 382, pp. 202– 211, 2011
[21] J. Jaafar, A. F. Ismail, and T. Matsuura0, "Preparation and barrier properties of SPEEK/Cloisite15A®/TAP nanocomposite membrane for DMFC application," Journal of membrane science, vol. 345, pp. 119-127, 2009
[22] M. Tsapatsis, J. K. Jeong, and S. Nair, "Layered silicate material and applications of layered materials with porous layers," US Patent 7, 087: 288 B2, 2006
[23] A. Mahreni, A. B. Mohamad, A. A. H. Kadhum, W. R. W. Daud, and S. E. Iyuke, "Nafion/silicon oxide/phosphotungstic acid nanocomposite membrane with enhanced proton conductivity," Journal of membrane science, vol. 327, pp. 32-40, 011
[24] S. S. Mohtar, A. F. Ismail, and T. Matsuura, "Preparation and characterization of SPEEK/MMT-STA composite membrane for DMFC application," Journal of membrane science, vol. 371, pp.10–19, 2011
[25] K. S Reolofs, T. Hirth, and T. Schiestel, "Dihydrogenimidazole modified sulfonated poly(ether ether ketone) hybrid materials as electrolyte membranes for direct ethanol fuel cells," Material of science engineering B, vol. 176, pp. 727-735, 2011
[26] S. Zhou, Q. Wu, and Z. Liu, "Preparation and conductivity of organic undecatungstochromic heteropoly acid hybrid material," Polymer Bulletin, vol. 56, pp. 95–99, 2006
[27] M. Wang, and S. Dong, "Enhanced electrochemical properties of nanocomposite polymer electrolyte based on copolymer with exfoliated clays," Journal of power sources, vol. 170, pp. 25–432, 2007
[28] K. D. Kreuer, "On the development of proton conducting polymer membranes for hydrogen and methanol fuel cells," Journal of membrane science, vol. 185, pp. 29-39, 2001
[29] J. R. Kim, S. Cheng, O. H. Sang-Eun, and B. E. Logan, "Power generation using different cation, anion, and ultrafiltration membranes in microbial fuel cells," Environmental Science and Technology, vol. 41, pp. 1004–1009, 2007

2

MATHEMATICAL MODEL TO PREDICT THE PERMEABILITY OF WATER TRANSPORT IN CONCRETE STRUCTURE

Eluozo, S. N

Department of Civil Engineering, Faculty of Engineering, University of Nigeria Nsukka

Corresponding E-mail : solomoneluozo2000@yahoo.com

Abstract

Coarse aggregate fine sand are component of concrete, the rate of macropores in concrete determine the rate porosity and void ratio in concrete structure, the influence of the permeability coefficient determine the rate of water transport in concrete. Mathematical model to predict the rate of permeability on water transport were mathematically developed, the model is to monitor the rate of water transport in concrete structure. Permeability established a relationship under the influence of macropores on the constituent that made of concrete, application of concrete placement determine the rate of permeability deposition in concrete structure, permeability establishment are under the influence of macropores between the mixture through the cement paste, considering the variables in the system, mathematical model were established to monitor the rate of water passing through concrete and also determine the rate of permeability coefficient on concrete structure.

Keywords: *Concrete Structures, Permeability and Mathematical model*

1.0 Introduction

The durability of concrete structure depends on the frequent rate of migration through the dissolved constituent. Such migration is influenced by permeability. This condition in terms of concrete mixture is through a continuous network of the micro pores that exist in the matrix of the concrete mix. Other influences are through the porosity that exists in the interfacial of the gradation structure of the aggregate. This study characterized the rapid and accuracy of measuring concrete permeability in a mix, this include establishment of theoretical model that describes the influence of permeability on concrete structure. Experiments are performed using transient permeability apparatus to monitor the measure coarse aggregate fine sand and water are the micropores between this material as a component of concrete known as porosity and void ratio in the concrete structure, the influence of the permeability coefficient determine the rate of water transport in concrete of water migration in concrete mixture, the apparatus allowed a rapid and accurate measure of water migration in concrete mix.

Concrete is a type of porous material and can be physically and chemically damaged due to its exposure to various environments from the placing of concrete to its to the service life. In particular, some external hazardous elements, such as sulfate, chloride ion, and carbon dioxide, permeate in concrete over a long-term period as a solution or a gaseous state and cause physical damage due to chemical reactions. These reactions affect the corrosion of steel bars applied in concrete and that decreases the durability life and strength of such steel bars. Thus, it is very important to insert

Corrosion inhibitors into steel bars in the case of a deterioration element that exceeds the critical amount of corrosion in the location of steel bars [1]. However, it is very difficult to guarantee corrosion resistance at the location of steel bars using conventional technology that applies corrosion inhibitors only on the surface of concrete [2-3]. This study attempts to develop a method that penetrates corrosion inhibitors up to the location of steel bars and investigate the penetration depth of corrosion inhibitors by verifying moisture migration in concrete under applied pressure. In the penetration of water in concrete, the penetration depth according to the passage of time can be estimated using the Darcy's law, which is also applicable to the penetration of sand stratums under low pressure conditions. Meanwhile, it is necessary to analyze the penetrative diffusion flow accompanied by internal deformation under high pressure conditions [4]. Under the circumstances, this study applied the experiment on the penetration depth of water for reinforced concrete structures using the water pressure applied to the holes in concrete as a variable. Based on the results of this experiment, this study also calculated coefficients of water penetration and diffusion and estimated the penetration depth for the concrete according to the water pressuring time and pressure. In addition, this study attempts to provide the basic data for the development of a diffusion method with high pressure penetration of corrosion inhibitors for penetrating the inhibitors to the location of steel bars through investigating the water penetration mechanism in concrete using a FEM analysis that reflects the interaction between solid and fluid [5-7].

To produce high quality long lasting concrete structures, cements of a high and consistent quality must be employed. Worldwide, the cement industry spends countless hours assuring the quality of its products, mainly based on laboratory tests. In the USA, most physical testing of cements is performed according to ASTM standards /1/; in Germany, testing is generally governed by the European Norm /2/. Technologies that could reduce the number of physical tests needed for cement production (and optimization) would clearly be a welcome addition. One such potential technology is the use of virtual testing. In virtual testing, starting materials are characterized and their performance predicted via the use of computer models. This should result in savings in both resources (materials, labor, etc.) and time, as 28 days of concrete performance can be simulated in just a few hours of computer time. An additional benefit of virtual testing is the capability to perform a large number of "what-if" type computations to explore new material systems and optimize existing ones, e.g., what is the optimum sulfate content and form for a particular cement or how will the performance of a cement change if its Blaine fineness is increased by 10 m2/kg.

Long-lasting structure is very important if it is to survive the harsh environment that it is often exposed to [8]. In Nigeria, where there is no maintenance culture, it is all the more imperative. Nigerian hot marine coastal waters, constitute an aggressive environment that has been found to be deleterious to concrete [9], leading to premature deterioration that affects the strength and durability characteristics of concrete structures. One of the major forms of chemical attack on concrete is the chloride ingress. This ingress leads to corrosion of reinforcement, reduction in strength, unserviceable structures, and structures that are aesthetically poor. According to Stanish [8], corrosion products put surrounding concrete in tension thereby causing tension cracking and spalling of the cover of concrete. The attendant adverse structural influences are: loss of bond between the reinforcement and concrete, loss of steel area, and loss of stiffness. The total effects of these are serious durability problems because of reductions in the strength, serviceability and aesthetics of concrete structures.

Result in early repair or premature replacement of the structures. Libby [10] and Gallegos & Quesada [11] listed many chloride-induced structural failures that required expensive rehabilitation work. Thus to increase the service life of coastal concrete structures, increasing the resistance of concrete to chloride penetration is very necessary. A common approach to prevent such deterioration is to prevent chloride penetration into the structure by using relatively impenetrable concrete. But impenetrability of chloride into concrete depends on its porosity which in itself is defined in terms of pore size, pore distribution and interconnectivity of the pore systems. And in increasing this resistance, it is necessary to take into cognizance the prevalence of elevated temperature curing conditions, lest it works against the durability. As previously observed by Detwiler [12], the effects of hot weather and/or accumulated heat of hydration can be mitigated by various measures, but only to a certain extent. Earlier works by Wee [13], Smith [14] and Kumar [15] with cement paste containing silica fume, fly ash, and granulated blast furnace slag suggested that supplementary materials could improve the performance of concrete cured at elevated temperatures against chloride intrusion. For this work, the author chose slag because it is available as by products from the steel rolling plants in Osogbo, Aladja, and Katsina in Nigeria.

And for the basis of comparison, degree of hydration, rather than the curing time was employed. Perenchio [16] had earlier pointed out that conclusions based on constant temperature curing would not necessarily apply to concrete cured under field conditions. And previous work by Detwiler [17] gave the number of days to reach 70% degree of hydration for Portland cement paste, cement paste containing silica fume and granulated blast furnace slag (at 30% replacement level). The numbers of days to reach this degree of hydration were employed for this work.

Presently natural resources are increasingly consumed due to rapid urbanization and thereafter human construction activities, so that various strategies are being investigated by engineers to protect and restore natural ecosystems all over the world. Permeable (porous/ pervious) pavement is termed as comprising materials that facilitate storm water infiltrate and transfer to the underlying subsoil [18-19].

In Australia, permeable pavement has been utilized as a potential tool of Water Sensitive Urban Design (WSUD) to manage natural water. From 1994 the University of New South Wales (UNSW) started to research into permeable concrete paving and more recently the University of South Australia (UniSA) is also involved. However, the previous studies conducted both in UNSW and UniSA mainly concentrated on water quality and pollution control through permeable pavements and, only the properties of base course materials in permeable pavement system and segmental paving have been studied. There is still a gap of optimizing the surface materials for permeable pavements pervious concrete. At ambient temperature conditions, a dosage of 5 fl oz/cwt of the HCA provides between 60 and 90 minutes of extra working time. Hydration controlling admixtures can eliminate inconsistencies and performance variability that may be brought on by the need to re-temper mixtures at the job sites [20]. Along with the HCA, VMA or viscosity modifying admixtures may be beneficial to the performance of pervious concrete. The use of VMAs results in better flow, quicker discharge time from a truck, and easier placement and compaction. Furthermore, VMAs prevent drain down, and may increase both compressive and flexural strength of pervious concrete. It should be noted that not all VMAs are made with pervious concrete in mind, and therefore, care should be taken when choosing the right VMA for pervious installation [20]. In California, Young's [21] reported that latex modifiers allowed harder surface finishing using Bunyan screeds, which in return produced "table-top" surface, and almost eradicated surface raveling. Latex modifiers assist in binding the cement paste to the aggregate. Mixtures with latex modifiers might allow utilization of pervious concrete in high speed pavement applications [21].

2.0 Theoretical Background

This model consider the cylinder sample depicted which produced total volume of $V = AL$ where A is the cross section area of the specimen and L is known to be the length, verification of this sample is through two pressurized reservoirs, this produced the upstream at ϕu and the downstream one at ϕd. These initial values can be expressed as upstream and downstream pressures are $\phi 1$ and $\phi 0$ respectively. In this condition application of partial differential expression which should govern pressure variation as a function of length and period $\phi(Z, t)$ within the sample which is expressed as but in this study concrete materials where considered, lack mix design application that has lead poor compressive strength in concrete, this has also developed to so many structural damage, the focus of this study are to determine the rate of permeability and there relation with the concrete constituent.

For the express of the parameter in the system are

ϕ	=	porosity
k	=	permeability
μ	=	Macropore in pore fluid viscosity
K	=	Sample permeability
Z	=	Distance with the region i.e. measure from the upstream reservoir.
t	=	Time

β = Lumped compressibility

3.0 Governing Equation

The model designing water transport is presented by Roy et al 1993 the model was to monitor the behaviour of water that is passing through a concrete structure.

$$\frac{\partial^2 \phi}{\partial Z^2} + \beta\left(\frac{\partial \phi}{\partial Z}\right)^2 = \frac{\beta u}{k}\frac{\partial \phi}{\partial t} \quad \text{...} \quad (1)$$

The mathematical expression of the parameter in the system where put into mathematical equation, the variables ϕ *denote* (Z, t) which is the function of Z distance and Time T, but for simplicity the equation where linearized so it will be easy to solve the problem under study. The concept of water forms the binding agent with cement paste fine sand coarse aggregate are mixed together to make up concrete structure, most time it reinforced either with mild steel or high tension steel depending on the imposed load, the concrete is placed on a form work, it is also vibrated for compaction, this compaction will only reduce the rate of void and decrease the rate permeability in the concrete structure, but it will not avoid permeability, the rate of permeability determine the rate of void ratio and porosity deposited in concrete structural component. Considering the equation by applying separation of variable; the variables in the system were represented, by the application of mathematical tools, a constant C_1 and C_2 where establish as the equation were derived where ϕ represent pressure, β and other parameters u K t where represented as λ^2 ,this condition express the entire variables.

$\frac{T^2}{T}\ \frac{\beta u}{K}$ Equal to λ^2. By splitting techniques this was applied to other variables. The parameters known as the variable in the system that produce the result of water transport in the concrete structure where all equal to λ^2 exception of pressure in the system denoted as ϕ. Permeability is the product of porosity, because of the pore fluid between the binding agents on the concrete structure, fine sand coarse aggregate reacting with the cement paste, are mixed together to produce concrete. The rate of permeability coefficient, are determine through structural characteristic from the concrete, where the permeability exist allows the fluid to pass through the concrete based on the rate of deposited micropores from permeability coefficient in the concrete structure. The model established a relationship within the variables in the system, where two constant were developed, this implies that the constant linearized the equation derived to solve water transport passing through the influence of permeability in concrete structure.

Substituting solution ϕ = Z, T into equation (1), we have

$$Z'T + \beta(Z'T)^2 = \frac{\beta u}{k}X'T \quad \text{...} \quad (2)$$

$$\frac{T''}{T} + \beta\left(\frac{Z''T}{Z}\right)^2 = \frac{\beta u}{k}\ \frac{Z'T}{T} \quad \text{...} \quad (3)$$

$$\frac{T''}{T} = \frac{\beta u}{k} + \beta - 2\frac{Z''}{Z} = \frac{\beta u}{k}\ \frac{Z''}{Z} - 2 \quad \text{....................................} \quad (4)$$

$$\frac{T''}{T} = \beta - 2\frac{Z'}{Z} - \beta - \frac{\beta u}{k}\frac{Z'}{Z} - \frac{\beta u}{k} - \beta - 2 \quad \text{..........................} \quad (5)$$

$$\frac{T''}{T} = \beta - 2 - \beta - \frac{\beta u}{k} - \frac{\beta u}{k} - \beta - 2 \quad \text{..........................} \quad (6)$$

Considering when $Ln\ \Delta Z \Rightarrow 0$

$$\frac{T''}{T} = \beta - \frac{\beta u}{k} - \frac{\beta u}{k} - \beta = \lambda^2 \quad \ldots\ldots\ldots\ldots \quad (7)$$

$$\frac{T''}{T} = \lambda^2 \quad \ldots\ldots\ldots\ldots \quad (8)$$

$$\frac{\beta u}{k} = \lambda^2 \quad \ldots\ldots\ldots\ldots \quad (9)$$

$$\beta = \lambda^2 \quad \ldots\ldots\ldots\ldots \quad (10)$$

This implies that the equation can be expressed as:

$$\frac{\beta u}{k}\frac{Z'}{Z} = \lambda^2 \quad \ldots\ldots\ldots\ldots \quad (11)$$

$$\frac{\partial^2 y}{\partial Z^2} = \lambda^2 \quad \ldots\ldots\ldots\ldots \quad (12)$$

$$\partial y = \frac{\lambda^2}{\beta} Z\partial y \quad \ldots\ldots\ldots\ldots \quad (13)$$

$$\int \partial y = \int \frac{\lambda^2}{\beta} Z \, \partial y + C_1 \quad \ldots\ldots\ldots\ldots \quad (14)$$

$$y = \frac{\lambda^2}{\beta} Z^2 \, \partial y + C_1 + C_2 \quad \ldots\ldots\ldots\ldots \quad (15)$$

$$y = 0 \quad \ldots\ldots\ldots\ldots \quad (16)$$

$$\boxed{y = \frac{\lambda^2}{\beta} + C_1 Z + C_2} \quad \ldots\ldots\ldots\ldots \quad (17)$$

$$\Rightarrow \frac{\lambda^2}{\beta} + Z^2 + C_1 Z + C_2 = 0 \ldots\ldots\ldots \quad (18)$$

Applying quadratic expression, we have

$$\frac{-b \pm \sqrt{b^2 - 4ac}}{2a} \quad \ldots\ldots\ldots\ldots \quad (19)$$

$$X = \frac{-(C_1) \pm \sqrt{(C_1) - 4C_1\left(\frac{\lambda^2}{\beta}\right)C_2}}{2C_1} \quad \ldots\ldots\ldots\ldots \quad (20)$$

$$= \frac{C_1 \sqrt{C_1^2 - 4C_2 \frac{\lambda^2}{\beta}}}{2C_1} \quad \ldots\ldots\ldots\ldots \quad (21)$$

$$X = \frac{-C_1 + \sqrt{C_1^2 - 4C_2 \frac{\lambda^2}{\beta}}}{2C_1} \quad \ldots\ldots\ldots\ldots \quad (22)$$

$$\frac{X - C_1 + \sqrt{C_1^2 - \frac{4C_2\lambda^2}{\beta}}}{2C_1} \qquad \text{...} \qquad (23)$$

Subject equation (23) to the following boundary conditions and initial values conditions

$$t = 0, \ \phi = Po \qquad \text{..} \qquad (24)$$

Therefore, $Z(z) = C_1\ell^{-M_1Z} + C_2\ell^{-M_2Z}$ .... (25)

$$= C_1 \, Cos \, M_1Z + C_2 \, Sin \, M_2Z \qquad \text{................................} \qquad (26)$$

Solving equation (17) gives

$$y = \frac{\lambda^2}{\beta} + C_1Z + C_2$$

Deriving the relationship between the variables where constant were integrated to linearized the equation, application of quadratic expression were introduced, this concept were applied to express the integrated constant, denoted as C_1 and C_2, the relationship of the parameter denotes as λ^2 were established considering the boundary values of t = 0, $\phi = \phi 0$ the boundary were integrated to produce the model. This will be applied to monitor the rate of water transport in concrete structure, for further simplicity application of Suncidal expression were established, where the constant introduced, where expression in equation (25) as $Z(z) = C_1\,\ell^{-M_{1z}} + C_2{}^{-M_{2z}}$ deriving it produce the equation (27) and (28).

The final model equation for water transport was established as

$$\phi(z,t) = \left(C_1 \, Cos \, M_1 \frac{\lambda^2}{\beta} Z + C_2 \, Sin \, M_2 \frac{\lambda^2}{\beta} Z \right) \qquad \text{.........................} \qquad (27)$$

But if $Z = \frac{V}{t}$

$$\phi(z,t) = \left(C_1 \, Cos \, M_1 \frac{\lambda^2}{\beta}\frac{V}{t} + C_2 \, Sin \, M_2 \frac{\lambda^2}{\beta}\frac{V}{t} \right) \qquad \text{.........................} \qquad (28)$$

4.0 Conclusion

The model established consider all the variables in the system that influence the transport of water in concrete structure, initially, concrete attained strength by curing the structure component, but when it has attained the required strength, the fluid passing through the macropores established is through the influence of permeability, this will decrease the structural compressive strength, water transport on the structure are caused by the variables on the model developed. Applying the mathematical expression will definitely monitor the rate of water migrating on concrete structure, the variable that allows the water transport has been expressed. The permeability influence that relate with the macropores between the constituent of the concrete is the mixture of fine. Sand, cement, water and coarse aggregate, including the mixture and compressibility of concrete placement through compaction are expressed base on various rates. The model can be simulated to develop the theoretical values that can be compared with the experimental values for validation of permeability coefficient in concrete structure.

References

[1] H. S. Lee, S. W. Shin, "Evaluation on the effect lithium nitrite corrosion inhibitor by the corrosion sensors embedded in mortar," Constr Build Mater, vol. 21, pp. 1–6, 2007

[2] M. Ormellese, "Corrosion inhibitor for chlorides induced corrosion in reinforced concrete structure," Cem Concr Res, vol. 36, pp. 536–47, 2006

[3] T. A. Soylev, and C. McNally, "Effectiveness of amino alcohol-based surface-applied corrosion inhibitor in chloride-contaminated concrete,", Cem Concr Res, vol. 37, pp. 972–7, 2007

[4] J, Murata, Y. Ogihara, S. Koshikawa, and Y. Itoh, "Study on watertightness of concrete," ACI Mater J, vol. 2, pp. 107–16, 2004

[5] V. G. Papadakis, C. G. Vayenas, and M. N. Fardis, "Physical and chemical characteristics affecting the durability of concrete,", ACI Mater J, vol. 8, pp. 186–96. 1991

[6] Automatic dynamic incremental nonlinear analysis (theory and modeling guide), vol. 3, ADINA CFD & FSI, 2008

[7] J. H. Yoo, H. S. Lee, and M. A. Ismail, "An analytical study on the water penetration and diffusion into concrete under water pressure," concrete and building materials, pp. 99-108, 2011

[8] K. D. Stanish, R. D. Hooton, and M. D. A. Thomas, "Testing the Chloride Penetration of Concrete," FHWA-Civil Engineering Department, University of Toronto, 1997

[9] D. K. Jain, J. Prasad, and A. K. Abu, "Ground Granulated Blast Furnace Slag Cement Concrete," NBM Media Construction Information, 2010

[10] J. R. Libby, "Three Chloride-Related Failures in Concrete Structures," Concrete International, vol. 9(6), pp. 29-31, 1987

[11] H. Gallegos, and G. A. Quesada, "Corosion Repair Procedure," Concrete International, vol. 9 (6), pp. 54-57, 1987

[12] R. J. Detwiler, K. O. Kjellsen, and O. E. Gjorv, "Resistance to Chloride Intrusion of Concrete Cured at Different Temperatures," ACI Materials Journal, vol. 88 91), pp. 19-24, 1991

[13] T. H. Wee, A. K. Suryananshi, and S. S. Tin, "Evaluation of Rapid Chloride Permeability Test Results for Concretes Containing Mineral Admixtures," India Concrete Journal, vol. 97 92), pp. 221-223

[14] B. G. Smith, "Durability of Silica Fume Concrete Exposed to Chloride in Hot Climates," Journal of Materials in Civil Engineering, pp. 41-45, 2001

[15] S. Kumar, B. K. Rao, and S. Mishra, "Chloride Penetration Resistance of Concrete Containing Blast Furnace Slag," The Indian Concrete Journal, 2002

[16] W. F. Perenchio, N. P. Rajamane, A. M. Neville, R. J. Detwiler, K. O. Kjellson, and O. E. Gjorv, "Discussion of Resistance to Chloride Intrusion of Cement Cured at Different Temperatures," ACI Materials Journal, vol. 88(6), pp. 676-679, 1991

[17] R. J. Detwiler, C. A. Fapohunda, and J. Natalie, "Use of Supplementary Cementing Materials to Increase the Resistance of Chloride Ion Penetration of Concretes Cured at Elevated Temperature," ACI Materials Journal, vol. 91 (1), 1994

[18] ARMCANZ & ANZECC: Australian and New Zealand Environment and Conservation Council and Agriculture and Resource Management Council of Australia and New Zealand, 2000

[19] C. Lain, and Y. Zhuge, "Investigation of the effect of aggregate on the performance of permeable concrete," Taylor & Francis Group, London, ISBN 978-0-415-56809-8, pp. 1, 2010

[20] M. A. Bury, A. M. Christine, and D. Fisher, "Making Pervious Concrete Placement Easy Using a Novel Admixture System," vol. 2, pp. 1-3, 2006

[21] Young's, Andy, "Pervious Concrete, The California Experience," 2006

3

DESIGN AND DEVELOPMENT OF SUSTAINABLE CONSTRUCTION STRATEGY FOR RESIDENTIAL BUILDINGS: A CASE STUDY FOR COMPOSITE CLIMATE

Chetan S. Dhanjode[1], R. V. Ralegaonkar[2], Vaidehi A. Dakwale [3*]

[1] M Tech. Construction Technology and Management, VNIT, Nagpur, India
[2] Associate Professor, Civil Engineering Department, VNIT, Nagpur, India
[3] Research Scholar, Civil Engineering Department, VNIT, Nagpur and Assistant Professor Civil Engineering Department, RCOEM, Nagpur, India

*Corresponding E-mail: vaidehi.dakwale@gmail.com

Abstract

Developing cities in India like Nagpur are growing exponentially in population due to industrialization. The ever increasing demand of the natural resources leads to depletion of limited resources and also affects local environment in terms of increasing pollution emissions especially carbon emission. With recourse to composite climatic condition (Nagpur, India) the sustainable construction strategy is developed in the present paper. The developed strategy includes conservation of soil, energy, resources, material and water. It provides a systematic approach towards sustainability of building through quantification of energy consumption. Analysis of conventional and non-conventional material and technology on cost, energy consumption and carbon emission parameters helps in highlighting suitable options for sustainable construction. Strategy is validated through a case study of new construction of the residential buildings for enhanced environmental performance. The amount of excavated soil and its reutilization on site is taken into consideration. Locally available sustainable construction materials are compared for material cost, and embodied energy for selection of appropriate construction material. In order to conserve the ground water various options for the low flow devices, roof top rainwater harvesting and gray water recycling and reuse are suggested. Application of sustainable construction strategy to case study building revealed that 97% of natural soil is conserved through backfilling. Estimated carbon emission reduction due to recommended construction materials (sustainable bricks, cement, and steel) with respect to conventional options is of the order of 60%. Recommended water conservation options resulted in 57% reduction in ground water demand. However suggested sustainable construction strategy options estimated in increase in project cost by 13% whose effect can subsequently be reduced over the design life span computations. The developed strategy can further be applied to the larger residential township areas with varying building types for conserving the natural resources as well as reducing the impact of environmental pollution.

Keywords: *Sustainable construction strategy, natural resource conservation, techno-economic feasibility, environmental pollution control, carbon emission reduction*

1.0 Introduction

Construction industry is a major consumer of natural resources. Construction related activities account for quite a large portion of energy consumption and CO_2 emissions. This energy is consumed at pre-occupancy state in manufacturing and transportation of building materials and components, which is termed as embodied energy (EE), and at occupancy state which corresponds to the running of the appliances when it is occupied. Construction sector in India emits about 22% of the total annual emission of CO_2 resulting from the Indian economy. Out of the emissions from the construction sector, 80% are resulting mainly from the products/industrial processes of energy intensive building materials, i.e. cement, lime, steel, bricks and aluminum [1]. Further, sustainability of the built form is affected by decisions to be taken at all the design stages. The design of built form with solar passive techniques includes shape and size of built form, orientation, site planning, design of building components such as roofs, walls, openings (doors and windows) and design of building elements such as windows and shading devices [2]. A sustainable construction strategy ideally should include all those parameters which reduce energy consumption and atmospheric carbon emission. Degree of sustainability of a building is judged by comparing it to a benchmark construction. Comparison benchmarks can be set according to building characteristics. Development of these reference building benchmarks requires some assumptions and data collection. While some factors such as activity, occupancy data, building area, number of pupils, age of building, etc. can be easy to obtain by means of questionnaires, other vital information for evaluation for energy performance of the building, such as construction details and type and efficiency of heating systems are often not known by respondents [3]. This information has to be collected from all stakeholders involved in planning and designing stage of building.

A strategy for sustainable construction is presented in this paper which takes in to account soil, resources, material, and water conservation. It helps in rational utilization of non-renewable resources with an increase in use of renewable resources, elimination of wastes, reduction of waste formation, customs management of consumption, utilization of equipments and materials, more efficiency and/or increase in quality of productive processes. Each parameter affects overall sustainability of building. Development and environmental performance of building is location specific. Proper control at all stages of building development such as construction, fitment, outdoor facility construction, transportation, operation, waste treatment, property management, demolition, and disposal, with related material, equipment, energy and manpower inputs improves environmental performance of building [4].

It is comparatively easy to construct a new building with an approach to sustainability. It is because in new buildings control over all actions can be easily planned than existing buildings. Quantification and comparison of resource consumption in building construction tends to minimize the consumption. Building energy labeling systems help in this. There are several voluntary energy labeling systems prevalent across the globe viz. BREEAM (Building Research Establishment's Environmental Assessment Method), GB Tool, LEED (Leadership in Energy and Environmental design), CASBEE (Comprehensive Assessment System for Building Environmental Efficiency) Green Globes, GRIHA (Green Rating for Integrated Habitat Assessment) [5]. Voluntary building-assessment scheme helps in encouraging owners greater in energy-reduction effort [6]. Researches carried in the area of improvement in environmental performance of building suggest that policy formulation, assessment and monitoring, life cycle analysis and simulation, and building material selection are major controlling factors [7]. In the strategy to improve overall performance of buildings, approach of authority [8-11], economy of nation [12, 13], availability of local energy sources [14], state of technology [15-17] and behavior of consumers [18, 19] have been identified as major parameters.

A sustainable building in its basic form is energy efficient and water efficient. In most of the sustainable construction practices take energy conservation in building occupancy state into account. Whereas research shows that the embodied energy of residential buildings can contribute up to 40% of the life-cycle energy use in residential buildings. The initial embodied energy of a building is the energy used in producing a building whereas the recurring embodied energy is the energy used in maintaining and repairing of the building over its effective life. [20]. EE data depends on 10 parameters viz. system boundaries, methods of EE analysis, geographic location of study area, primary and delivered energy, age of data sources, source of data, completeness of data, technology of manufacturing processes, feedstock energy consideration and temporal representativeness [21]. Present study takes in to account embodied energy of building material as one of the strategic parameter.

Uncontrolled exploitation of ground water resources results in to scarcity of potable water. Rain water harvesting is an apt answer to such situation. Even though initial investment in such technology is considerable it proves to be cost effective in long run. Efficiency of rainwater harvesting depends on the materials used, design and construction, maintenance and the total amount of rainfall. Cement tiles are used as a roofing material, the year-round roof runoff coefficient is about 75%, whereas clay tiles collect usually less than 50% depending on the harvesting technology. Plastic and metal sheets are best with an efficiency of 80-90% [22]. A combination of water management options results in decrease in water consumption up to 50%, thereby reducing the pressure on the scarce water resources. This reduction can be achieved because the water is recycled for different usage reducing the use of fresh groundwater. Environmental and financial impacts can be reduced similarly; the annual environmental impact of the in-house water use can be reduced by 8% when using low-flow shower heads, and up to 38% when using rainwater harvesting systems. These options are financially viable in long term use such as faucet aerators, low-flow shower heads and dual flush toilets [23].

Similarly during a building's life cycle, operational energy services including Heating ventilation and air conditioning (HVAC), lighting, equipment and appliances etc. contribute to approximately 80% of total building energy consumption [24]. Improvement in thermal performance of building envelope also leads to reduction in carbon emission. Retrofitting, reconstruction and selection of appropriate U-factors for building materials lead to a saving of carbon emission up to 31– 36%. Reuse, recycling and regeneration energy through combustion together can save up to 10% of total energy and subsequently emission [25]. However, energy consumption pattern in domestic application varies over geographic location with occupant's behavior. In order to design the sustainable construction strategy for the residential building in composite climate various labeling and rating systems are reviewed. Different parameters associated with sustainable construction practices are identified. The data is collected by field survey. The developed strategy for the sustainable construction is applied. The economic viability of the developed sustainable construction strategy for the considered case study of the residential building is estimated on the annual basis.

2.0 Developed sustainable construction strategy

Strategic planning is a decision-making process involving diverse objectives which requires a systematic approach. In the present paper the design process for new buildings is addressed. To validate the effect of strategy a case study of cluster of eighteen residential buildings located in composite climate is taken up and results are quantified. Present strategy includes;

1. Conservation of excavated soil by various means such as backfilling and landscaping.
2. Use of low embodied energy, natural materials, and recycled building materials.
3. Reduction in per capita consumption of water in the home through various measures such as rain water harvesting, use of low flow devices.
4. Reuse of waste water and reducing consumption of municipal water.

3.0 Design Data

Proposed site is located at Nagpur city, Maharashtra in India. Case study building is a cluster of 18 individual residential bungalows. The building is located at nonagricultural land and is sanctioned by local town planning authority. Each bungalow is a four bedrooms, hall, and kitchen house with approximate total occupancy of 5 to 6 persons. Plot area of each proposed building is 11.15 m x 19.10 m i.e. 212.96 m^2. Proposed area of construction of one bungalow is 237.00 m^2. Thus, for 18 bungalows total area is 3833.28 m^2. Orientation of building is 0° towards North and the short wall of the building is facing East–West. Sizes of openings are 20% of wall area. The longer window provides better light with better penetration and spread, so as to reduce lighting load. Outer walls are 0.23 m thick & inner walls are 0.115 m thick. Reinforce concrete roof is proposed for the construction of building. Roof insulation is proposed with the aid of false ceiling to reduce the heat gain through roof. Around 28% of the plot area is covered by landscaping on front and rear side of building. Soil used for landscaping is partially excavated dump and garden soil. The building is designed as a framed reinforced concrete structure which has M-20/M-25 concrete as primary building material

3.1 Reuse of excavated soil

Soil it is collected in process of excavation for foundation pits and is used for landscaping, refilling and leveling of low level areas. Total Quantity of excavation is estimated 36.50 m^3 for each bungalow. Excavated soil for total 18 buildings is 657.00 m^3. Out of this soil 430.20 m^3 is used in filling in plinth and floor, 74.68 m^3 is utilized in garden top soil replacement in landscaping and remaining 131.40 m^3 is used for leveling internal roads surface

3.2 Cost analysis of building materials

To calculate embodied energy of buildings estimate of material quantity has been prepared. Construction cost is calculated on the basis of quantity estimate. Rates have been taken from current schedule rates of Central Public Work Department (CPWD), India 2009-2010. For non-conventional material cost data has been collected at source. Major conventional building materials have been compared to non-conventional building materials. Several building material options for each conventional building material have been proposed. For masonry clay bricks have been compared to Fly ash bricks, Autoclaved aerated concrete blocks (A A C blocks), and Stabilized Mud Block (SMB). Ordinary Portland cement (OPC) is compared with Portland Pozzolana cement (PPC). Cost of conventional building material has been considered as base value and percentage saving in cost has been estimated for non- conventional building material. Details of cost of various options have been given in Table 1 below.

3.3 Estimation of embodied energy of building materials

Embodied energy of basic building materials has been calculated. Unit embodied energy data has been derived from various literatures [1], form source calculations and market survey. Fly ash bricks, Autoclaved aerated concrete blocks (A A C blocks), and Stabilized Mud Block (SMB) are compared with clay bricks. These materials use waste material as raw material and least energy in manufacturing process and thus have lower embodied energy. Portland Pozzolana cement uses fly ash in the process of manufacturing in as a replacement to costly and energy intensive cement clinkers. Thus PPC has lower embodied energy as compared to OPC. Steel being energy intensive material is compared with recycled steel. Table 2 gives comparative analysis of embodied energy for basic building materials

Table 1: Comparative analysis of cost of conventional and energy efficient materials

Description	Quantity	Unit	Rate (Rs.)	Cost (Thousand Rs.)	Cost Saving %
Clay Brick	520895	No.	3.90	2031.490	.....
Fly Ash Brick	520895	No.	3.75	1953.356	3.85
AAC block	73800	No.	28.00	2066.400	-1.72
SMB	520895	No.	3.80	1979.401	2.56
OPC	15120	Bag	290.00	4384.800	.....
PPC	15120	Bag	260.00	3931.200	10.34
TMT steel	109.930	Ton	46000.00	5056.780	.....
Recycled steel	109.930	Ton	42000.00	4617.060	8.70

Source: (Quantity of construction materials has been worked out and rates are taken from current scheduled rates of PWD 2009)

Table 2: Comparative analysis of embodied energy for basic building materials

Description	Quantity	Unit	energy (MJ/ unit)	Total energy in MJ	% Savings
Clay Brick	1132.38	Cum	4050.00	4586139.00	.....
Fly Ash Brick	1132.38	Cum	1100.00	1245618.00	72.84
AAC block	1132.38	Cum	1920.00	2174169.60	52.59
S M B	9846.78	Sqm	110.00	1083145.80	76.38
OPC	524.66	Cum	852.00	447010.32	.....
PPC	524.66	Cum	507.00	266002.62	40.49
TMT steel	109930.00	Kg	32.00	3517760.00	.....
Recycled steel	109930.00	Kg	10.10	1110293.00	68.44

3.4 Estimation of carbon emission of building materials

Carbon emission for bricks, cement and steel has been calculated. Unit carbon emission data has been collected from various literatures, form source calculations and market survey. For each construction material above options have been compared and saving in emission is highlighted. Table 3 gives comparative analysis of carbon emission for building materials. Carbon emission data has been taken from Environmental Protection Agency (EPA) report and for steel it is taken from M/s Kamboj Ispat Pvt. Ltd., Nagpur, India

Table 3: Comparative analysis of carbon emission for building materials

Description	Quantity (No.)	CO_2 emission per brick (Kg CO_2 /bricks)	Total CO_2 emission (Kg CO_2)	% Savings
Clay Brick	520895	0.59	307328.05	.....
Fly Ash Brick	520895	0.11	57298.45	81.36
AAC block	73800	0.22	16236.00	94.72
S M B	520895	0.22	114596.90	62.71
OPC	15120	0.89	13456.80	.....
PPC	15120	0.60	9072.00	32.58
TMT steel	109.930	1.987	218.43	.....
Recycled steel	109.930	0.357	39.25	82.03

3.5 Water conservation through low flow devices

Nagpur lies in central part of India where average intensity of annual rainfall is 800 mm. Rest of the need of water is catered by municipal water supply and withdrawal of ground water through wells and sumps. Water is a scarce natural resource in this part and thus essentially needs to be conserved. There are two methods of water conservation reducing water use on sully side and restoring and recycling water on demand side. In this case study water conservation is achieved through deploying low water fixture which is commercially available in market, implementing rainwater harvesting and installing waste water recycling system for case study buildings. Various systems for Low flow devices (LFD) have been studied. Low-flow toilet flushing device use a maximum of 1.6 gallons of water per flush compared with about 3.5 gallons of water used by a conventional toilet. Low-flow shower heads use about 2½ gallons of water per minute compared to between four and five gallons per minute used by conventional heads. Low-flow faucet aerators can cut the water usage of faucets by as much as 40% from 4 gallons per minute to 2½ gallons. As far as cost is concerned low flow fixtures are high in cost as compare to conventional fixtures. Thus an increase in initial installation cost is there but they save considerable amount of water and thus pay off in long run of time. Table 4 gives reduction in water usage by low flow fixtures.

Table 4: Comparative analysis of conventional fixtures and low water fixture

Water consumption liter/capita/day	Conventional system	Low flow Device
Per day per person consumption of water (l)	180	108
Annually Consumption for 18 buildings (l)	7123204.8	4758004.8
Cost of water (annually) (Rs.)	427392.29	285480.28
Cost of Fixture & Fittings (Rs.)	1291770.00	2251170
Total Cost (Rs.)	1719162.29	2536650.288
Excess in cost (%)	......	47.55
Saving in water (%)	......	33.20

Rates of water supply is taken from Nagpur Municipal Corporation water tariff as 5.0 Rs/unit (1 unit=1000 L

3.6 Rainwater Harvesting

For landscaping area of 80.04 m^2 water requirement is taken as 3.5 liters /m^2 per day. Yearly water requirement with 30% evaporation losses for 18 buildings is computed as 1210204.80 liters /annum. To cater this need it is propose to have rainwater harvesting system. Roof top rainwater harvesting is analyzed for total 18 buildings. Water collected annually for 18 Building is 1243584 liters/annum. Thus harvested water totally caters the need of water required for landscaping with a 2.68% excess storage.

3.7 Recycling & reuse of wastewater

Decentralised Wastewater Treatment Systems are effective, reliable, cost-efficient and custom-built wastewater treatment systems which are perfectly suited for small to medium-size systems. The technical options within decentralised wastewater treatment system are based on a modular and partly standardized design. Decentralised wastewater treatment system is based on basic technical treatment processes: mechanical treatment (sedimentation and flotation) and biological (anaerobic and aerobic) treatment. The most common decentralised wastewater treatment system modules are septic tanks, biogas digesters, anaerobic baffled reactors (ABR), anaerobic filters (AF), planted gravel filters (PGF), and (if needed) polishing ponds. The systems can be designed to specific individual site-specific needs. In the present study a Decentralised Wastewater Treatment Systems is designed collectively for 18 buildings. Water required for washing & cleaning of house & flushing of toilets is 35 litres per capita per day thus annual water requirement for 18 buildings is 1.14 million litres. Whereas grey water collected out of bathing and cloths washing is 48 litres per capita per day. Assuming that 80% of recycling water will be available after treatment total quantity of water available is computed as 1.26 million litres. Thus requirement for washing cleaning is 100% catered with grey water recycling with a saving of 10% water.

3.8 Operating energy and related carbon emission

Household lighting typically accounts for 10% of electric bill. Alternative sources of energy are not carbon based and therefore do not release greenhouse gas (carbon dioxide) into the atmosphere when consumed. They can be designed to partially or fully satisfy the building / township's energy needs. Increasing utilization of these commercially available energy resources, including solar, wind, geothermal, and hydroelectric, is an important component in a municipal initiative to improve energy efficiency.

Compact Fluorescent Lights (CFLS) and Light Emitting Diodes (LEDs) have risen in popularity in recent years due to their longevity and energy-efficient design. These bright, compact lights are designed to reduce energy consumption without sacrificing light or productivity. Replacing CFL by LED saves cost and energy considerably.

All figures must carry numbers in the text (e.g. Fig. 1) and captions. Captions should be complete enough to allow understanding of the illustration without referring to the text. In addition, a source of the image other than the author's own archive should be placed directly under the image (author, date) and the font size by 1pt smaller than the caption. Use single blank lines before and after the image.

4.0 Results and Discussions

Selection of site is a major criterion in conservation of natural resources. Site must be a non agricultural land. Top soil must be preserved. In present study excavated stuff has been used for landscaping. Remaining soil is used for backfilling and leveling. No soil is disposed out of site.

The comparative study of construction material such as bricks, cement, and steel has been done. For bricks four options have been considered viz. Clay Brick, Fly Ash Brick, AAC (Autoclaved Aerated Concrete), and Stabilized Mud Block. These bricks are locally available within a periphery of 30-50 km. Various locally available options for above building materials have been compared for embodied energy and cost. It is observed that these options when compared to clay brick consume less embodied energy viz. Fly Ash Brick consumes 73%, AAC (Autoclaved Aerated Concrete) consumes 53% and Stabilized Mud Block (m^2) consumes 76% less embodied energy. More over Fly ash brick utilizes fly ash which is waste material and land filling. Stabilized Mud Block do not used any manufacturing energy and these are only compressed blocks. Autoclaved concrete blocks uses less concrete for given volume and thus, save scarce natural resources such as cement and aggregates. Cost of these blocks fly ash bricks and stabilized mud blocks are 4% and 3% less as compared to conventional clay bricks respectively whereas, AAC blocks are 2% costlier than conventional clay blocks.

Similarly in the comparison of Ordinary Portland cement and Portland Pozzolana Cement it is found that later is 11% cheaper than former one and saves energy to an extent of 41%. Thermo Mechanically Treated (TMT) steel is being used for the building under study and it has a potential of recycling the scrap as well as use of recycled steel as reinforcement in construction. It is found that recycled steel saves 9% energy and 68% cost as compared to conventional steel. When conventional materials are replaced by non-conventional material such as fly ash bricks or PPC cement it helps in reducing carbon emission as well as gives a solution to waste utilization. Fly Ash Brick, AAC (Autoclaved Aerated Concrete), and Stabilized Mud Block emit respectively 82%, 95%, 63% less CO_2 to atmosphere as compared to clay Bricks. Pozzolana Portland Cement (PPC) emits 33% less CO_2 as compared to Ordinary Portland Cement (OPC). It is found that recycled steel saves 82% as compared to conventional steel.

Comparative study of conventional and low flow devices shows that water consumption has been reduced to 108 liter per capita per day than 180 liter per capita per day by using the low flow device. This helps to reducing ground water load as well as municipal water bills. For the present study building cost of the RWH system is found as Rs. 113640.00. The annual collected water from the RWH system is 69088 Lit/year. The requirement of water for the landscaping and gardening is 67233 Lit/year. Hence, required water for the landscaping and gardening can be 100% fulfilled by the RWH system. More over cost of the system is mere 1% of the total cost of the building so it can be easily constructed with building itself. Similarly treated grey water caters the need of household washing and flushing. Lighting, home appliances and cooling system is the important component for human beings. Now a day's consumption of electricity is more as per the living standards, hence it is necessary to use the alternative option for the like solar, geothermal cooling system or use latest low conserving lights and star rated appliances.

Case study helps in elaborating the strategy and indicating its application. If implemented it results in to a systematic approach towards the sustainable construction. Analysis of case study clearly shows that application of the proposed construction strategy to a building results in to potential energy and emission saving. The strategy guides through each stages of construction process for quantifiable reduction in use of non-renewable resources and helps in using resources efficiently and minimizing waste by closing cycles. It further extends solution towards pollution reduction. This strategy can be implemented to a larger buildings and townships for reduction in energy and emission.

5.0 Conclusions

The conclusion of this study can be summarized as follows:

a. Conservation of natural resources by utilizing non-agricultural land for developing into residential or commercial building helps in conserving natural vegetation and reducing environmental pollution.
b. Utilization of waste material in production of construction material and minimizing conventional energy sources in material manufacture such as fly ash and stabilized mud blocks reduces energy in a range of 70-80% and saves cost up to 2-4%. Saving in energy in cement is in a range of 41% and 11% respectively
c. Water being scarce resource it is essential to reduce use of ground water and treated municipal water. Low flow fixtures save water and also reduce utility bill. Even if cost of low flow devices is quite high as compared to the conventional devices but in long turn they pay off by saving.
d. Total water calculated in the study taking standard water consumption 180 litres per capita per day which has been reduced to 108 litres per capita per day by using the low flow device.
e. Similarly sanitary water requirement may be reduced 80-90% by deploying decentralised Wastewater Treatment System and landscaping water requirement may be cut down by almost 70-80% by the rain water harvesting

Acknowledgement

Authors are thankful to Department of Science and Technology, New Delhi, India for encouraging and extending the facilities for the on-going research project (DST SERC FAST TRACK ROJECT SR/FTP/ETA-067/2009)

References

[1] B. V. V. Reddy, and K. S. Jagadish, "Embodied energy of common and alternative building materials and technologies," Journal of Energy and Buildings, vol. 35, pp.129-137, 2003.

[2] G. Patle, V. A. Dakwale, and R. V. Ralegaonkar, "Design of Green Building: A Case Study for Composite Climate," International Journal of Engineering Research and Applications, vol. 1(2), pp.388-393, 2011.
[3] P. Hernandez, K. Burke, and O. Lewis, "Development of energy performance benchmarks and building energy ratings for non-domestic buildings: An example for Irish primary schools," Journal of Energy and Buildings, vol.40, pp. 249-254, 2008.
[4] G. Q. Chen, H. Chen, Z. M. Chen, B. Zhang, L. Shao, and S. Guo, "Low carbon building assessment and multi-scale input–output analysis," Communications in Nonlinear Science and Numerical Simulation, vol. 19, pp. 583–595, 2011.
[5] K. M. Fowler, and E. M. Rauch, "Summary report on sustainable building rating systems," Pacific Northwest National Laboratory, U.S. Department of Energy, 2006.
[6] R. Zmeureanu, "Development of an energy rating system for existing houses." Journal of Energy and Buildings, vol. 42, pp. 1238–1247, 2010.
[7] V. A. Dakwale, R. V. Ralegaonkar, and S. Mandavgane, "Improving environmental Performance of building through increased energy efficiency: a review," International Journal of sustainable cities and society, vol. 1, pp. 211-218, 2011.
[8] P. Mazzarachio, A. Raggi, and G. barbiroli, "New Method for assessing global quality of energy system," Journal of applied Energy, vol. 53, pp. 315-324, 1996.
[9] S. Kumar, R. Taylor, and D. Probert, "Reassessing a proposed low-energy strategy for the UK," Journal of Applied Energy, vol. 62, pp. 213-239, 1999.
[10] W. L. Lee, C. K. Chau, and F.W.H. Yik, "On the study of the credit-weighting scale in a building environmental assessment scheme," Journal of building and Environment, vol. 37, pp. 1385 – 1396, 2002.
[11] M. S. Su, J. Deng, and Zhao C., "Interaction of Renewable Energy Policy and CO_2 Emission Control Policy: Case Study," Journal of energy engineering, vol. 63, pp. 63-70, 2008.
[12] S. Kadoshin, T. Nishiyama, and T. Ito, "The trend in current and near future energy consumption from a statistical perspective," Journal of Applied Energy, vol. 67, pp. 407-417, 2000.
[13] O. Osama, and R. Kouhy, "Future environmental regulation issues to remote energy efficiency", Journal of Energy engineering," vol. 132 (2), pp. 67-73, 2006.
[14] W. Zongxin, and W. Zhihong, "Mitigation assessment results and priorities for China's energy sector," Journal of Applied Energy, vol. 56, pp. 237-251, 1997.
[15] V. Nikolay, "Assessment of GHG mitigation measures in Ukraine," Journal of Applied Energy, vol. 56, pp. 367-380, 1997.
[16] H. Lund H, "Implementation of energy-conservation policies: The case of electric heating conversion in Denmark," Journal of Applied Energy, vol. 64, pp. 117-127, 1999.
[17] L. Wang, J. G. william, and P. Jones, "Case study of zero energy house design in UK," Journal of Energy and Buildings, vol. 41, pp. 1215–1222, 2009.
[18] D. Castro, J. A. Sefair, and F. R. Laura, "Optimization model for the selection of materials using a LEED-based green building rating system in Colombia," Journal of building and Environment, vol. 44, pp. 1162-1170, 2009.
[19] O. Santin, L. Itard, and H. Visscher, "The effect of occupancy and building characteristics on energy use for space and water heating in Dutch residential stock," Journal of Energy and Buildings, vol. 41, pp.1223-1232, 2009.
[20] T.Y. Chen, J. Burnett, and C. K Chau, "Analysis of embodied energy use in the residential building of Hong Kong," Energy, vol. 26, pp. 323–340, 2001.
[21] M. K. Dixit, J. L. Fernandez, and S. Lavy, "Identification of parameter for embodied energy measurement: a literature review," Journal of Energy and Buildings vol. 42, pp. 1238–1247, 2010.
[22] D. Mandloi, D. Khare, and T. Pareek, "Rain water harvesting in Indore City: A demanding need for sustainable development," Journal of Chemical, Biological and Physical Sciences, vol. 1, pp. 88-101, 2011.
[23] D. W. Nazer, M. A. Siebel, P. Zaag, Z. Mimi, and H. Gijze, "A Financial, Environmental and Social Evaluation of Domestic Water Management Options in the West Bank, Palestine," Water Resource Management, vol. 24, pp. 4445–4467, 2010.
[24] Y. K. Juan, and P. Gao, "A hybrid decision support system for sustainable office Building renovation and Energy performance improvement," Journal of Energy and Buildings, vol. 42, pp. 290–297, 2010.
[25] V. A. Dakwale, and R. V. Ralegaonkar, "Review of carbon emission through buildings: threats, causes and solution," International Journal of Low-Carbon Technologies, vol. 7, 143–148, 2012.

4

VULNERABILITY OF MOTIVATION SCHEMES IN ENHANCING SITE WORKERS PRODUCTIVITY FOR CONSTRUCTION INDUSTRY'S SUSTAINABILITY IN NIGERIA

Olufisayo Adewumi Adedokun*, Olajide Timothy Ibironke and Ayokunle Olubunmi Olanipekun

Department of Quantity Surveying, Federal University of Technology, Akure, Nigeria

*Corresponding E-mail: fisayoadedokun@gmail.com; oaadedokun@futa.edu.ng

Abstract

The construction industry provides appreciable employment for the unskilled and semi-skilled labour considering the nature of most activities carried out on construction sites. However, the productivity of this categories of labour have trended poor overtime. This study therefore assessed the financial and non financial motivational schemes; in light of their preference to construction site workers so as to assure on their vulnerability in enhancing productivity. The data collected through the administration of questionnaires to construction site operatives were analysed using the relative importance index (RII). The research instrument was also legitimized with content validity index (CVI) for items ≥ 0.80 and cronbach alpha value, as a measure of internal consistency, ≥ 0.70. Allowances, wages stock option and bonuses are the financial motivators while delegation, independence and autonomy, facilities for transportation are the non financial motivations that are important to site workers on construction sites. Giving praises, encouragement to make contributions, free medical care, availability of relaxation centres and first aid provision are the motivation schemes that site operatives believed to enhance their productivity. This paper concludes that both financial and non financial motivational schemes are important to enhancing the productivity of operatives on construction sites. The vulnerable motivational schemes to the enhancement of productivity, as revealed in this study is recommended to the construction and subcontracting firms, foremen, clerk of works, site supervisors, site engineers and all other stakeholders in ensuring a maximum output for a unit input from site operatives.

Keywords: *Construction Industry, Motivation, Nigeria, Productivity, Site Operative*

1.0 Introduction

Ashiru [1] opines that construction industry is responsible for the provision of capital projects which comprises the building and civil engineering sub-sector. The construction industry stands out as the most conspicuous contributor to capital formation in Nigeria [2]. In the pre-independence decade it accounted for about 40% of the total capital formation and in the immediate post independence decade its contribution was a little above 50% [3]. It is therefore no exaggeration to assert that the development of the construction industry is the base for all industrial activities in the country [1]. Government, according to [4] being the largest client, still invests in capital projects as a way of revamping the depressed economy. The largest share of investment capital in all the developing countries finds its way into construction. Unfortunately, the Nigerian construction industry is least developed in terms of mechanization, productivity and the benaefits derived from industrial progress. The construction industry in Nigeria is labour intensive; and it is the highest employer of the nation's work force. The industry accounts for over 50% of the country's gross fixed capital formation. Unfortunately most empirical studies have revealed that output of the industry is quite low when compared with many developed countries [5]. Workers' productivity on construction sites has been shown to be very poor and this has been the trend for a long time [6, 7].

2.0 Literature Review

Ashiru [1] opines that construction projects are more complex than ever before, there are new constructional methods, materials, components [4] and different procurement options [8] thereby compounding the risky and complex nature of the industry [9]. Despite all these, thousands of tasks have to be performed precisely if a project is to run smoothly, on time and within the budget allocated [10]. The completion of a construction project requires the judicious scheduling and allocation of available resources. Manpower, equipment and materials are the most important project resources that require close management attention. Labour (manpower) is usually the resources most affected by changes in project structure such as extension of time or compression of time frame which may require the offering of incentives for overtime [11].

2.1 Motivation schemes in construction industry

Construction industry according to [12] possesses a strong and particular structure all over the world as compared with manufacturing industry. Variations in the construction labour productivity can naturally make a great impact on national economy and productivity, since the sector also augments production capacity of its dependent sub-sectors that are many [13].

However, construction workforce especially in developing countries is not seen as an important input, although project labour generally make up the most variable and the largest percentage of total project costs [14]. Due to this high cost of labour, labour cost control is a very important function for profitability in the industry. Despite the fact that these costs change with the type of projects in a broad range, site workers according to [15] typically account for up to 40% of the direct capital cost in large construction projects. The labour cost components of electrical and mechanical works according to [15] also represent 40% -60% of their final cost.

The present situation of the industry makes productivity improvement imperative to the construction sector. Industry is better characterized by traditional or manual methods and hard labour condition, besides a low productivity. Labour intensive technology, on the other hand, requires much more diligence and insight than equipment. Intensive construction processes, and thus labour resource becomes a more important input in the production phase. Moreover, production-related inactivity obvious in a labour-intensive project and the solutions of the inactivity are not easily found, it means that it is easy for a contractor to be lulled into a false sense of accomplishment, when the crews are all working. Residential and commercial construction is labour-intensive as compared to industrial and heavy construction tending to be capital-intensive. In the same way, renovation or retrofitting work, by its own nature, according to [16] requires a higher labour capital ratio than the new construction.

2.2 Motivation as a tool for improving construction labour productivity

Motivation according to [17] has been defined as the characteristic of an individual willing to expend effort towards a particular set of behaviours. It imparts incentives that require a response on part of someone else to achieve a defined goal. In business, motivation is not synonymous with salaries; money is a means for accommodating the economic needs of workers [18]. Motivation according to [19] means an inner wholesome desire to exert effort without the external stimulus of money. Also, Motivating according to [20] is the ability of indoctrinating the personnel with a unity of purpose and maintaining a continuing, harmonious relationship among all people. It is a force which encourages and promotes a willingness of every employee to cooperate with every member of the team [21]. To maintain it is to create and perpetuate the climate which brings harmony and equilibrium into the entire work group for the benefit of all who are involved [21].

Motivation according to [22] is the work managers perform to inspire, encourage and impel people to take action and its efforts must be directed towards improving company operations [23].

To be effective, however, they must also be designed to show benefits to the employee. In fact, motivation can best be accomplished when workers are able to merge their personal ambitions with those of the company. Motivation is defined by [24] as a driver to organization's performance, conditioned by the effort's ability to satisfy some individual need" Motivation as a set of independent and dependant relationships that explains the direction, amplitude and persistence of an individual's behaviour holding constant the effects of aptitude, skills, understanding of a task and the constraints operating in the work environment [21].

The relationship between the employer and employee must be one of understanding in order for the employee to identify himself with his work and with the business he is working for. Lack of motivation in return affects productivity [25]. A number of symptoms according to [26] may point to low morale: declining productivity; high employee turnover; increasing number of grievances; higher incidence of absenteeism and tardiness; increasing number of defective products; higher number of accidents or a higher level of waste materials and scrap. A motivated employee according to [27] is a loyal employee and to be loyal implies that the employee supports the actions and objectives of the firm. The appearance of the job as a whole has, in fact a bearing on the willingness and quality of an employee's performance.

Individuals tend to develop certain motivational drives on the cultural environment in which they live and these drives affect the way people view their jobs. Marti, Gil and Barrasa [28] asserts that achievement, affiliation, competence and power are four types of motivational drives that are found in individuals that are self-motivated and this may be the case for many construction workers. Motivation plays a part in enhancing construction labour productivity and forms the basis for identification of the work environment factors. Prawitz, Garman, Sorhaindo, O'Neill, Kim and Drentea [29] advocated the use of financial incentive programmes to improve construction labour productivity, reinforcing [30] theory of driving forces that led to productivity improvements. Autonomy and comradeship are also found to be important aspects that add to the way construction workers are self-motivated about their work. However, much work in linking motivation and productivity relied on Hertzberg's sample involving mainly white-collar professionals [31]. Bowen, Cattell, Distiller and Edwards [32] discovered that productive job creates high job satisfaction while non-productive job (one which fall behind schedule) produce dissatisfaction at all levels of the management/worker chain. The relationship is believed to be due to the very nature of construction, thus different from the one found in an office or factory setting which states that high job satisfaction leads to greater productivity. In construction, a worker, through his own efforts produces a highly visible, physical structure in which great satisfaction comes from completion.

2.3 Incentives

In order to keep construction workers motivated their needs must be addressed as project goals are reached. Satisfying workers' needs according to [33] can be viewed as distributing incentives when certain objectives are achieved. Employees have needs that they want met and employers have goals that they reach and they can work together as a team to satisfy the wants of both the employees and their employers. Workers who are motivated to help reach the goal of the employer and do so should be recognized with an incentive/reward [34]. When considering what type of incentives to use there are two types to be aware of, extrinsic and intrinsic. Extrinsic rewards according to [22, 35] are external rewards that occur apart from work, such as money and other material things. On the other hand, intrinsic rewards according to [36, 37] are internal rewards that a person feels when performing a job, so that there is a direct and immediate connection between work and reward.

The power of incentives is immense and pervasive, which is all the more reason they require careful management [38]. Goldman, Thompson and Daily [39] have summarized a list of these advantages and disadvantages associated with financial incentives. Many construction companies have already considered that there can be advantages and disadvantages of developing an incentive program. A study by [40] showed that those companies that keep their program simple with the main objective of the program in mind (to benefit the project in reference to cost, schedule, customer service, environment and quality) are also deemed success of any incentive program.

Incentives are usually defined as tangible rewards that are given to those who perform at a given level. Such rewards may be available to workers, supervisors, or top managers. Whether the incentive is linked directly to such items as safety, quality, the reward follows successful performance [41]. Many companies feel that pocket money is a good motivator. Others contend that small rewards such as toasters and blenders do not motivate. Many companies therefore offer profit sharing plans; or companies have abandoned monetary rewards and instead offer lavish trips to such places as Europe and some Caribbean islands.

Because of the expense, these programs require careful monitoring. Some companies merely reward good producers with an extra day off with pay. Other concerns reward top performers with better working conditions. Since incentive programs aim to increase workers' performance levels, the measure used to decide if a reward has been earned should be carefully set [42]. The performance level must be attainable or workers won't try to reach the goal. That fact underscores the usefulness of having workers themselves contribute their ideas about what constitutes a reasonable level of performance. An incentive scheme may also fail if the measure of success ignores quality or safety [43]. An obvious problem exists when an incentive is applied to work that is machine paced. Incentives according to [44] should be clearly linked to performance, but not all incentives can be clearly tied to objective criteria. Some incentive rewards are issued on the basis of a subjective assessment by a superior on the merit of particular workers. Moreover, an incentive scheme is set up to motivate employees to perform better.

3.0 Methodology

This study adopted the use of data which were collected through administration of questionnaires to the site operatives. Two operatives were randomly selected from five trades in 10 different busy sites of Lagos state. A total number of 75 structured Questionnaires were distributed, 49 received, were valid for analysis thus representing 65% of population. The responses to questions were presented in tables and analyzed using the percentile method. Responses to multiple choice type questions that required weighting were analyzed using the relative index for ranking and the values ranked in descending order.

3.1 Research rigour

Researches can be classified into positivist or the interpretative paradigm or philosophy and if a research reflects the principles of positivism, then it has taken the stance that there is a reality out there waiting to be discovered and this reality could be measured and assessed objectively [45]. This research has adopted the positivism paradigm by assessing and measuring the vulnerability of motivation schemes in enhancing sustainable productivity in the Nigerian construction industry. Rigour; which ensures that scientific results stems from standard and accurate means in research [46] needs to be established [47]. Tobin and Begley [48] relate rigour to legitimising research process; where validity and reliability are the tests of rigour in positivist (quantitative) research [46]. Wong, Ong and Kuek [45] added a third criterion (sensitivity) to validity and reliability as means of ensuring good measuring instrument. Sensitivity is the ability of research instrument to capture the variability in responses and 5-point likert scale was advocated because it allows for optimum choice of response [45]; and thus adopted in this research work.

Validity concerns with the extent to which a measure reflects the social phenomena being observed [49]. Face and content validity are used to evaluate the degree to which elements of the research instrument are relevant to and representative of the targeted construct in this research.

Face validity for the research instrument used in this research was carried out according to [50]. Sushil and Verma [50] state that face validity is assessed by having expert researchers to review the contents of the test to see if the items seem appropriate, therefore, two randomly selected experts each in the field of Human Resource Management and Construction Management carried out the face validity test of the questionnaire and; was adjudged face valid. Content validity index (CVI); which quantifies the extent of agreement between experts [51] was used in evaluating the content validity of the questionnaire for this research and was carried out in accordance to [52]. Parsian, and Dunning [52] require that experts in the area of study be asked to rate the relevance of the items in the questionnaire draft to the study concept using Likert scale. This was carried out on a 3-point Likert scale (1=not relevant, 2=somewhat relevant, 3=relevant) by four experts in the area of study and the mean aggregate CVIs for all constructs presented on Table 1 as follows:

Table 1: Content Validity Index (CVI) test

Contents	**Mean Content Validity Index (CVI)**
Financial motivational schemes	0.81
Non-financial motivational scheme	0.89
Vulnerability of motivational schemes	0.80

For financial motivation schemes, the four experts agreed on the 'allowance' while three out of four agreed on wages, stock option and bonuses. Where there's a total agreement, the CVI is 1 (4/4) and where one of the experts differ, the CVI is 0.75 (3/4). The mean aggregate of the experts rating is 0.81. The item 'option' in the financial motivation schemes was not judged by the experts. In similar manner, experts had total agreement in the rating of items delegation, independence and autonomy, flexible hour and opportunities to contribute for non-financial motivation schemes. Three out of the experts agreed on the other items in the rating of the non-financial motivation schemes and the mean aggregate of experts rating is 0.89. The items representing the vulnerability of the motivation schemes in enhancing productivity have a mean aggregate expert rating of 0.80. Polit, and Beck [53] state that a CVI of .80 or better indicates good content validity, therefore, the questionnaire used in this research is content valid.

In further validating the questionnaire, the test of internal consistency was carried out using Cronbach's alpha. Cronbach's alpha is often considered a measure of item homogeneity where large alpha values indicate that the items are tapping a common domain [54]. The internal consistency of the measured attributes in this study as perceived among the respondents within the Likert scale (0 to 5) was explained by the reliability coefficient that is based on the average correlation among the attributes and the total number of attributes in the sample. Using the statistical Package for Social Sciences (SPSS), the Cronbach's alpha (α) was computed and presented on Table 2 below:

Table 2: Test of Reliability

Scale of measure	**α-value**
Financial motivational schemes	0.73
Non-financial motivational scheme	0.79
Vulnerability of motivational schemes	0.70

In order to examine the internal consistency of the scale, the Cronbach's alpha (α) test was performed on the attributes under each factor and on all the attributes. Sushil, and N. Verma [50] reiterate that if a test has a strong internal consistency, it should show only moderate correlation among items (.70 to 0.90). Too low values means unreliability and too high values revealing some items are redundant and should be removed from the test. Hence, for this study value of α as 0.7 was used as cut-off value as adopted by [55]. The Table 2 above presents Cronbach's alpha (α) values which are 0.73, 0.79 and 0.70 for financial motivational schemes, non-financial motivational schemes and the vulnerability of motivational schemes respectively. The Cronbach's alpha (α) values indicate that the research instrument is reliable.

4.0 Results and Discussions

This section presents the analysis results of the collected data. The findings of the analysis are presented in tables and discussed under the various headings captured below.

Table 3: Years of Experience of Respondents (Site Operatives)

Years of Experience	Frequency	Percentage
Less than 5yrs	2	4
5-10yrs	4	8
11-15yrs	14	29
16-20yrs	8	16
Above 20 years	22	43
Total	**49**	**100**

From the Table 3 shown above, respondents with more than 20 years of experience representing 43% responded to the questionnaire while those representing 29% and 16% have range of years of experience between 11- 20 years and thus meaning that reasonably high percentage of respondents have requisite experience in their respective companies and are qualified to respond to the questions.

Table 4: Importance of financial motivation schemes to construction site workers

Schemes	R. I.	Rank
Allowances	0.776	1
Wages	0.696	2
Stock option	0.68	3
Bonuses	0.676	4
Others	0.608	5

Tables 4 reveals that allowances under financial scheme ranked 1st with relative index of 0.776, this is in line with the statement of [56] that such monetary rewards may psychologically appease the employees when they are asked to work longer hours and finally [57] that giving such allowances create in employees a feeling that they are being cared by their firms and thus, may be willing to work harder.

Table 5: Importance of non-financial motivation schemes to construction site workers

Schemes	R. I.	Rank
Delegation	0.704	1
Independence and autonomy	0.704	1
Facilities for transportation	0.688	3
Others	0.660	4
Training and development	0.632	5
Flexible hour	0.608	6
Opportunity to contribute	0.556	7
Recreational programmes	0.584	8

Tables 5 reveals that independence and autonomy and delegation under non financial scheme also ranked 1st on the list and this as well is in consonance with [58] that from industrial sociology, we know that blue-collar workers put emphasis on independence and autonomy and by expressing independence and autonomy in various ways, workers differentiate themselves from one another; they become individualised and individuated as e.g. skilful, trustworthy or productive. The least as shown on the table is recreational programmes that ranked 8th under non financial incentive schemes with relative index of 0.584.

Table 6: Vulnerability of motivation schemes in enhancing productivity of site workers

Schemes	R.I	Rank
Giving of praises for work or responsibility well discharged	0.872	1
Encouragement to make contributions	0.728	2
Provision of free medical care for the operatives	0.700	3
Availability of relaxation centre for operatives to relax after work	0.700	3
Provision of first aid as a measure of safety	0.696	5
Organizing end of the year party for operatives to interact with top officers	0.696	5
Job security	0.684	7
Provision of protective devices	0.684	7
Avoidances of delay after day's work	0.676	9
Giving of awards for meritorious services	0.636	10
Provision of working implements and tools	0.628	11
Availability of transport facilities	0.620	12
Provision of insurance scheme against unforeseen circumstances	0.600	13
Availability of training opportunities and development of employees	0.600	13

Table 6 shows that giving of praises for work or responsibility well discharged contributes immensely towards increasing productivity while availability of training opportunities and provision of insurance schemes against unforeseen circumstances with relative index of 0.600, have least effect in boosting productivity on construction sites.

Various motivation schemes are highlighted and allowances was ranked to be the most prominent in all the sites visited as part of the financial motivation schemes while the most predominant under non financial schemes was to allow employees to be independent and giving them the autonomy to perform, as in table 4. Motivation schemes that is most vulnerable to enhancing productivity in the industry and of all listed, the most susceptible is giving of praises to work or responsibility well discharged which is in table 6.

5.0 Conclusions

With reference to data collected from the field and duly analysed, it can be concluded that productivity in construction industry can largely be enhanced by paying adequate attention to the motivation schemes available and that giving of praises for work or responsibility well discharged can boost the morale of operatives on site. It should also be noted that combination of two or more of the motivation schemes such as encouragement to make contributions and availability of relaxation centre for operatives to relax after work can as well yield same result of maximising the output of the workers such that high degree of productivity can be attained.

From the foregoing, it can be recommended that the construction and subcontracting firms, foremen, clerk of works, site supervisors, site engineers and all other stakeholders should motivate their workers by giving them praises for work or responsibility well discharged and also, allowing them to make contributions, provisions of free medical care and making available relaxation centres after work, make them work to their maximum capacity (maximum output) for a unit input from site operatives towards the construction industry's sustainability in Nigeria.

References

[1] S. A. Ashiru, "Impact of motivation on productivity in Nigeria construction industry", An unpublished Postgraduate Diploma dissertation submitted to Department of Quantity Surveying, Federal University of Technology, Akure, Nigeria, 2011.

[2] O. Aboyade, "Study of investment and growth in Nigeria", Preager Publishers, New York, 1996.

[3] J. U. Umo, "Capital Formation: Structure of the Nigeria Economy", Macmillan and University of Lagos Press, Nigeria, 1984.

[4] I. H. Seeley, "Building Economics: appraisal and control of building design cost and efficiency", Macmillan press limited, London, 2nd Edition, 1984.

[5] O. Ortiz, F. Castells, and G. Sonnemann, "Sustainability in the construction industry: A review of recent developments based on LCA", Construction and Building Materials, vol. 23, pp. 28-39, 2009.

[6] O. Olubodun, "Framework for Construction Costs Minimization in Southern Nigeria", Unpublished M.Sc. Thesis, Department of Building, Obafemi Awolowo University, Ile-Ife, Nigeria, pp. 15-20, 1986.

[7] J. O. Ayandele, "Evaluation of factors that affect labour productivity of some selected building trades on Nigeria construction sites". An unpublished Ph.D Thesis submitted to Department of Building, University of Lagos, Nigeria, 1996.

[8] H. O. Adebiyi, "An appraisal of financial risks in Build Operate-Transfer projects in Lagos, Nigeria", An unpublished Postgraduate Diploma dissertation submitted to Department of Quantity Surveying, Federal University of Technology, Akure, Nigeria, 2009.

[9] O. A. Adedokun, "Evaluation of Risk Analysis Techniques in Selected Large Construction Companies in Nigeria", Published M.Tech Thesis submitted to School of Postgraduate Studies, Federal University of Technology, Akure, Nigeria, 2012.

[10] A. H. Memon, and R. M. Zin, "Resource-Driven Scheduling Implementation in Malaysian Construction Industry", International Journal of Sustainable Construction Engineering & Technology, vol. 1 (2), December, pp. 2010.

[11] P. Janes, and M. Wisnom, "Changes in Tourism Industry Quality of Work Life Practices", Journal of Tourism Insights, vol. 1 (1), pp. 106-113, 2010.

[12] A. Dubois, and L, Gadde, "Supply strategy and network effects – purchasing behaviour in the construction industry", European Journal of Purchasing & Supply Management, vol. 6, pp. 207- 215, 2000.

[13] K. Manley, "Innovate now; Improving performance in the building and construction industry", CRC for Construction Innovation, Brisbane, 2006.

[14] E. L. Glaeser, J. Gyourko, and R. E. Saks, "Why Have Housing Prices Gone Up"? Harvard Institute of Economic Research Discussion Paper Number 2061, Retrieved on January 2, from http://post.economics.harvard.edu/hier/2005list.html, 2005.

[15] W. Ibbs, L. D. Nguyen, and S. Lee, "Quantified Impacts of Project Change", Journal of professional issues in engineering education and practice @ ASCE, vol. 133 (1), pp. 45-52, 2007.

[16] A. Kazaz, E. Manisali, and S. Ulubeyli, "Effect of basic motivational factors on construction workforce productivity in Turkey", Journal of Civil Engineering and Management, vol. 14 (2), pp. 95-106, 2008.

[17] A. A. Tabassi, and A. H. Abu Bakar, "Training, motivation, and performance: The case of human resource management in construction projects in Mashhad, Iran", International Journal of Project Management, vol. 27, pp. 471-480, 2009.

[18] M. M. Alajloni, Z. M. S. Almashaqba, and M. A. N. Al-Qeed, "The Classical Theory of Organisation and its Relevance", International Research Journal of Finance and Economics, vol. 41, pp. 60-67, 2010.

[19] O. T. Ibironke, O. A. Adedokun, and A. A. Hungbo, "Drivers and Challenges of Motivation for Casual Workers on Construction Sites", Journal of Emerging Trends in Economics and Management Sciences, vol. 2 (5), pp. 413-417, 2011.

[20] K. Hallahan, "Enhancing Motivation, Ability, and Opportunity to Process Public Relations Messages", Public Relations Review, vol. 26 (4), pp. 463-480, 2000.

[21] W. D. Thwala, and L. N. Monese, "Motivation as a tool to improve productivity on the construction site", Paper presented at 5th Post Graduate Conference on Construction Industry Development with a theme: "construction as a cornerstone for economic growth and development" on 16-18 March at Bloemfontein, South Africa, pp. 139-145, 2008.

[22] A. M. M. Liu, and Z. Fang, "A power-based leadership approach to project management", Construction Management and Economics, vol. 24, pp. 497-507, 2006.

[23] D. K. Ahadzie, D. G. Proverbs, and P. Olomolaiye, "Towards developing competency-based measures for construction project managers": Should contextual behaviours be distinguished from task behaviours? International Journal of Project Management, vol. 26, pp. 631-645, 2008.
[24] R. Dwivedula, and C. N. Bredillet, "Profiling work motivation of project workers", International Journal of Project Management, vol. 28, pp. 158-165, 2010.
[25] O. A. Shadare, and T. A. Hammed, "Influence of work motivation, leadership effectiveness and time management on employees' performance in some selected industries in Ibadan, Oyo State, Nigeria", European Journal of Economics, Finance and Administrative Sciences, vol. 16, pp. 7-17, 2009.
[26] K. Appleton, A. House, and A. Dowell, "A survey of job satisfaction, sources of stress and psychological symptoms among general practitioners in Leeds", British Journal of General Practice, vol. 48, pp. 1059-1063, 1998.
[27] J. D. Cesare, and G. Sadri, "Do all carrots look the same? Examining the impact of culture on employee motivation", Management Research News, vol. 26 (1), pp. 29-40, 2003.
[28] M. Marti, F. Gil, and A. Barrasa, "Organizational leadership: Motives and Behaviours of leaders in current organizations", The Spanish Journal of Psychology, vol. 12 (1), pp. 267-274, 2009.
[29] A. D. Prawitz, E. T. Garman, B. Sorhaindo, B. O'Neill, J. Kim, and P. Drentea, "Incharge Finance Distress/Financial Well-Being Scale: Development, Administration, and Score Interpretation", Financial Counseling and Planning, vol. 17 (1), pp. 34-50, 2006.
[30] S. Tangen, "Professional Practice Demystifying productivity and performance", International Journal of Productivity and Performance Management, vol. 54 (1), pp. 34-46, 2005.
[31] N. Bassett-Jones, and G. C. Lloyd, "Does Herzberg's motivation theory have staying power"? Journal of Management Development, vol. 24 (10), pp. 929-943, 2005.
[32] P. Bowen, K. Cattell, G. Distiller, and P. J. Edwards, "Job satisfaction of South African quantity surveyors: an empirical study", Construction Management and Economics, vol 26, pp. 765-780, 2008.
[33] Y. A. Dauda, and W. A. Akingbade, "Employee incentive management and financial participation in the Nigerian banking industry: Problems and prospects", European Journal of Economics, Finance and Administrative Sciences, vol. 24, pp. 100-110, 2010.
[34] T. R. Tudor, "Motivating employees with limited pay incentives using equity theory and the fast food industry as a model", International Journal of Business and Social Science, vol. 2 (23), pp. 95-101, 2011.
[35] P. Xiang, A. Chen, and A. Bruene, "Interactive impact of intrinsic motivators and extrinsic rewards on behaviour and motivation outcomes", Journal of teaching in Physical Education, vol. 24, pp. 179-197, 2005.
[36] R. C. Mahaney, and A. L. Lederer, "The effect of intrinsic and extrinsic rewards for developers on information systems project success", Project Management Journal, vol. 37 (4), pp. 42-54, 2006.
[37] J. Wang, and H. Yuan, "Factors affecting contractors' risk attitudes in construction projects: Case study from China", International Journal of Project Management, vol 29, pp. 209-219, 2011.
[38] A. Dixit, "Incentives and organizations in the public sector: An interpretative review", The Journal of Human Resources, vol. 37 (4), pp. 696-727, 2002.
[39] R. L. Goldman, B. H. Thompson, and G. C. Daily, "Institutional incentives for managing the landscape: Inducing cooperation for the production of ecosystem services", Ecological Economics, pages: 1-11, 2007.
[40] R. F. Cox, R. R. A. Issa, and A. Frey, "Proposed Subcontractor-Based Employee Motivational Model", Journal of Construction Engineering and Management @ ASCE, vol. 132 (2), pp. 152-163, 2006.
[41] K. R. Molenaar, J. Park, and S. Washington, "Framework for measuring corporate safety culture and its impact on construction safety performance", Journal of Construction Engineering and Management @ ASCE, vol. 135 (6), pp. 488-496, 2009.
[42] S. H. Han, S. H. Park, E. J. Jin, Y. K. Kim, and Y. K. Seong, "Critical issues and possible solutions for motivating foreign construction workers", Journal of Management in Engineering @ ASCE, vol. 24 (4), pp. 217-226, 2008.
[43] G. Manso, "Motivating Innovation", The Journal of Finance, vol. 66 (5), pp. 1823-1860, 2011.
[44] V. Agarwal, N. D. Daniel, and N. Y. Naik, "Role of managerial incentives and discretion in hedge fund performance", The Journal of Finance, LXIV (5), pp. 2221-2256, 2009.
[45] K. Wong, S. Ong, and T. Kuek, "Constructing a Survey Questionnaire to Collect Data on Service Quality of Business Academics", European Journal of Social Sciences, vol. 29 (2), pp. 209 – 221, 2012.
[46] S. L. T. McGregor, and J. A. Murnane, "Paradigm, methodology and method: Intellectual integrity in consumer scholarship", International Journal of Consumer Studies,vol. 34 (4), pp. 419 – 427, 2010.
[47] P. Darke, S. Graeme, and M. Broadbent, "Successfully completing case study research: combining rigour, relevance and pragmatism", Information System Journal, vol. 8, pp. 273-289, 1998.

[48] G. A. Tobin, and C. M. Begley, "Methodological rigour within a workforce framework", Journal of Advanced Nursing, vol. 48 (4), pp. 388 – 396, 2004.

[49] D. Wahyuni, "The Research Design Maze: Understanding Paradigms, Cases, Methods and Methodologies", JAMAR, vol. 10 (1), pp. 69 – 80, 2012.

[50] S. Sushil, and N. Verma, "Questionnaire Validation Made Easy", European Journal of Scientific Research, vol. 46 (2), pp. 172 – 178, 2010.

[51] C. Malmgreen, P. Graham, L. M. Shortridge-Baggett, M. D. Courtney, and A. M. Walsh, "Establishing content validity of a survey research instrument : the older patients in acute care survey - United States", Journal for Nurses in Staff Development, vol. 25 (6), pp. E14-E18, 2009.

[52] N. Parsian, and T. Dunning, "Developing and Validating a Questionnaire to Measure Spirituality: A Psychometric Process", Global Journal of Health Science, vol. 1 (1), pp. 2 – 11, 2009.

[53] D. F. Polit, and C. T. Beck, "Nursing Research: Generating and Assessing for Nursing Practice", (8th ed.). Philadelphia: Lippincott, Williams & Wilkins, 2008.

[54] C. S. Wells, and J. A. Pollack, "An Instructor's Guide to Understanding Test Reliability", Testing and Evaluation Services, University of Wisconsin, USA, 2003.

[55] I. A. Rahman, A. H. Memon, and A. T. A. Karim, "Relationship between Factors of Construction Resources Affecting Project Cost", Modern Applied Science; vol. 7 (1), pp. 67-75, 2013

[56] S. Werner, and S. G. Ward, "Recent Compensation Research: An Electric Review", Human Resource Management Review, vol. 14 (2), pp. 201 – 227, 2004.

[57] J. W. L. Lewis, and Y. Y. L. Florence, "Human resource practices of Contractors that lead to job Satisfaction of professional Staff", ECAM, vol. 19 (1), pp. 101 – 118, 2012.

[58] B. Charlotte, "Collective individualism: the informal and emergent dynamics of practising safety in a high-risk work environment", Construction Management and Economics, vol. 27, pp. 949 – 957, 2009.

5

QUANTITATIVE ASSESSMENT OF COST AND TIME IMPLICATION OF SUSCEPTIBILITY OF BUILDING ELEMENTS TO VARIATION IN NIGERIA

Solomon Olusola Babatunde

Department of Quantity Surveying, Obafemi Awolowo University, Ile-Ife, Nigeria

*Corresponding E-mail : Sholly_intl@yahoo.com

Abstract

A number of research studies have been carried out on the causes and effects of variation on construction project delivery, thereby taking for granted the susceptibility of building elements to variations. This formed the basis of this paper with a view to assessing the cost and time implications of the susceptibility of each building element to variation during construction process in Nigeria. Archival record comprises contract drawings, original bill of quantities, addendum and reduction bill of quantities, and minutes of site meetings among others were used to extract data relating to initial cost, final construction cost, estimated period, final completion period of each building element attributed to variations. The data obtained were analyzed using statistical methods of average, percentage, regression analysis, and analysis of variance (ANOVA). The study identified the building elements having greater than 20% of cost overrun due to variation as earthwork and fillings, frame, windows and external doors, fittings and furnishings, water installation, and external services. The study further identified the building elements having greater than 25% time overrun due to variation as earthwork and fillings, block work (at substructure), upper floors, external walls, wall finishing among others. The results of ANOVA and regression analysis on the building elements cost and time were used to establish models. Thus, the established models are: AFC= 981690 + 1.033AIC; and AFCO= 608390.865 +1.310AIC to predict the average final cost of each building element, and the average final cost overrun of each building element due to variation respectively, where AFC= Average Final Cost, AIC= Average Initial Cost, and AFCO= Average Final Cost Overrun. Also, the study established the model: Y = 1.379(X) – 0.251 for predicting the average actual completion period of each building element, where Y= Average Actual Completion Period, and X= Average Estimated Period (week). The study provides information that enables the clients, consultants/professionals, and other construction stakeholders to understand the building elements that demands specific attention in terms of cost and time implications of variations. The study also provides building clients a foreknowledge of cost variance that is likely to occur on each building element during construction process.

Keywords: *Building elements, construction, cost overrun, time overrun, Nigeria*

1.0 Introduction

It is a fact that very few of building construction project executed is completed within the project sponsors' originally estimated budget and time frame. However, the problem of cost and time overrun in the construction industry is a worldwide phenomenon especially in large transport infrastructure projects are worrisome [1]. For instance, Halloum and Bajracharya,[2] state that 93% of the projects experiencing cost overrun, and more than 90% of the projects witnessing time overun in Abu Dhabi, UAE. In Malaysia, Memon, Rahman and Azis, [3] found that 92% of construction projects witnessing time overrun, and 89% of projects experiencing cost overrun. In such cases where final cost turned out to be several times higher than original estimate, the situation is unsuitable and always affects the sponsors' original plan in terms of cost, time and quality [4]. This affects construction process and also frustrates the client's requirement. Great concern has been expressed in recent years regarding the adverse impact of variations to the construction projects. Construction variations are identified as a major source of conflicts and disputes in the construction industries of many countries [5]. Building projects are liable to variations due to change of mind on the part of the clients, the consultants, or unforeseen problems raised by the main contractor or sub-contractor [6]. Mokhtar, Bedard and Fazio, [7] assert that variations are inevitable in any construction project. Needs of the client may change in the course of design or construction, market conditions may impose changes to the parameters of the project, and technological developments may alter the design and the choice of the engineer [8]. The engineer's or architect's review of the design may bring about changes to improve or optimize the design and the operations of the project. Arain, [9] states that errors and omissions in construction may force a change. Variations are common in all types of construction projects [10 - 12]. The nature and frequency of variations occurrence vary from one project to another depending on various factors [13-14]. The impact of variations varies from one construction project to another. It is generally accepted that variations affect the construction projects with unpalatable consequences in time and cost [15-17]. However, severe criticisms of the industry are generated when projects take far longer than planned [18]. These effects usually undermine the objectives of a client (sponsor) who devotes his/her scarce fund on a developmental project.

Variations in construction had been studied by a number of researchers in the light of its causes and effects on project delivery, thereby taking for granted the susceptibility of building elements to variation. For instance, Ssegawa, Mfolwe, Makuke and Kutua [19] study construction variations on building projects in Botswana; and concluded that variations are unavoidable consequence. Arain and Low [20] investigate developer's views on potential causes of variation order on institutional buildings in Singapore. The authors established that errors and omissions in design; change in specifications by client and consultants are the most significant causes of variations in institutional buildings. Arain and Low [21] study consultants' related variations in school building projects. Their study revealed that non-compliance design with government regulations; design discrepancies; and change in design by consultants are the most frequent causes of consultant related variation in school building projects. Therefore, this study becomes necessary to assess the cost and time implications of the susceptibility of each building element to variation during construction process in Nigeria. The study identified the building elements that demands specific attention in terms of cost and time implications of variations with a view of providing clients; construction professionals; and other stakeholders in the construction industry foreknowledge of cost variance that is likely to occur as result of variations on each building element during construction process. This study also provides useful information to the clients in planning for finance and helps the consultants in controlling construction costs and time.

2.0 Brief Literature Review

2.1 Cost and Time Implication of Variations on Construction Projects Delivery

The construction project delivery system has suffered considerable set back due to variations. For the fact that a project was completed without being abandoned does not mean it has been successful in its entirety. The table 1 shows few examples of cost and time overruns on construction projects by prior researchers.

Table 1: Identified percentage of cost and time overruns on construction projects

S/n	Author and Year	Focus	Country/Study area	Findings
1	Kumaraswamy, Miller and Yogeswaran [22]	Extension of time due to excusable delays	Hong Kong	The authors identified 15-20% time overrun in civil engineering projects.
2	Bordat, McCullouch, Labi and Sinha [23]	Analysis of cost overruns and time delays of INDOT projects	United States	The authors found 55% cost overrun and 12% time overrun in all Indiana Department of Transportation (INDOT) contracts.

S/n	Author and Year	Focus	Country/Study area	Findings
3	Oladapo, [24]	Quantitative assessment of the cost and time impact of variation orders on construction projects	Nigeria	The author found that variations accounted for about 79% cost overrun and 68% time overrun.
4	Priyantha, Karunasena and Rodrigo [25]	Causes, nature and effects of variations in highways	Sri Lanka	The authors identified 9.9% cost overruns of the initial contract sum
5	Halloum and Bajracharya, [2]	Cost and time overrun on Infrastructure construction projects	Abu Dhabi, UAE	The authors identified 8.7% cost overrun and 8.3% time overrun in physical infrastructure
6	Cantarelli, Flyvbjerg and Buhl [1]	Geographical variation in project cost performance	Netherlands, North West Europe, and other geographical areas	The authors identified 11% cost overrun in rail project in Netherlands, 27% cost overrun in North West European countries, and 44% cost overrun in other geographical areas. The authors further identified 7% cost overrun in bridge project in Netherlands, 45% cost overrun in other NW European countries and 27% cost overrun in other geographical areas
7	Memon Rahman and Azis [3]	Time and cost performance in construction projects	Malaysia	The authors found between 5-10% cost overrun and about 5-10% time overrun in construction projects.

It reflects from the table that cost and time overruns on construction projects are global phenomenon, and the magnitude varies from one project to another. Even, the impact also varies from location to another. However, it is widely believed that the consequence of variations undermine the objectives of client.

2.2 Building Elements and Classifications

A building element is a component of the building that fulfils specific function(s) irrespective of its design, construction or specification. Robert and Harrold [26] define building elements as components common to most buildings that usually perform a given function, regardless of the design, specification, construction method, or materials used. Element classification ensures consistency in the economic evaluation of building projects over time and from project to project, and it enhances project management and reporting at all stages of the building life cycle planning, programming, design, construction, operations, and disposal [26]. However, building elements classifications include: The Building Cost Information System of the Royal Institute of Chartered Surveyors (RICS); The Canadian Institute of Quantity Surveyors (CIQS); UNIFORMAT II; Construction Economics European Committee (CEEC) among others. Therefore, the Royal Institute of Chartered Surveyors (RICS) Classification of building element provides the comprehensive list of building elements typical of any building projects. Thus, this study adopted the RICS building elements classification.

2.3 Nature of Variations

Variation involves not only changes to the work or matters relating to the work in accordance with the conditions of the contract but also changes to the working conditions. The nature of variations is usually defined by variation clauses in the conditions of contract. For instance, a variation is defined in clause 5.1 of the JCT' [27] Standard Building Contract as follows:

- an addition, omission or substitution of any work;
- the alteration of contractual standards for the materials or goods;
- the removal from site of work executed or materials that are in accordance with contract;
- any alteration of working restrictions such as access to the site, site hours, working areas
- or the order of work sequencing or;
- the expenditure of provisional sums.

FIDIC [28] clause 13 states that variations may be initiated by the Engineer at any time prior to issuing the taking-over certificate for the works, either by an instruction or by a request for the contractor to submit a proposal. Each variation may include the followings:

(a) Changes to the quantities of any item of work included in the Contract (however, such changes do not necessarily constitute a Variation).
(b) Changes to the quality and other characteristics of any item of work.
(c) Changes to the levels, positions and/or dimensions of any part of the works.
(d) Omission of any work unless it is to be carried out by others.
(e) Any additional work, plant, materials or services necessary for the permanent works including any associated tests on completion, boreholes and other testing and exploratory work.
(f) Changes to the sequence or timing of the execution of the Works.

Therefore, variation can be described as a deviation from an outlined planned process of construction work vis-à-vis alteration/deviation from the basis upon which the contract was awarded. The term does not only embrace changes to the work or matters appertaining to the work in accordance with the provision of the contract, but also changes the contract conditions.

2.4 Factors Predispose Building Elements to Variations

Susceptibility of building elements to variation does not just happen on building elements it is due to various factors that made the elements to be susceptible. These factors affect each element differently. Arain, [29] groups the factors into four, this includes: owner related factors; consultant related factors; contractor related factors; and natural related factors. Babatunde, Babalola, Jagboro and Opawole [30] identify nine factors that predispose building elements to variations. These include consultant related factor; contractor related factor; client related factor; workmanship related factor; government related factor; economic related factor; technological related factor; safety related factor; and natural related factor. The predisposing factors and other causes of variation orders on construction projects have also been identified by many researchers [31-35].

3.0 Methodology

The study adopted documentary reports/archival records, this includes: the contract drawings; original bill of quantities; addendum; and reduction bill of quantities, minutes of site meetings among others of completed educational building projects attributed to variations on each building element in Lagos metropolis, Nigeria. Educational building projects were sampled because there is always proper contract documentations, and all the professionals within the built environment are involved which led to the award of contract to the contractor within a time frame, and also there were huge construction of educational building projects in the study area because the Lagos State Government initiated a plan tagged "Millennium Model Schools" an effort to build modern educational building projects across the study area. Thus, the educational building projects were limited to the one awarded and completed between year 2002 and 2012 because the periods experienced similar economic conditions in Nigeria. However, there was no official list stipulating the number of educational building projects completed in the study area. Therefore, as a result of researcher's knowledge of the area and the help of Lagos State ministry of educational coupled with the assistance of colleagues working in the construction industry within the study area. Fifty (50) completed educational building projects were identified together with the names and addresses of consultancy firms involved on each project. It was on this premise, that the list of the consultancy firms, particularly architectural and quantity surveying firms were generated. However, the identified firms were visited. In view of this, archival records of Thirty Nine (39) out of Fifty (50) identified and completed educational projects were reviewed. This was due to the fact that few of the firms considered the information classified and confidential while others could not retrieve them due to poor data storage. Thus, in order to achieve the purpose of this study the information relating to initial cost, final construction cost, estimated period in week(s), and final completion period in week(s) on each building element attributed to variations were securitized to eliminate those that fail to provide adequate information. Therefore, Thirty (30) educational building projects were found suitable for the analysis. The data obtained were analyzed using statistical methods of average, percentage, regression analysis, and analysis of variance (ANOVA).

4.0 Results and Discussions

Table 2 reveals the cost implications of each building element to variation during construction process. The table shows the percentage of cost overruns due to variation on each building element and it classified as follows: cost overrun greater than '>' 20% include the following building elements: earthwork and fillings; frame; windows and external doors; fittings and furnishings; water installation; and external services having the percentage of cost overruns of 20.08%; 28.84%; 39.09%; 31.71%; 21.55% and 20.69% respectively. The table also reveals the percentage range of cost overrun between 10 to 20%; this includes the following building elements: wall finishing; floor finishing; service equipment; disposal installation; electrical installation; and drainage with the percentage of cost overruns of 11.18%; 19.63%; 17.66%; 16.15%; 18.63%; and 10.05% respectively.

Table 2: Cost Implication of Variation on each Building Elements

Building elements	Average Initial Cost (Naira)	Average Final Cost (Naira)	Difference (Average Cost Overrun) (Naira)	Percentage of Cost Overrun
A. Substructure				
i. Earthwork and Fillings	5,948,594.96	7,442,828.79	1,494,233.83	20.08 %
ii. Concrete works	9,153,882.85	9,926,950.46	773,067.61	7.79%
iii. Block works (if any)	2,935,140.54	2,762,889.12	172,251.42	5.87%
B. Superstructure				
i. Frame	8,762,897.75	12,315,182.85	3,552,285.1	28.84 %
ii. Upper floors	14,174,208.14	14,929,252.3	755,044.16	5.06 %
iii. Roof	6,672,276.48	7,395,181.59	722,905.11	9.78 %
iv. Stairs	2,464,121.08	2,521,205.88	57,084.8	2.26%
v. External walls	4,524,693.47	4,737,670.41	212,976.94	4.50 %

vi. Windows and external doors	6,161,368.81	10,115,674.19	3,954,305.38	39.09 %
vii. Internal wall and internal partitions	4,695,491.02	5102386.73	406895.71	7.97%
viii. Internal doors	3,623,466.73	3,858,414.44	234,947.71	6.09%
C. Internal Finishes				
i. Wall finishing	5,191,418.05	5,844,816.12	653,398.07	11.18%
ii. Floor finishing	7,605,887.59	9,463,098.77	1,857,211.18	19.63 %
iii. Ceiling finishing	3,492,507.56	3,637,182.89	144,675.33	3.98%
D. Fittings and Furniture				
i. fittings and furnishings	3,742,969.9	5,480,876.17	1,737,906.27	31.71 %
E. Services				
i. Sanitary appliances	8,447,568.3	8,507,460.22	59,891.92	0.70%
ii. Service equipment	7,321,909.22	8,891,796.5	1,569,887.28	17.66%
iii. Disposal installation	4,742,581.65	5,655,982.89	913,401.24	16.15%
iv. Water installation	2,568,899.88	3,274,462.51	705,562.63	21.55 %
v. Ventilation system	3,228,467.28	3,279,998.48	51,531.209	1.57%
vi. Electrical installation	10,549,346.2	12,964,360.28	2,415,014.08	18.63%
F. External works				
i. Drainage	3,881,252.67	4,314,958.28	433,705.61	10.05%
ii. External services	3,472,856.12	4,378,978.43	906,122.31	20.69%

1US \$ =150 Nigerian Naira; 1 £ =250 Nigerian Naira.

Table 2 finally shows the percentage range of cost overruns between 0 to 10%; these include the following building elements: concrete works; block work (at substructure); upper floors; roof; stairs; external walls; internal wall and internal partitions; internal doors; ceiling finishing; sanitary appliances; and ventilation system having the percentage of cost overruns of 7.79%; 5.87%; 5.06%; 9.78%; 2.26%; 4.50%; 7.97%; 6.09%; 3.98%; 0.70%; and 1.57% respectively. Table 3 and 4 were derived from the table 2, thus, the following relationships were established.

Table 3 shows a strong linear correlation between Average Final Cost (AFC) and Average Initial Cost (AIC) of each building element. The value of R= 0.908 and coefficient of determination R^2=0.825 which implies that 82.5% of the variation in final cost of each building elements is explained by initial cost of each building elements during construction process. The P value of 0.000 indicates that the relationship is statistically significant. Also, Table 3 indicates the relationship between 'AIC' and 'AFC' illustrated using a model of this form: AFC= 981690 + 1.033AIC. This model can be used to predict Average Final Cost of each building element. Table 2 further reveals a strong linear correlation between Average Initial Cost (AIC) and Average Final Cost Overrun (AFCO) on each building element. The value of R= 0.962 and coefficient of determination R^2=0.926 implies that 92.6% of the variation in Average Final Cost Overrun of each building element is explained by Average Initial Cost (AIC) of each building element during construction process. The P value of 0.000 reveals that the relationship is statistically significant. Moreover, Table 3 further indicates the relationship between 'AIC' and 'AFCO' due to variation on each building element illustrated using a model of this form: AFCO= 608390.865 +1.310AIC. Therefore, the model is most appropriate in predicting Average Final Cost Overrun of each building element.

Table 3 also reveals a strong linear correlation between Average Initial Cost (AIC); Average Final Cost Overrun (AFCO) and Percentage of Cost Overrun (PCO) on each building element. The value of R= 0.966 and coefficient of determination R^2=0.933 which implies that 93.3% of the variation in Percentage of Cost Overrun of each building element is explained by initial cost of each building element during construction process.

Table 3: Model Summary of Cost Implication on Building Elements to Variation

Model	R	R Square	Adjusted R Square	Std. Error of the Estimate	Change Statistics				
					R Square Change	F Change	df1	df2	Sig. F Change
1	.908[a]	.825	.818	1.69395E6	.825	113.266	1	24	.000
2	.962[b]	.926	.920	1.12319E6	.101	31.589	1	23	.000
3	.966[c]	.933	.924	1.09183E6	.007	2.340	1	22	.140

a. Predictors: (Constant), Average Initial Cost (Naira)
b. Predictors: (Constant), Average Initial Cost(Naira), Average Cost Overrun (Naira)
c. Predictors: (Constant), Average Initial Cost(Naira), Average Cost Overrun (Naira), Percentage of Cost Overrun

In Table 4, the relationship between 'AIC,' 'AFCO' and 'PCO' due to variation was illustrated using a model of this form: PCO= 1.482E6 + 2.59 - 58400AIC. Hence, introducing Percentage of Cost Overrun (PCO) into the model, although produces a significant model but its contribution to prediction of the Average Final Cost Overrun is not significant (t =-1.53, p>0.05).

Table 4: Coefficients of the Developed Model for Cost Implication on Building Elements to Variation

Model		Unstandardized Coefficients		Standardized Coefficients	T	Sig.
		B	Std. Error	Beta		
1	(Constant)	981690.454	579684.569		1.693	.103
	Average Initial Cost (Naira)	1.033	.097	.908	10.643	.000
2	(Constant)	608390.612	390060.833		1.560	.132
	Average Initial Cost (Naira)	.865	.071	.760	12.177	.000
	Average Cost Overrun Due to Variation Only (Naira)	1.310	.233	.351	5.620	.000
3	(Constant)	1.482E6	685765.124		2.162	.042
	Average Initial Cost(Naira)	.767	.094	.674	8.159	.000
	Average Cost Overrun Due to Variation Only (Naira)	1.823	.405	.489	4.501	.000
	Percentage of Cost Overrun Due to Variation Only	-58400.241	38176.423	-.153	-1.530	.140

a. Dependent Variable: Average Final Cost Overrun Due to Variation only (Naira)

Table 5 indicates the time implications of each building element to variation during construction process. The table shows the percentage of time overruns due to variation on each building element as follows: time overruns greater than '>'20% include the following building elements: earthwork and fillings; block work (at substructure); frame; upper floors; roof; stairs;

external walls; wall finishing; ceiling finishing; fittings and furnishings; sanitary appliances; service equipment; disposal installation; water installation; ventilation system; electrical installation; and drainage with the percentage of cost overruns of 40.65%; 37.48%; 29.31%; 25.74%; 22.80%; 21.64%; 29.94%; 30.04%; 20.63%; 26.82%; 23.11%; 23.79%; 25.82%; 25.91%; 26.89%; 23.48%; and 26.81%. It can be deduced from the finding that almost all the building elements experienced time overruns greater than 20%. The table also reveals the building elements suffered time overruns between 10 to 20%. The building elements are: external services; concrete works; internal wall and internal partitions; internal doors; and floor finishing having the percentage of cost overruns of 19.97%; 14.08%; 10.27%; 13.94%; and 14.98%. The table finally indicates percentage of time overruns between 0 to 10% the only building element in this category is windows and external doors with percentage time overrun of 9.32%.

Table 5: Time Implication on each Building Element to Variation

Building Elements	**Average Estimated Period (week)**	**Average Actual Completion Period (week)**	**Difference (Period Overrun) (Week)**	**Percentage of Period Overrun**
A. Substructure				
i. Earthwork and Fillings	6.25	10.53	4.28	40.65%
ii. Concrete works	6.53	7.60	1.07	14.08 %
iii. Block works (if any)	3.87	6.19	2.32	37.48%
B. Superstructure				
i. Frame	9.31	13.17	3.86	29.31%
ii. Upper floors	8.77	11.81	3.04	25.74%
iii. Roof	3.42	4.43	1.01	22.80%
iv. Stairs	4.20	5.36	1.16	21.64%
v. External walls	5.43	7.75	2.32	29.94 %
vi. Windows and external doors	5.84	6.44	0.60	9.32 %
vii. Internal wall and internal partitions	5.33	5.94	0.61	10.27%
viii. Internal doors	3.52	4.09	0.57	13.94%
C. Internal Finishes				
i. Wall finishing	3.75	5.36	1.61	30.04 %
ii. Floor finishing	5.11	6.01	0.90	14.98 %
iii. Ceiling finishing	3.27	4.12	0.85	20.63%
D. Fittings and Furniture				
i. fittings and furnishings	3.22	4.40	1.18	26.82%
E. Services				
i. Sanitary appliances	3.16	4.11	0.95	23.11%
ii. Service equipment	2.98	3.91	0.93	23.79%
iii. Disposal installation	2.73	3.68	0.95	25.82 %
iv. Water installation	3.46	4.67	1.21	25.91 %
v. Ventilation system	3.29	4.50	1.21	26.89 %
vi. Electrical installation	5.15	6.73	1.58	23.48%
F. External works				

i. Drainage	5.05	6.90	1.85	26.81 %
ii. External services	4.65	5.81	1.16	19.97%

Table 6 and 7 were derived from the table 5. The result from table 6 shows a strong linear relationship between average estimated period and average actual completion period due to variation on each building element. The value of R= 0.955 and coefficient of determination R^2=0.912 which implies that 91.2% of the variation in actual completion period of each building element is explained by average estimated period of each building element during construction process. The P value of 0.000 reveals that the relationship is statistically significant. Moreover, table 7 indicates the model as Average Actual Completion Period (Y) = 1.379(X) – 0.251. The model is statistically significant, hence the model can be adopted in predicting the average actual completion period of each building element in building projects.

Table 6: Model Summary of Time Implication of Variation on Building Elements

Model	R	R Square	Adjusted R Square	Std. Error of the Estimate	Change Statistics				
					R Square Change	F Change	df1	df2	Sig. F Change
1	.955[a]	.912	908	.77065	.912	218.042	1	21	.000

a. Predictors: (Constant), Average estimated period (week)

Table 7: Coefficients of the Developed Model for Time Implication of Variation on Building Elements

Model		Unstandardized Coefficients		Standardized Coefficients	t	Sig.
		B	Std. Error	Beta		
1	(Constant)	-.251	.468		-.537	.597
	Average estimated period (week)	1.379	.093	.955	14.766	.000

a. Dependent Variable: Average actual completion period (week)

Y = ax +b

If Y =Average Actual Completion Period

X = Average Estimated Period (week)

b = Slope or Relationship between X and Y

5.0 Conclusions

The study concluded that variation on each building element had a significant impact on educational building projects delivery in terms of cost and time. The building elements to variations have varying percentage of cost overruns and time overruns with earthwork and fillings; frame; windows and external doors; fittings and furnishings; water installation; and external services have higher percentage of cost overruns greater 20%. On the other hand concrete works; block work (at substructure); upper floors; stairs; external walls; internal wall and internal partitions; internal doors; ceiling finishing; sanitary appliances; and ventilation system have the least cost overruns ranges between 0 to 10%. The study further revealed time overruns on each building element to variation with earthwork and fillings; block work (at substructure); frame; upper floors; roof; stairs; external walls; wall finishing; ceiling finishing; fittings and furnishings; sanitary appliances; service equipment; disposal installation; water installation; ventilation system; electrical installation; drainage; and external services have higher percentage of time overruns greater than 20%. While windows and external doors only have the least time overruns ranges between 0 to 10%. The study established the models: AFC= 981690 + 1.033AIC and AFCO= 608390.865 +1.310AIC to predict the average final cost of each building element and the average final cost overrun of each building element respectively, where AFC= Average Final Cost; AIC= Average Initial Cost; and AFCO= Average Final Cost Overrun. The model for predicting the average actual completion period of each building element due to variation was established as average actual completion period(Y) = 1.379(X) – 0.251, where X= Average Estimated Period (week). The study recommends identified building elements that have higher percentage of cost and time overruns due to variations are to be given outmost consideration during the design and construction processes in order to minimize its effects and contribution to variations. Also, the clients should be encouraged by the various professionals to be to be more comprehensive and firm in their briefs to facilitate adequate planning and reduce variations.

References

[1] C. C. Cantarelli, B. Flyvbjerg and S. L. Buhl, "Geographical Variation in Project Cost Performance: The Netherlands versus Worldwide," Journal of Transport Geography, vol.24, pp. 324–331, 2012.
[2] M. A. Halloum, and A. Bajracharya, "Cost and Time Overrun Revisited: A Study on the Infrastructure Construction Projects in Abu Dhabi, UAE," Third International Conference on Construction in Developing Countries (ICCIDC–III), 'Advancing Civil, Architectural and Construction Engineering & Management', July 4-6, Bangkok, Thailand, 2012.
[3] A. H. Memon, I. A. Rahman and A. A. A. Azis, "Time and Cost Performance in Construction Projects in Southern and Central Regions of Peninsular Malaysia," International Journal of Advances in Applied Sciences, vol.1 (1), pp. 45-52, 2012.
[4] O. E. Akinsiku, "Assessment of Elements' Proneness to Variation and its Effects on Construction Cost Overruns," An unpublished BSc Thesis, Department of Quantity Surveying, Federal University of Technology, Akure, 2006.
[5] D. J. Yates and C. Hardcastle, "The Causes of Conflict and Dispute in the Hong Kong Construction Industry," RICS Research Paper Series, vol. 22, pp. 1-50, 2003.
[6] J. K. Ssegawa, K. M. Mfolwe, B. Makuke and B. Kutua, "Construction variations: A Scourge or a Necessity?" 2002, Accessed at http://buildnet.csir.co.za/cdcproc/docs/3rd/ssegawa_jk_mfolwe_km.pdf on 4 February 2011.
[7] A. Mokhtar, C. Bedard and P. Fazio, "Collaborative Planning and Scheduling of Interrelated Design Changes," Journal of Architectural Engineering, ASCE, vol. 6, pp. 66-75, 2000.
[8] F. M. Arain, S. A. Assaf and S. P. Low, "Causes of Discrepancies between Design and Construction," Architectural Science Review, vol.47, pp. 237-249, 2004.
[9] F. M. Arain, (2005) "Strategic Management of Variation Orders for Institutional Buildings: Leveraging on Information Technology," 2005, Accessed at http://www.pmi.org/pmief/scholarship/documents/BNS041.pdf on 4 February 2011.
[10] Construction Industry Institute (CII), "Pre-project Planning: Beginning a Project the Right Way," Construction Industry Institute, University of Texas at Austin, TX, 1994.
[11] E. R. Fisk, "Construction Project Administration," Prentice Hall, New Jersey, 1997.
[12] C. W. Ibbs, C. K. Wong and Y. H. Kwak, "Project Change Management System," Journal of Management in Engineering, ASCE, vol. 17, pp. 159-165, 2001.
[13] Construction Industry Institute (CII), "Constructability: A Primer," Construction Industry Institute, University of Texas at Austin, TX, 1986.
[14] P. F. Kaming, P. O. Olomolaiye, G. D. Holt and F. C. Harris, "Factors Influencing Construction Time and Cost Overruns on High Rise Projects in Indonesia", Construction Management and Economics, vol. 15, pp. 83-94, 1997.
[15] W. Hester, J. A. Kuprenas and T. C. Chang, "Construction Changes and Change Orders: Their Magnitude and Impact," CII Source Document 66, University of California- Berkeley, 1991.
[16] D. S. Barrie and B. C. Paulson, "Professional Construction Management: Including CM, Design-Construct, and General Contracting," McGraw Hill, New York, 1992.
[17] Construction Industry Institute (CII), "Project Change Management," Special Publication, Construction Industry Institute, University of Texas at Austin, TX, 1994.
[18] D. W. M. Chan and M. M. Kumaraswamy, "An Evaluation of Construction Time Performance in the Building Industry," Building and Environment Journal, vol. 31, pp. 569–578, 1996.
[19] J. K Ssegawa, K. M. Mfolwe, B. Makuke and B. Kutua, "Construction variations: A Scourge or a Necessity?" 2002, Accessed at

http://buildnet.csir.co.za/cdcproc/docs/3rd/ssegawa_jk_mfolwe_km.pdf on 4 February 2011.
[20] F. M. Arain and S. P. Low, "Developers' Views of Potential Causes of Variation Orders by Institutional Buildings in Singapore," 2006, Accessed at http://goliath.ecnext.com/coms2/gi_0199-5445784/Developers-views-of-potential-causes.html on 4 February 2011.
[21] F. M. Arain and S. P. Low, "PSSCOC and Variations: An Analysis of Consultant Related Variations in School Building Projects," International Construction Law Review, vol. 24, pp. 366-381, 2007.
[22] M. M. Kumaraswamy, D. R. A. Miller and K. Yogeswaran, "Claims for Extensions of Time in Civil Engineering Projects," Construction Management and Economics, vol.16 (3), pp. 283-294, 1998.
[23] C. Bordat, B. G. McCullouch, S. Labi, and K. C. Sinha, "An Analysis of Cost Overruns and Time Delays of INDOT Projects," Publication FHWA/IN/JTRP-2004/07, Joint Transportation Research Program, Indiana Department of Transportation and Purdue University, West Lafayette, Indiana, 2004, Accessed at http://docs.lib.purdue.edu/cgi/viewcontent.cgi?article=1482&context=jtrp on 11 November 2013.
[24] A. A. Oladapo, "A Quantitative Assessment of the Cost and Time Impact of Variation Orders on Construction Projects," Journal of Engineering, Design and Technology, vol. 5(1), pp. 35-48, 2007.
[25] T.H.S. Priyantha, G. Karunasena and V.A.K. Rodrigo, "Causes, Nature and Effects of variations in Highways," Built - Environment - Sri Lanka - Vol. 9 (1), pp.14-20, 2011.
[26] P. C. Robert and E. M. Harold, "UNIFORMAT II Elemental Classification for Building Specifications, Cost Estimating, and Cost Analysis," 1999, Accessed at http://www.bfrl.nist.gov/oae/publications/nistirs/6389.pdf on 20 January 2010.
[27] Joint Contract Tribunal, "Standard Form of Building Contract," 2011.
[28] FIDIC, "Conditions of Contract for Building and Engineering Works- Designed by the Employer," Fédération Internationale des Ingénieurs-Conseils, Geneva, 2005.
[29] F. M. Arain, "Potential Barriers in Management of Refurbishment Projects," Journal of Independent Studies and Research, vol. 3, pp. 22-31, 2005.
[30] S. O. Babatunde, O. Babalola, G. O. Jagboro and A. Opawole, "An Evaluation of Factors Predisposing Building Elements to Variation in Nigeria," Civil Engineering Dimension, vol. 15(1), pp. 11-17, 2013.
[31] H. R. Thomas and C. L. Napolitan, "The Effects of Changes on Labour Productivity: Why and How Much," CII Document 99, The Pennsylvania State University, USA, 1994.
[32] R. H. Clough and G. A. Sears, "Construction contracting," John Wiley & Sons Inc., New York, 1994.
[33] C. W. Ibbs, S. A. Lee and M. I. Li, "Fast tracking's Impact on Project Change," Project Management Journal, vol. 29, pp. 35-41, 1998.
[34] J. J. O'Brien, "Construction Change Orders," McGraw Hill, New York, 1998.
[35] C. Gray and W. Hughes, "Building design management", Butterworth-Heinemann, Oxford, UK, 2001.

6

MICROSURFACING – AN ECO-EFFICIENT TOOL FOR ROAD SAFETY AND PAVEMENT MAINTENANCE

Rajesh S. Gujar[1], Krupesh A. Chauhan[2], Gautam Dadhich[3]

[1] Research Scholar, SVNIT, Surat, Gujarat, India
[2] Associate Professor, Civil Engineering Department, SVNIT, Surat, India
[3] UG Student, PanditDeendayal Petroleum University, Gandhinagar, Gujarat, India

*Corresponding E-mail : gautamdadhich3392@gmail.com

Abstract

With the development of infrastructure, in rapidly developing modern era, frequent maintenance of existing infrastructure, especially roads is of prime importance. Microsurfacing is a preventive maintenance technique (preventative maintenance strategy aims at producing the most cost-effective improvements in pavement quality and life) involving three to six millimeter sized bitumen aggregate treated with a special emulsion. The layer is neatly laid down on road section requiring maintenance. This paper aims at focusing on the contributions made by researchers on Microsurfacing as a tool for pavement maintenance and repair. Although the technique is finding its place in Indian sub-continents little late, but nevertheless it is universally considered as an inevitable tool as far as preventive pavement maintenance is concerned and should be used to extend life of the underlying pavement.

Keywords: *Microsurfacing, Skid Resistance, Surface Texture, roughness, Pavement maintenance*

1.0 Introduction

Microsurfacing is a mixture of polymerized bitumen emulsion, specially graded fine aggregates, cement, water and necessary additives, mixed homogeneously on site in a special Microsurfacing machine, uniformly spread immediately over a properly prepared surface evenly by means of a spreader box attached behind the machine. It is an environment friendly cold mix application treatment applied over an existing pavement surface which is structurally sound, but the surface shows signs of premature ageing, aggregate loss, cracking, high degree of polishing etc. Microsurfacing helps in preservation of pavement strength and can be used both for preventive and periodic renewal treatment. Factors that reportedly limit the application of microsurfacing consist of established limitations and debated limitations that may act more as suggested guidelines for use. Microsurfacing is shown to be most effective under certain conditions [1]. Microsurfacing treatments can be expected to last at least seven years when placed on medium to high volume roads [2]. Microsurfacing is very successful on both low and high volume roadways and is recommended for night applications on heavy-traffic streets [3].

2.0 Literature Review

One of the early studies done on microsurfacing in the United States foreshadowed the favorable conclusions of further research by recommending that microsurfacing be approved for routine use in restoring flexible pavements to fill surface ruts and cracks, seal the surface and restore skid resistance [5]. Hicks et al. documented that microsurfacing as an appropriate maintenance strategy for more types of pavement distress than any other commonly used strategy, as well as having a longer life expectancy than all but thin HMA overlays which cost 30% more. Temple et al. observed the median Pavement Condition Index (PCI) of microsurface sections is approximately 85 after 60 months of service with significantly fewer cracks detected after treatment. Microsurfacing leads to reduction in initial rut after treatment [7]. The friction numbers

(FN) for smooth tire are in the low teens to low thirties on 50% of the microsurface section. Microsurfacing can be a highly cost-effective safety treatment for both intersections and road segments that warrant skid resistance improvement because of a high frequency of wet road accidents and low friction numbers [4]. Correlation is developed between International roughness index values to the amount of material placed in the surface preparation course [2]. Microsurfacing provide a tighter surface texture (improving noise) and require less weight of aggregate per square area to provide adequate coverage (reducing cost) [6].

3.0 Environmental Impact of Microsurfacing

Microsurfacing is an environmental friendly technique which is used for the preventive maintenance of road. As environmental footprint according Mr. Takamura energy used is almost zero because it cold mix asphalt as presented in table 1. Numbers of accidents also get reduced due to increases in skid resistance after microsurfacing. A comparison between hot mix asphalt and cold mix asphalt is described in figure 1.

Table 1: comparisons between hot mix asphalt and cold mix asphalt

CHARACTERISTICS	HOT ASPHALT	MICROSURFACING	SAVINGS
Materials (tons/ lane km)	46.875	6.09	87
Fuel consumption (kg/ton)	8	1	88
Fuel consumption (kg/lane km)	3,750	400	89
CO2 emission (Kg/lane km)	1,078	13.40	99
NO2 emission (Kg/lane km)	5.1	1.3	75
Accidents (estimated ratio)	1.0	0.5	50
Health risk (estimated ratio)	1.0	0.2	80

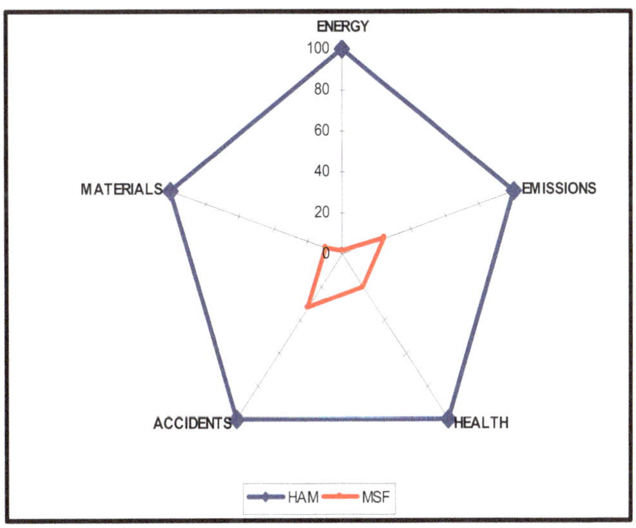

Figure 1: Footprint according to Mr.Takamura(Courtesy: ELSAMAX)

4.0 A Case Study of Ahmedabad - Vadodara Expressway

Microsurfacing, preventive maintenance techniques is first time used in Ahmedabad-Vadodara expressway. The Ahmedabad Vadodara Expressway also known as National Expressway 1 is a major artery of public transport connecting the cities of Ahmedabad and Vadodara in the state of Gujarat, India. This 95 Km long expressway has 2 lanes each side and one service lane. It is a part of the Golden Quadrilateral Project by National Highway Authority of India (NHAI). Two-wheeler vehicles are forbidden on the road due to safety concerns. Microsurfacing (Type II) used as a preventive maintenance solution on Ahmedabad – Vadodara Expressway. In this expressway, a study of microsurfacing is done in a section between CH. 17.00 to CH. 22.

Pavement Performance evaluation is done to determine potential of microsurfacing as a preventive maintenance. For the measurement of performance of microsurfacing as preventive maintenance tool following performance measure are taken in to consideration:

- Skid resistance
- Roughness
- Surface Texture

Skid resistance is the force developed when a tire that is prevented from rotating slides along the pavement surface (Highway Research Board, 1972). Skid resistance is an important pavement evaluation parameter because inadequate skid resistance will lead to higher incidences of skid related accidents. Skid resistance depends on pavement surface texture. Skid resistance changes over time. Typically it increases in the first two years following construction as the roadway is worn away by traffic and rough aggregate surfaces become exposed, and then decreases over the remaining pavement life as aggregates become more polished. Skid Resistance measurement on Ahmedabad – Vadodara Expressway is done using Portable Pendulum Skid tester (ASTM-E-303-93-2003). Skid resistance is measured before laying microsurfacing and after 6 months of laying microsurfacing on highway between sections of CH. 17.00 to CH. 22. Figure 2 shows Results obtained from Portable Pendulum Skid tester before and after lying of microsurfacing.

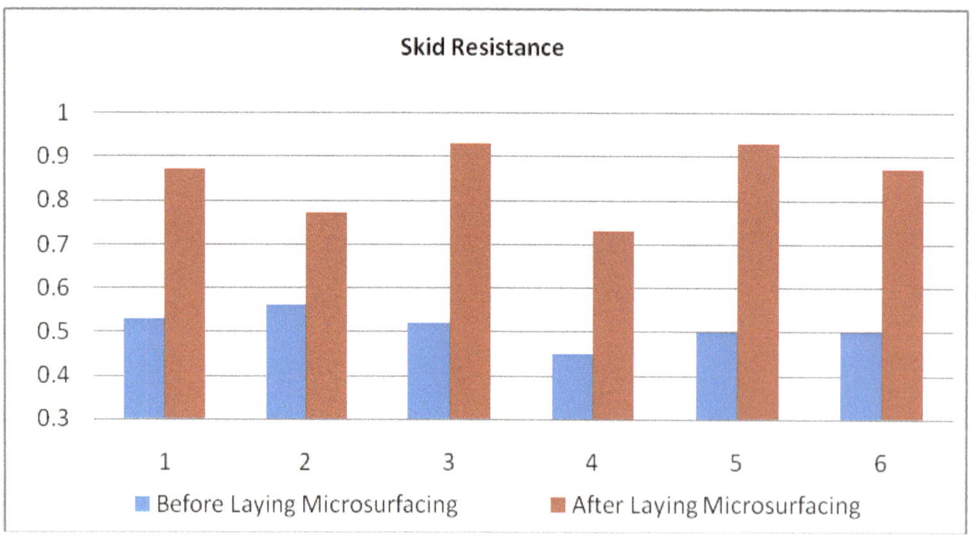

Figure 2: Graphical Results of Improved skid resistance

Pavement roughness is generally defined as an expression of irregularities in the pavement surface that adversely affect the ride quality of a vehicle (and thus the user). Roughness is an important pavement characteristic because it affects not only ride quality but also vehicle delay costs, fuel consumption and maintenance costs. Roughness measurement on Ahmedabad – Vadodara Expressway is done using fifth wheel bump integrator (developed by CRRI) and is reported as Unevenness Index (UI) in mm/km. Figure 3 shows Unevenness index before and after lying of microsurfacing.

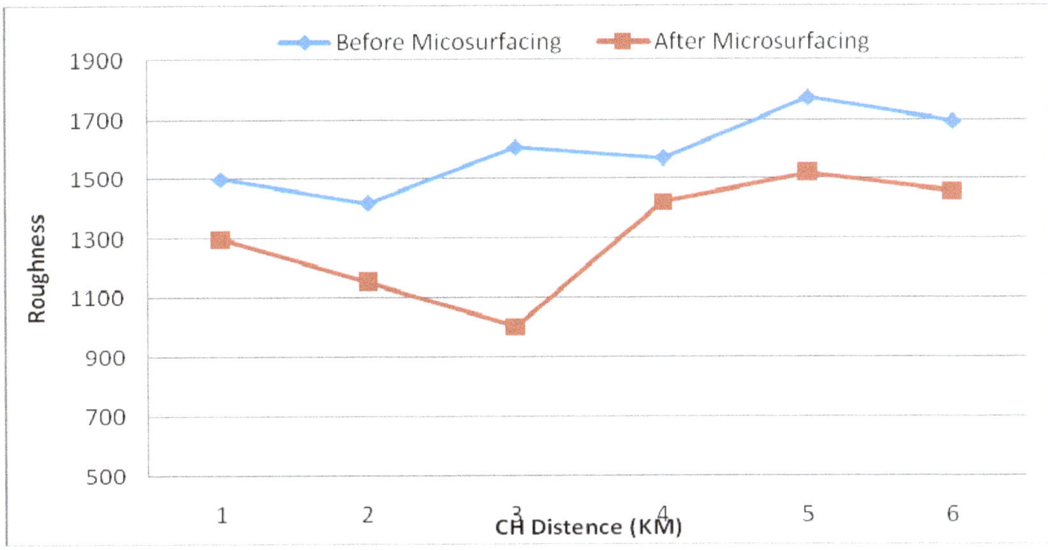

Figure 3: Graphical Results of Roughness

Texture depth test procedure covers the determination of the average texture depth of a paved surface using sand to give the volume of voids. The method is suitable for the measurement of surfaces with average texture depths greater than 0.45 mm (less than 350mm sand circle diameter). Figure4 Results of surface texture before and after lying of microsurfacing.

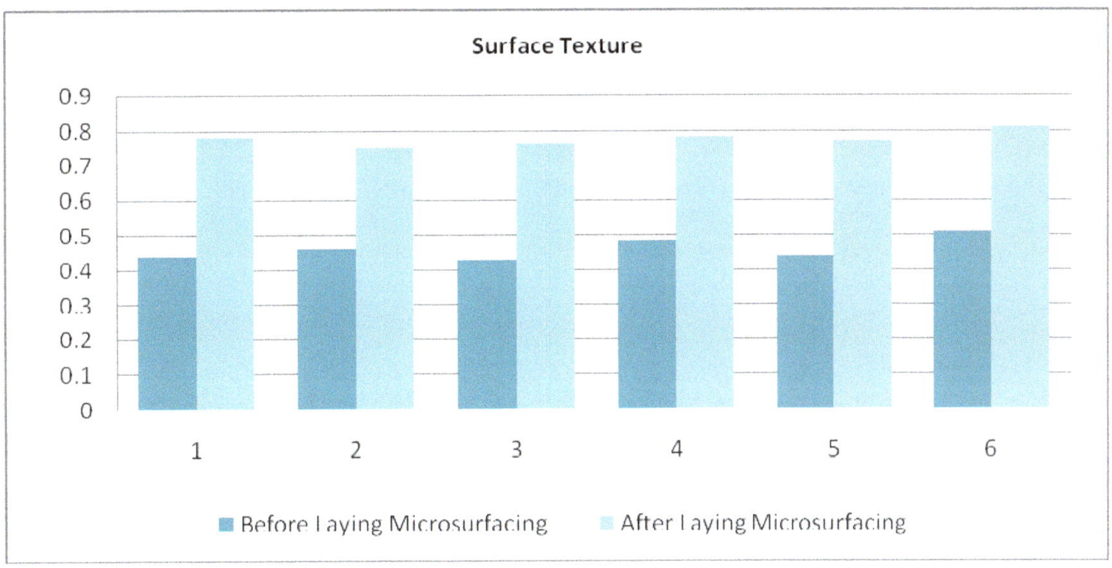

Figure 4: Graphical Results of Improved Surface Texture

In this study roughness is determined by Bump Indicator and skid resistance is measured. The roughness before lying of microsurfacing is 1461(mm/Km) to1772 (mm/Km) and roughness after microsurfacing is 1000 to 1517.The skid resistance is also improved from 0.2 to 0.6. Which shows that microsurfacing improve riding quality. The skid resistance is also improved by using microsurfacing.

5.0 Conclusion

Microsurfacing technology, which is applied to Ahmadabad-Vadodara expressway for the first time in Gujarat, shows outstanding results. Based on these results following conclusions are drawn :

a) Carbon Footprint of Microsurfacing shows that this technology is environmental friendly and economical technology.

b) Post results on Microsurfacing technology at expressway the shows that this method is not only an economical solution for preventive maintenance but also it's also provide safety and riding quality by improving skid resistance and reducing roughness respectively.

c) The roughness before lying of microsurfacing is 1461(mm/Km) to1772 (mm/Km) and roughness after microsurfacing is 1000 to 1517.

d) The skid resistance is also improved from 0.2 to 0.6 and surface Textures improved from 0.4 to 0.8. Which shows that microsurfacing contributes in Road Safety by reducing accidents.

References

[1] L. Samuel, G. Lamptey, and S. H. Kong, "Effectiveness of microsurfacing treatments," Journal of transportation engineering, vol. 133 (5), pp. 298-307, 2007.

[2] E. N. Johnson, T. J. Wood, and R. C. Olson, "Flexible slurry-microsurfacing system for overlay preparation: Construction and seasonal monitoring at minnesota road research project," Transportation Research Record: Journal of the Transportation Research Board, vol. 1989 (1), pp. 321-326, 2007.

[3] K. L. Smith, and D. G. Peshkin, "Pavement Preservation on High-Traffic-Volume Roadways." Transportation Research Record: Journal of the Transportation Research Board, vol. 2235 (1), pp. 54-65, 2011.

[4] L. Craig, and B. Persaud, "Safety effects of targeted program to improve skid resistance." Transportation Research Record: Journal of the Transportation Research Board, vol. 2068 (1), pp. 135-140, 2008.

[5] C. M, Pederson, W. J. Schuller, and C. D. Hixon, "Microsurfacing with natural latex modified asphalt emulsion: a field evaluation," Transportation Research Record, pp. 108- 112, 1988.

[6] R. G. Hicks, K. Dunn, and J. Moulthrop, "Framework for selecting effective preventative maintenance treatments for flexible pavements," Transportation Research Record, vol. 1597, pp. 1- 10, 1997.

[7] T. William, "Performance of Louisiana's chip seal and microsurfacing program," Transportation Research Record: Journal of the Transportation Research Board, vol. 1795 (1), pp. 3-16, 2002.

7

Assessment of Cost Management Functions of Quantity Surveyors with Lean Methodology

Maroof Opeyemi Anifowose[1], Wasiu Adeniran Ola-awo[2], Tanimu U Mohammed[3]

Department of Quantity Surveying, Federal University of Technology Minna, Nigeria

*Corresponding E-mail: anifowosemo@futminna.edu.ng, anny4yemi2000@yahoo.com

Abstract

The Construction industry in Nigeria is made up of a wide variety of activities which include the provision of professional and technical services to clients in the built environment. Despite the provision of these services to a large number of clients worldwide, the construction industry is still awash by the chronic problems of low productivity, insufficient quality, time over-runs, and poor safety, which hinder customer delivered value. The Just-In-Time phenomenon is a characteristic of lean production systems which operate with very little "fat" (e.g. excess inventory extra workers, wasted space).This study aimed at assessing the construction management function of the quantity surveyor in line with the principle of lean methodology (Just-In-Time). This was achieved by exploring the cost management function of the quantity surveyor, to investigate the current practice of cost management by quantity surveying firms. Data for the study were sourced primarily with the use of questionnaire and the subsequent data analysis, which employed the use of descriptive analysis of presenting the data as obtained on tables during the field survey and attempts a rudimentary establishment of patterns using percentages. The study concluded amongst others, that: all activities involved in the cost management function of the quantity surveyor are important, and value adding, corresponding to conversion activities in line with the Just-In-time/lean methodology.

Keywords: *Just In Time/Lean Methodology, Cost, Management, Quantity Surveyor*

1.0 Introduction

The construction industry has achieved a significant impact in term of contribution to nation's economy when compared with other sectors of the economy worldwide. Ref [1] confirmed construction industries as the major productive sector any nation. In Nigeria Construction industry is made up of a wide variety of activities that includes the provision of professional and technical services to clients within the built environment. In spite of the provision of these services to a large number of clients locally and internationally, the construction industries are still overflowing with the chronic problems of low productivity, unsatisfactory quality, time over-runs, and poor safety, which adversely affect customer delivered value [2]. However, for a better performance, to provides an excellence services and to avoid the issues stated above, is the Quantity Surveyors that plays the major role among the professionals involves in construction process within the built environment [3].

The Quantity Surveyor, often refer to as a cost economist, is the recognized professional as cost and value consultant within the construction industry [4]. He has the required knowledge and expertise; therefore shoulder with the responsibility of advising the client on the cost implication that may arise during the design, the subsequent construction process and with the overall control of the cost of a construction project.

The Royal Institute of Chartered Surveyors (RICS) report titled "The Challenge of Change" [5], has identified the current and potential changes in the external factors affecting the services that is being provided by the Quantity Surveyors and the way it is being provided by them. Such factors were identified as follows: (i) the structural changes that occur in the construction sector, (ii) the continuing Information Technology (IT) revolution, and (iii) changes in both the client base and its growing anticipation. However, the services provided by the Quantity Surveyors has transcended from the traditional role of preparing bills of quantities, valuation of variation, interim valuation, and final account etc., to a more advanced service delivery such as risk management, value management, project management etc., all these were as a result of the changes in external factors as stated above.

The Just-In-Time phenomenon is a characteristic of lean production systems which operate with very little "fat" (e.g. excess inventory extra workers, wasted space) [6]. The principle JIT was originated and first used in a Japanese Toyota car manufacturing company in the late 1940s. Although predominantly used in the manufacturing industry and the practice not limited to but include the following: (i) pull system, (ii) cellular manufacturing, (iii) agile manufacturing strategy, and (iv) bottleneck removal [7]. The adoption and application of JIT into service based industry most especially the construction industry were discussed by [8].

The importance or main objective of JIT is to improve performance by avoiding, eliminating, or reducing the waste. While others objectives are to eliminate disruption, reduce lead and set up time, system flexibility and most importantly value generation [8]. In addition recent study by [9-11] emphasized the importance of JIT principles to competitive capability and firm performance in a present day time-based competitive environment [7].However, the idea has been summarized into conversion, flow and value generation. Thus conversion was further explained by [8]as the transforming of the client's requirement into a desired finished product and flow activities binding the conversion activities. Several tools and techniques have been devised for implementation of lean thinking (JIT) aimed at increasing the performance of the professionals in the field/area of quality assurance, computerized integration of design, procurement, and electronic data interchange, which can be implemented to the profession of Quantity Surveying.

This study is therefore aimed at assessing the cost management function of the Quantity surveyor in line with the principle of JIT /lean methodology. The justification of this study is that there is no single studied of this nature previously carried out on the application of the JIT/lean process to cost management function of the Quantity Surveyors, although some studies has been carried out in other field/areas such manufacturing, operational management etc., as it was revealed by the previous literatures.

2.0 Literature Review

2.1 Lean Methodology: Just-in-time

Lean methodology is a repetitive production system in which processing and movement of materials or goods occurs just as they are needed, usually in a bit by bit or in small batches. It is used to refer to a production system in which both the movement of materials or goods during production and deliveries from suppliers are carefully timed so that at each step of the process, the next batch of the materials arrives for processing just as the preceding batch is completed. According to [12] the concept of JIT is aimed at facilitating the production of mass customized end products in cost efficient way, with lower inventory, faster and less amount of working capital tied up in the process. However, the outcome of this is a system with no idle items waiting to be processed and no idle worker or equipment waiting for items to be processed thus it is a routine operation [6]. Ref [12, 13], describes a system of lean methodology that delivered products free from defects to customers in zero time, and with nothing left in the inventory. The ultimate goal of JIT is a balanced system, in the sense that it achieves a smooth and rapid flow of materials throughout the system. The idea of lean methodology JIT is to make the processing time as short as possible by using resources in the best possible way. The degree to which the overall goal is achieved strongly relies on how well certain supporting goals are achieved. Those goals include: (a) Avoid disruption, (b) System flexibility, (c) Reduce set up and lead times to minimal, (d)less inventory, and (e) Eliminate waste.

Ref [14], adopted a performance criteria and defined waste for a particular production system, stated that failure to meet the unique requirement of the client is a waste. He goes further in outlying this criterion by defining waste as time, space, or materials used in the discharge of an activity that does not directly add value to the end product. Waste according to the lean methodology[15], include; (a) excess use of materials leading to overproduction, (b) inadequate space resulting to waiting time that adds no value to the production, (c) unnecessary transporting increases handling charges, and work-in-process inventory, (d) processing waste cause unnecessary production steps and scrap, (e)inventory cause idle resources, hides quality problem and production inefficiencies, (f) inefficient work methods indicates poor layout, material movement patterns, and increases work-in-process inventory, and (g) product defects requires rework, increases costs and possible sales loss as a result of customer dissatisfaction.

2.2 Application of JIT/Lean Methodology to Service Based Industries

Despite the enormous success recorded by the manufacturing industry in it application of lean methodology, little achievement has been made in Nigeria construction industry as a result of low level of acceptance of its application in the industry. This according to [14] is as a result of uniqueness and the complexity of construction projects. He stressed further on the uncertainties that associated with construction projects in respect to time and schedule pressure, these factors however differentiate it from the manufacturing industry. [16], backing up by [14] stated that the critical factors responsible and making the implementation of the lean methodology a difficult task is largely due to the slow response to changes in the construction industry. In view this, [16] further said that is due to the numerous decisions that has to be made over years with numerous interdependence under uncertainties.

Moreover the complex nature of construction projects has demand for different professionals with different backgrounds and cultures in its actualization [16]. These professionals are experts in the various disciplines with different learning styles which make it difficult in understanding each other and reaching agreements in making relevant decisions that are necessary for the accomplishment of the project. Thus the interactions between the various participants accrue a lot of flow waste [17], some of which cannot be easily identified. While [17] puts forward that people do not really understand the concept and principles of lean methodology. On this note [18] confirmed that the implementation of methodology involves a lot of money. These factors make industrial practitioners weary of implementing lean methodology and other new techniques on large and complex projects [12, 19]; rather, simulations of the techniques are presently developed to test the applicability of the philosophy to construction activities.

In other to achieve successful implementation of the lean process, [3] brought out the following points: (a) there should be proper and careful selection of the right people (management) who would see to the overall processes involved, (b) the culture of the construction industry has to be changed. There should be closer cross relation and staff training. This view was also supported by [20], and (c) there should be strategic procurement methods which would emphasis on mutual trust, effective communication and long term commitment to add value along design and construction process. [16]Also proposed the stages needed for improvement in achieving lean design process. The stages are: (a) Evaluation and diagnosis of the problems involved in the process, (b) Implementation of changes occurring from the diagnosis done, (c) Control of the implementation and processes involved, and (d) Standardization.

Notwithstanding the problems encountered in its adoption to construction as mentioned above, the lean methodology creates a better avenue for value generation in the Construction industry than the various project management structures already existing [21]. They pointed out the fact that Flow, the movement of materials and information through networks of interdependent specialists is almost invincible to those who see through the eyes of traditional project management. They further stated that traditional project management assumes that variability in work flow is outside management control and does not attempt to systematically reduce variability. Furthermore, the existing project management thinking is not based on theory. This is due to the fact that traditionalist are of the opinion that project management cannot be founded on theory, as the decisions of project managers are so much conditioned by context. The identification of flow wastes and value generation which the traditional thinking ignores, [22], gives lean production an upper hand and thus making it suitable for improving the activities and services provided by construction professionals, the quantity surveyor inclusive.

2.3 Factors Affecting the Service Delivery Process of the Quantity Surveyor

Reports produced by the Davis, Langdon, and Everest consultancy group QS 2000 [23], and the challenge of change [14], identified current and potential changes in the external factors affecting the services that QS's provide and the way in which they provide them. These include:

i. The structural changes that are occurring in the construction sector, via the changing nature of contracting and management of construction project.
ii. Continuing IT revolution
iii. Changes in both the client base itself and its increasing expectations
iv. Changes in markets
v. Changes in the profession.

The structural changes such as those outlined above have resulted in QSs working in private practice or consultancy, and to a lesser extent those working in a commercial capacity, being subject to changes in: identity and types of client; client objectives and expectations; type, size and location of projects; sources and structure of finance; and skills and services demanded by clients.

3.0 Research Methodology

This paper employs a quantitative approach to the study of cost management function of quantity surveyors in line with the principle of lean methodology. Data for this study was sourced from quantity surveyors using questionnaire. Questionnaires were prepared and designed in a simplified and closed ended format. The sampling frame of the study covers both the public and the private quantity surveyors in consulting and contracting firms in Kaduna and Niger state. A total number of fifty five sets of questionnaires were distributed to quantity surveyors within Kaduna and Minna, in a random manner, out of which thirty seven sets were retrieved. According to Ref [24] cited by [25], a sample size larger than 30 and less than 500 are recommended for most research study. The research instrument employed compared with those adopted by [26] in his study of cost comparison between built-in security components and some physical characteristics of buildings in Nigeria and [27] in her study of internationally comparable indicators of violence, which relied on a questionnaire to elicit relevant information.

The data obtained from the questionnaires were analyzed using the descriptive and inferential analysis and respondent's opinion was graded on a Likert scale. In addition graphical illustration were produced which provides visual evidence of patterns within the data. The results are presented in the subsequent figures and tables below.

4.0 Result and Discussion

Data was collected in two states of Nigeria among the practitioners with various levels of expertise which were also analyzed and are presented in this section. Figure 1 shows the two states from which data was collected for this study. 65% of the overall number of the respondents was recorded from Kaduna in Kaduna state, while the remaining 35% of the respondents were from Minna in Niger state. The spread of the respondents within these two locations shows a good representation of the practice of quantity surveying within the region.

Figure 2 shows the academic qualification of the respondents. 59.5% of the respondents were BSc holders in quantity surveying from various universities, 35.1% had higher national diploma in quantity surveying, while 4.5% owned ordinary national diploma in quantity surveying.

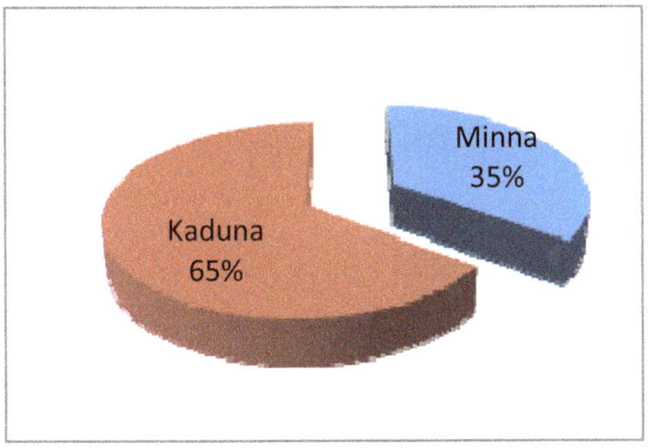

Figure 1: Frequency of the Data collected

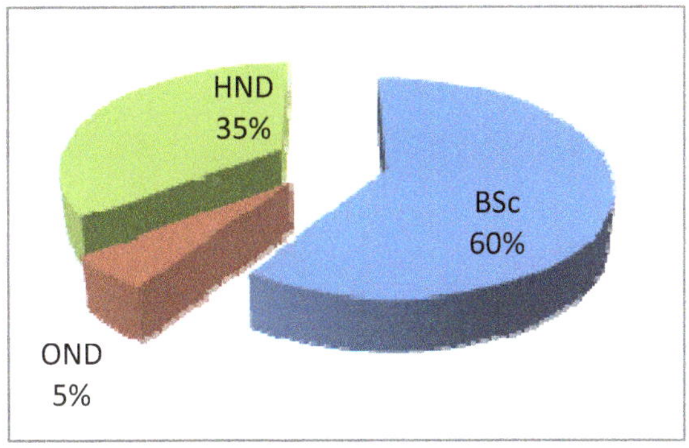

Figure 2: Qualification of respondents

Figure 3 above shows that majority of the respondents with 51%, had experience of between one to four years. 22% of the respondents had working experience of between five to nine years. 13% of the respondents had experience of between ten to fourteen years. While the remaining 14% covered respondents with working experience of fifteen years and above.

Figure 4 depicts information about respondent's place of work. Majority of the respondents with 65% work in quantity surveying consultancy firms. Contracting firms had minimum number of 13%, while the respondents working with the ministry of works covered the remaining 22%. However, the spread of the respondents within the various sectors is an indication that the data obtained is reliable for making analysis.

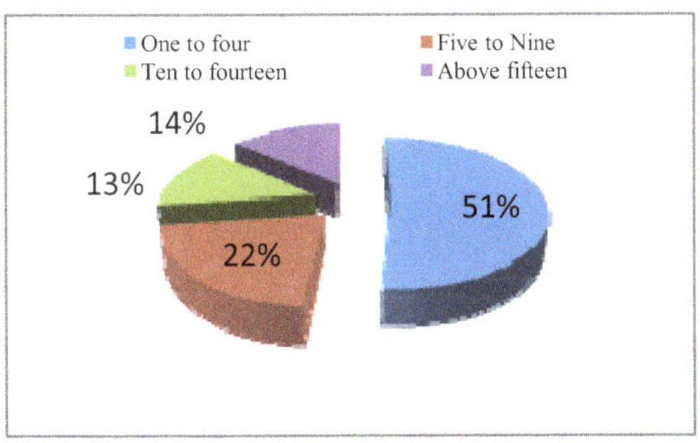

Figure 3: Working experience of respondents

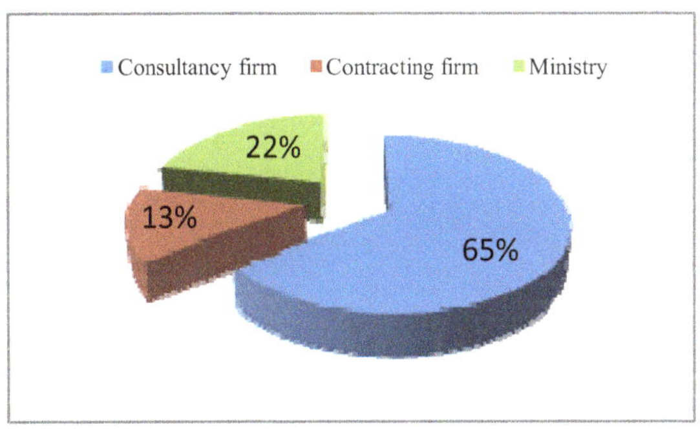

Figure 4: Place of work of respondents

Figure 5 below shows how often the respondents carry out the various cost management activities highlighted while performing their cost management functions. It was observed from the figure that over 95% of the respondents carry out all the cost management activities frequently with taking off quantities for works being done by them. Though, it can also be seen from the figure that some of the respondents only carry out the activity occasionally, but when compare the results, those who do it occasionally were being over shadowed. In the area of cost planning(group b),it was an exceptional result in which the selection of suitable cost analyses, and analyzing the chosen analysis with approximate estimate produced is only being occasionally done by most of the respondents. This can be seen from the difference in percentages obtained from the respondents on comparison, having 55% to 45% in the case of obtaining and selecting suitable cost analysis, and 59% to 35% in the case of analyzing approximate estimate with suitable cost analysis chosen for occasionally done and regularly done respectively. A very negligible percentage of the respondents have indicated that they have never carried out some of the cost management activities. This could be as a result of the level of working experience of such respondents. Their percentages though very minute, the lowest being 4%. While the highest were recorded in the case of developing budgets, and determination of design and price risk with percentages of 20% and 21.1% respectively, to those who do not carry out these activities at all as presented in table 1.

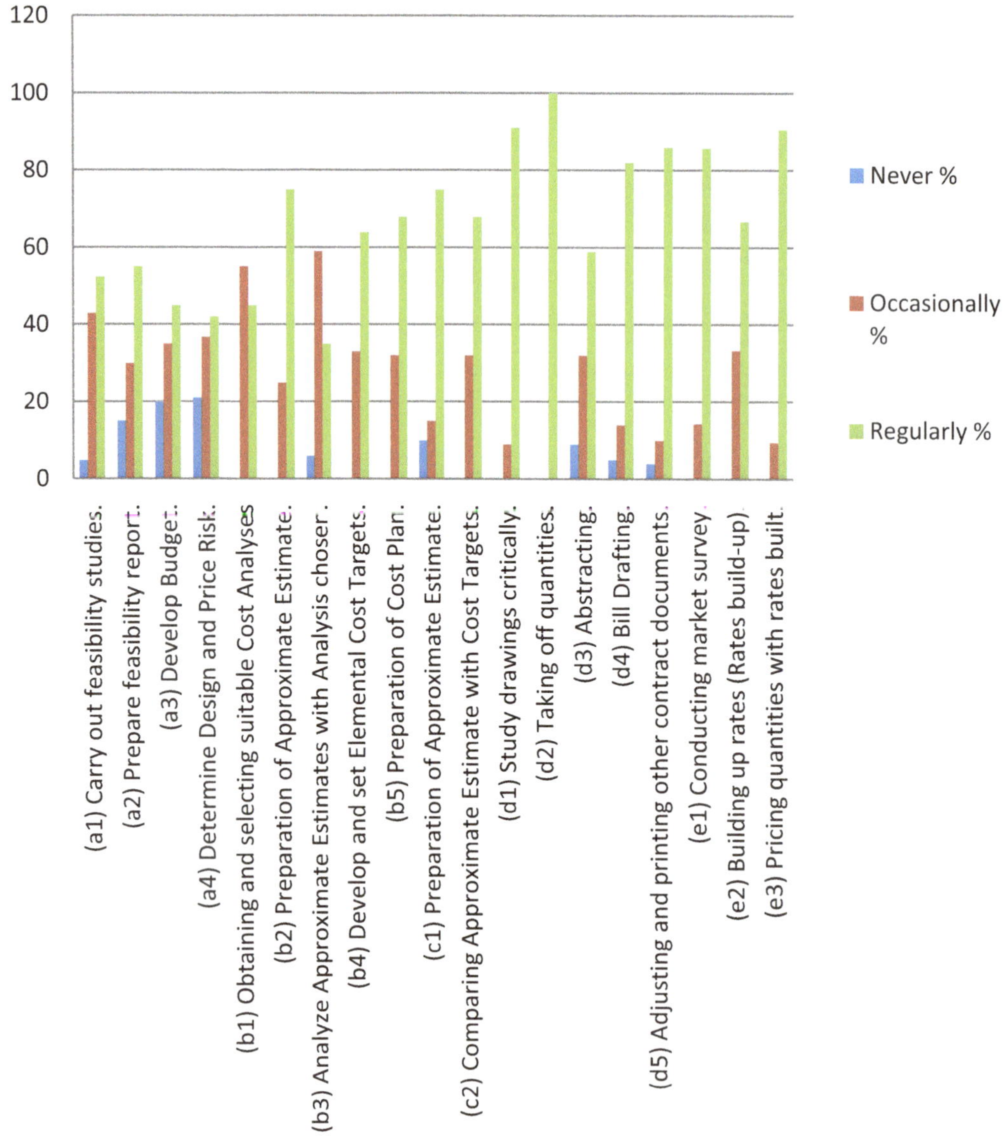

Note: a- establishing the budget, b-cost planning, c- cost control, d- contract documentation, and e- pricing contract bills

Figure 5: Frequency of conducting activities

Table 1: Frequency of conducting activities

Frequency of conducting activities	Never %	Occasionally %	Regularly %	Mean Score	SD
ESTABLISHING THE BUDGET.					
Carry out feasibility studies.	4.8	42.9	52.4	1.48	0.6
Prepare feasibility report.	15	30	55	1.4	0.75
Develop Budget.	20	35	45	1.25	0.79
Determine Design and Price Risk.	21.1	36.8	42.1	1.21	0.79
COST PLANNING.					
Obtaining and selecting suitable Cost Analyses	0	55	45	1.45	0.51
Preparation of Approximate Estimate.	0	25	75	1.75	0.44
Analyze Approximate Estimates with Analysis chosen.	6	59	35	1.29	0.59
Develop and set Elemental Cost Targets.	0	33	64	1.67	0.49
Preparation of Cost Plan.	0	32	68	1.68	0.48
COST CONTROL					
Preparation of Approximate Estimate.	10	15	75	1.65	0.67
Comparing Approximate Estimate with Cost Targets.	0	32	68	1.68	0.48
Frequency of conducting activities	**Never %**	**Occasionally %**	**Regularly %**	**Mean Score**	**SD**
CONTRACT DOCUMENTATION					
Study drawings critically.	0	9	91	1.91	0.29
Taking off quantities.	0	0	100	2	0
Abstracting.	9	32	59	1.5	0.67
Bill Drafting.	5	14	82	1.77	0.53
Adjusting and printing other contract documents.	4	10	86	1.81	0.51
PRICING CONTRACT BILLS					
Conducting market survey.	0	14.3	85.7	1.86	0.36
Building up rates (Rates build-up).	0	33.3	66.7	1.67	0.48
Pricing quantities with rates built.	0	9.5	90.5	1.9	0.3

Figure 6 shows the level of importance of activities for each of the various cost management activities previously outlined.

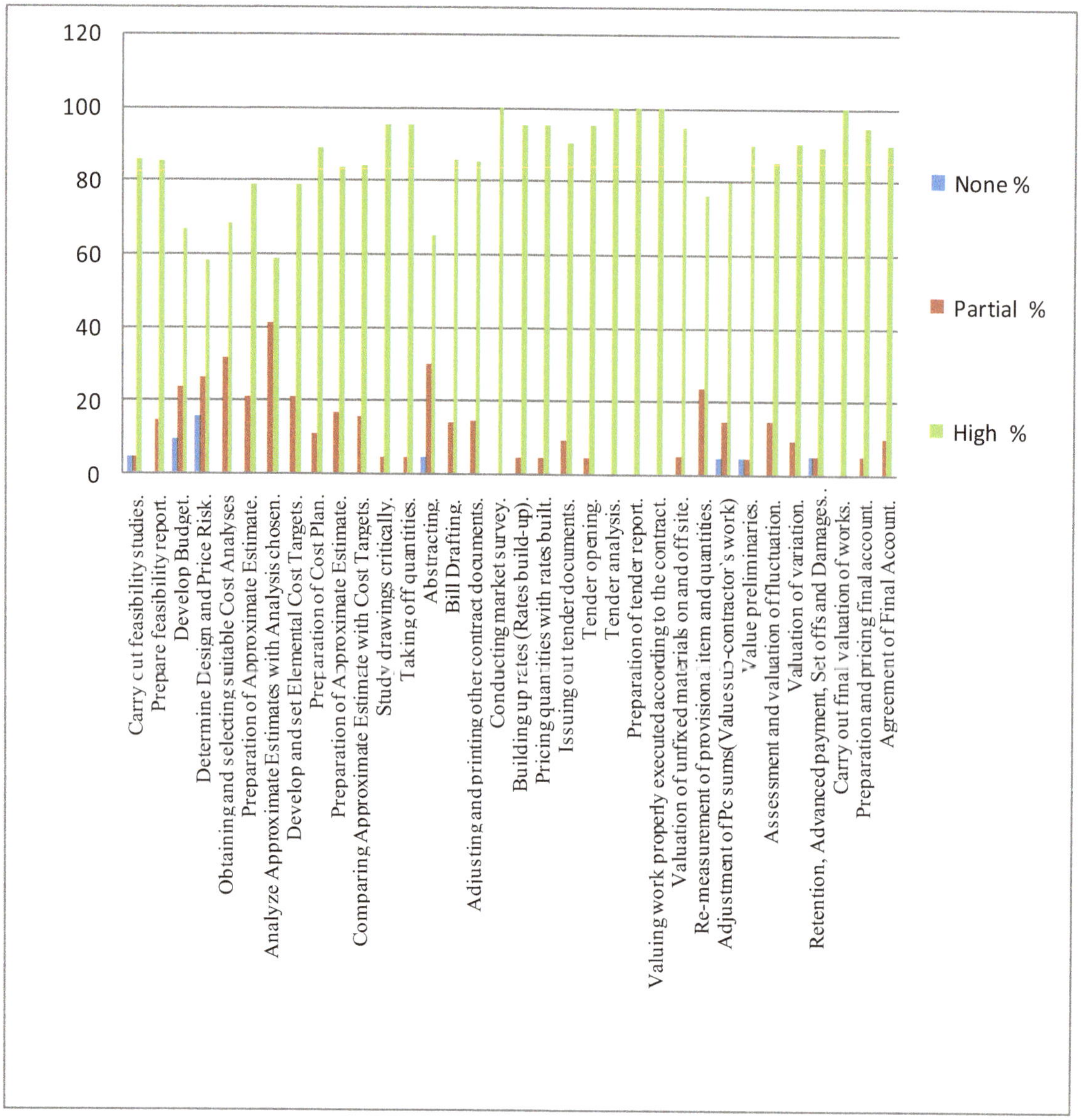

Figure 6: Level of importance of Activities

Results in figure 6 shows that majority of the respondent strongly believe that all the activities outlined are of paramount importance in performing the cost management functions of the quantity surveyor. This can clearly be seen from the figure above, as the activity which tagged lowest in the category of those with high importance had the support of about 57.9% of the respondent corresponding to the determination of design and price risk which is relatively high enough to determine the level of importance of such activity. Activities such as conducting market survey, and the valuations for both the interim and final valuation of work were shown to be of very high importance by all the respondents (100%). Other activities have percentages ranging from about 60% to 96% from the respondents as being activities of very high importance as shown in table 2. Level of difficulty encountered during the execution of works is summarized in figure 7.

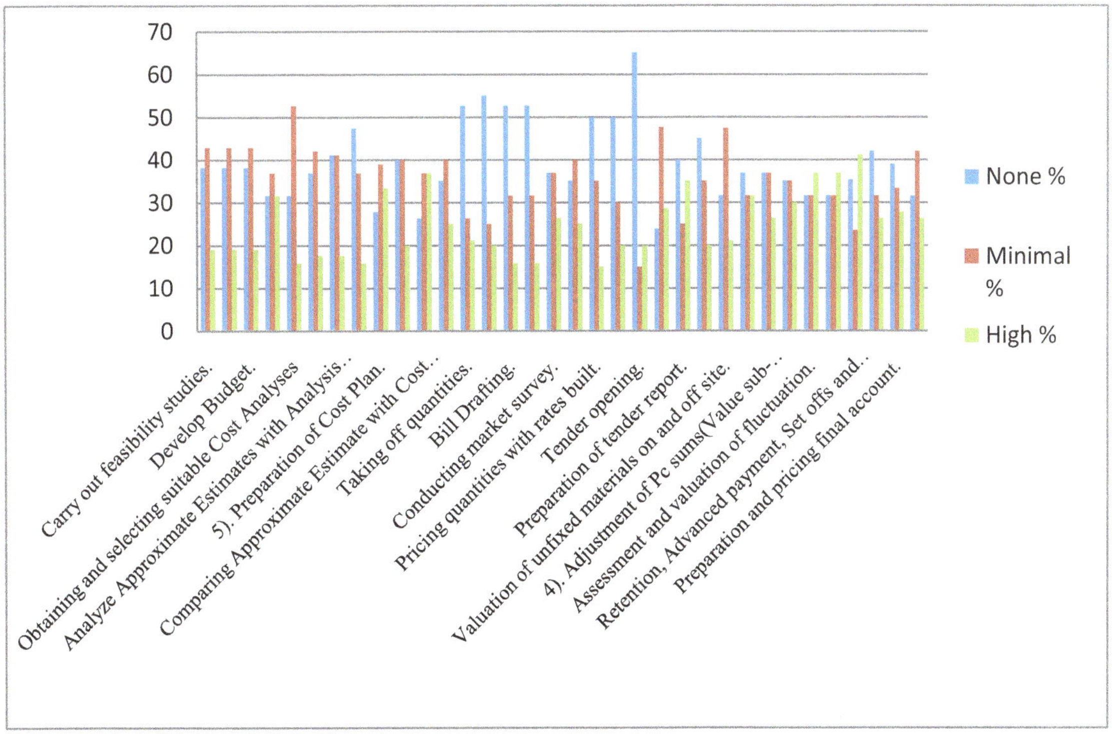

Figure 7: Level of Difficulty encountered while performing Activity

Table 2: Level of importance of Activities

IMPORTANCE OF ACTIVITY	None %	Partial %	High %	Mean Score	SD
ESTABLISHING THE BUDGET.					
1). Carry out feasibility studies.	4.8	4.8	85.7	2.81	.423
2). Prepare feasibility report.	0	15	85	1.85	.366
3). Develop Budget.	9.5	23.8	66.7	1.57	.676
4). Determine Design and Price Risk.	15.8	26.3	57.9	1.42	.769
COST PLANNING.					
1). Obtaining and selecting suitable Cost Analyses	0	31.6	68.4	1.68	.478
2). Preparation of Approximate Estimate.	0	21.1	78.9	1.79	.419
3). Analyze Approximate Estimates with Analysis chosen.	0	41.2	58.8	1.59	.507
4). Develop and set Elemental Cost Targets.	0	21.1	78.9	1.79	.419
5). Preparation of Cost Plan.	0	11.1	88.9	1.89	.323
COST CONTROL					
1). Preparation of Approximate Estimate.	0	16.7	83.3	1.83	.383
2). Comparing Approximate Estimate with Cost Targets.	0	15.8	84.2	1.84	.375
CONTRACT DOCUMENTATION					
1). Study drawings critically.	0	4.8	95.2	1.95	.218
2). Taking off quantities.	0	4.8	95.2	1.95	.218
3). Abstracting.	5	30.0	65.0	1.60	.598
4). Bill Drafting.	0	14.3	85.7	1.86	.359
5). Adjusting and printing other contract documents.	0	15.0	85.0	1.85	.366

IMPORTANCE OF ACTIVITY	None %	Partial %	High %	Mean Score	SD
PRICING CONTRACT BILLS					
1). Conducting market survey.	0	.0	100.0	2.00	.000
2). Building up rates (Rates build-up).	0	5.0	95.0	1.95	.224
3). Pricing quantities with rates built.	0	5.0	95.0	1.95	.224
TENDERING					
1). Issuing out tender documents.	0	9.5	90.5	1.90	.301
2). Tender opening.	0	4.8	95.2	1.95	.218
3). Tender analysis.	0	0	100.0	2.00	.000
4). Preparation of tender report.	0	0	100.0	2.00	.000
INTERIM VALUATION					
1). Valuing work properly executed according to the contract.	0	0	100.0	2	0
2). Valuation of unfixed materials on and off site.	0	5.3	94.7	1.95	0.229
3). Re-measurement of provisional item and quantities.	0	23.8	76.2	1.76	0.436
4). Adjustment of Pc sums(Value sub-contractor`s work)	5	15	80.0	2.8	4.549
5). Value preliminaries.	5	5	90.0	1.85	0.489
6). Assessment and valuation of fluctuation.	0	15	85.0	1.85	0.366
7). Valuation of variation.	0	9.5	90.5	1.9	0.301
8). Retention, Advanced payment, Set offs and Damages management.	5.2	5.3	89.5	1.84	0.501
FINAL ACCOUNT					
1). Carry out final valuation of works.	0	0	100.0	2.00	.000
2). Preparation and pricing final account.	0	5.3	94.7	1.95	.229
3). Agreement of Final Account.	0	10	90.0	1.90	.308

Results from the figure 7 shows that the level of difficulty faced is at the extreme, it is about 11.76% for all the activities. These activities include selection of suitable cost analysis, determination of design and price risk, preparation of cost plan, comparing approximate estimate with cost plan, tender analysis, Assessment and valuation of fluctuation, Valuation of variation, and Retention, Advanced payment, Set offs and Damages management with percentages from respondents as 52.6%, 31.6%, 33.3%, 36.8%, 26.8%, 36.8%, 36.8% and 41.2% respectively as shown in table 3 while figure 8 shows the results of cost indices.

Table 3: Level of Difficulty encountered while performing Activity

Level of Difficulty encountered while performing Activity	None %	Minimal %	High %	Mean Score	SD
ESTABLISHING THE BUDGET					
1). Carry out feasibility studies.	38.1	42.9	19	.81	.750
2). Prepare feasibility report.	38.1	42.9	19	.81	.750
3). Develop Budget.	38.1	42.9	19	.81	.750
4). Determine Design and Price Risk.	31.6	36.8	31.6	1.00	.816
COST PLANNING					
1). Obtaining and selecting suitable Cost Analyses	31.6	52.6	15.8	.84	.688
2). Preparation of Approximate Estimate.	36.8	42.1	17.6	.84	.765
3). Analyze Approximate Estimates with Analysis chosen.	41.2	41.2	17.6	.76	.752
4). Develop and set Elemental Cost Targets.	47.4	36.8	15.8	.68	.749
5). Preparation of Cost Plan.	27.8	38.9	33.3	.62	.744

Level of Difficulty encountered while performing Activity	None %	Minimal %	High %	Mean Score	SD
COST CONTROL					
1). Preparation of Approximate Estimate.	40	40.0	20	1.06	.802
2). Comparing Approximate Estimate with Cost Targets.	26.3	36.8	36.8	.80	.768
CONTRACT DOCUMENTATION					
1). Study drawings critically.	35	40.0	25.0	0.90	0.788
2). Taking off quantities.	52.6	26.3	21.1	0.68	0.820
3). Abstracting.	55	25.0	20.0	0.65	0.813
4). Bill Drafting.	52.6	31.6	15.8	0.63	0.761
5). Adjusting and printing other contract documents.	52.6	31.6	15.8	0.63	0.761
PRICING CONTRACT BILLS					
1). Conducting market survey.	36.8	36.8	26.3	0.89	0.809
2). Building up rates (Rates build-up).	35	40.0	25.0	0.9	0.788
3). Pricing quantities with rates built.	50	35.0	15.0	0.65	0.745
TENDERING					
1). Issuing out tender documents.	50	30	20.0	0.7	0.801
2). Tender opening.	65	15	20.0	0.55	0.826
3). Tender analysis.	23.8	47.6	28.6	1.05	0.74
4). Preparation of tender report.	40	25	35.0	0.95	0.887
INTERIM VALUATION					
1). Valuing work properly executed according to the contract.	45	35	20.0	0.75	0.786
2). Valuation of unfixed materials on and off site.	31.6	47.4	21.1	0.89	0.737
3). Re-measurement of provisional item and quantities.	36.8	31.6	31.6	0.95	0.848
4). Adjustment of Pc sums(Value sub-contractor`s work)	36.8	36.8	26.3	0.89	0.809
5). Value preliminaries.	35	35	30.0	0.95	0.826
6). Assessment and valuation of fluctuation.	31.6	31.6	36.8	1.05	0.848
7). Valuation of variation.	31.6	31.6	36.8	1.05	0.848
8). Retention, Advanced payment, Set offs and Damages management.	35.3	23.5	41.2	1.06	0.899
FINAL ACCOUNT					
1). Carry out final valuation of works.	42.1	31.6	26.3	.84	.83
2). Preparation and pricing final account.	38.9	33.3	27.8	0.89	0.832
3). Agreement of Final Account.	31.6	42.1	26.3	0.95	0.78

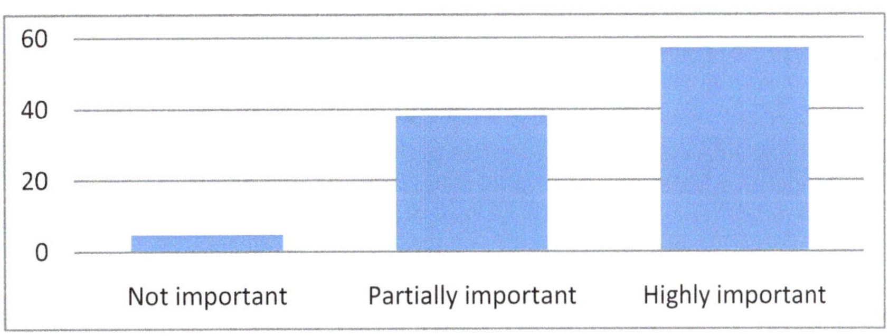

Figure 8: Establishing the Budget Using Cost Indices

Figure 8 shows the importance of cost indices for establishing the budget, were shown to be highly importance by 57% of the respondents and 38% of the respondents were partially in support of using cost indices to establishing the budget while about 5% of were for non-important this result was very low and insignificant when compare with respondents with highly and partially important. Cost management functions were also indentified and results are presented in table 4.

Table 4: Cost management functions performed for the project

COST MANAGEMENT FUNCTIONS	Client Type		Nature of project		Magnitude of Project in Million (Naira)		
	Public %	Private %	Building %	Civil %	50-100 %	100-150 %	Above 200 %
Establishing the Budget	40	29	40	60	33	33	33
Cost planning	60	86	80	50	67	67	83
Cost control	60	86	80	50	67	67	83
Contract Documentation	60	100	90	50	67	67	100
Pricing contract bills	80	29	50	50	33	100	50
Tendering	60	29	50	0	67	33	50
Interim valuation	40	100	80	50	67	33	100
Final Account	60	86	80	50	67	67	83

The table 4 above shows the extent to which cost management functions were conducted on some selected projects which the respondents have handled very recently or are currently being handled. It was observed from the table that budgets were established for 40% of the projects belonging to the public client, while 29% of projects owned by private clients had their budgets established by the respondents. 40% of project related to building works had the same function performed on them, while the same goes for 50% of projects relating to civil engineering works. Still on establishing the budget, a uniform percentage, 33% of the projects under the various magnitudes of project defined also had the function performed on them. With regards to cost planning and cost control, 86% of the projects belonging to the private client had these functions carried out on them by the respondents, while 60% for those belonging to the public client. As regards to the nature of works, 80% of building the projects went through cost planning and cost control, while 50% for those of civil engineering works. Based on magnitude of the project, projects over 200 million had the highest percentage of 83%. For the other two categories, 67% of the projects had cost planning and cost control conducted on them. The high percentage on the side of the private client shows that the private clients are in for value for money and minimization of wastages which is one of the main goals of JIT.

Contract documentation appeared as one of the most important cost management function of the QS. Contract documentation was conducted on all the projects belonging to the private client, equivalent to 100%, while 60% of those belonging to the public client had the same privilege. Also 90% and 60% of building and civil engineering projects went through the process of contract documentation. Based on magnitude of project, all the projects above 200 million went through the process of contract documentation, while 67% of the other two categories went through the same process as well. Projects over 200 million are definitely of large magnitude involving numerous consultants, sub-contractors and suppliers. Proper contract documentation is thus needed for proper administration of the project.

In the case of interim valuation, all projects belonging to the private client had this function performed on them, while 40% of projects owned by public clients went through the process. This again shows the differences in the orientation of both client types. The private client is more thorough in cost monitoring than the public client, and thus follows up the construction process strictly step by step making sure that payment is only made for what is due unlike what occurs in projects of public clients. With respect to nature of project, interim valuation was conducted for

80% of building projects and to 50% of civil engineering works. Likewise all the projects above 200 million went through interim valuation, 33% for projects within the range of 100-200 million ,and 67% for those between 50-100 million. The high percentage obtained for projects above 200 million could be as a result of various stake holders who form time to time demand for payment of work done, and also to finance the contractor and sub-contractors for work to continue without stoppage.

Moreover, 86% of the projects by private clients had their final accounts prepared after the execution of the projects. On the other hand, 60% of public projects had their final accounts prepared also. For building projects, 80% had their final account prepared, while 50% of civil engineering projects also had theirs prepared. 83% of projects above 200 million had their final account prepared on completion while 67% of projects between 50-100 million, and 100-200 million had their final accounts prepared.

5.0 Conclusions

All activities involved in the cost management function of the quantity surveyor are important, and value adding, corresponding to conversion activities in line with the Just-In-time/lean methodology. The activities are also being carried out on a frequent basis and supported by [12]. There are however extreme difficulties encountered while performing some of these activities especially in the area of cost planning, cost control and interim valuation. Since the activities are important and frequently done, they cannot therefore be removed from the cost management process in line with the principles of lean methodology as seen in figures 5,6,7,8, and tables 1, under data analysis. Improvement strategies only need to be sought in those grey areas for effective and improved service delivery in line with [12, 13].

The study indicates that all the documents highlighted used by the quantity surveyor are of paramount importance for effective service delivery. The study however revealed the unavailability of cost indices and standard guides which are needed to enhance the service delivery process of the cost management function of the quantity surveyor. The documents in question are also not sufficient enough for use by the Quantity Surveyor.

The study also revealed that some areas of information flow between the quantity surveyor and other stake holders need to be addressed especially in respect of the clients and designers (architects and engineers).

It was also discovered that majority of the cost management functions are performed under the private sector client and on projects ranging from two hundred million and above. This could be as a result of the demand for value for money by the private clients who are out on cost minimization and profit maximization.

6.0 Recommendations

The following recommendations are proposed based on the study

1. Designers should endeavor to service the quantity surveyor with all relevant contract drawings in good time, to avoid time wasted in no gainful venture, so as to meet up with the just-in-time principle of prompt service delivery.
2. Clients should endeavor to reply the Quantity Surveyor with the details of budget approval and confirmation of cost plan in good time, to void time wasted in no gainful venture, so as to meet up with the just-in-time principle of prompt service delivery.
3. Quantity Surveyors under the umbrella of the Nigeria Institute of Quantity Surveyor (NIQS) should come together and devise ways in developing an information system

like the Building Cost Information System (BCIS) which would go a long way in solving the problem of getting suitable cost analysis and cost indices.

4. The Nigeria Institute of Quantity Surveyor (NIQS) should also strive to see that quantity surveyors are highly engaged in both civil engineering work and Building work, so that value for money can fully be achieved on construction projects.
5. The public sector client should try and engage the Quantity Surveyor on time in their projects to perform his cost management functions.
6. The use of internet facilities in information should be explored to the fullest.

References

[1] A. S. Ali and K. H. Wen, "Building defects: Possible Solution for Poor construction Workmanship," Journal of Building Performance, vol. 2, pp. 59-69, 2011.
[2] D. Cartlidge, New Aspect of Quantity Surveying Practice. Oxford: Butterworth Heinemann, 2002.
[3] N. Nissanka and S. Sepani, "Acceptability of Lean Concepts to Functions of Quantity Surveyors in Sri Lanka," Lean Construction Journal, vol. 1, pp. 1-11, 2007.
[4] A. Ashworth and K. Hogg, Willis's Practices and Procedures for the Quantity Surveyors. Oxford: Blackwell Publishing Ltd, 2007.
[5] RICS, " QS Think Tank: The Challenge of Change," RICS, 1998.
[6] B. M. Testa, "Lean Manufacturing: processing buzzword or operational livesaver," Engineered wood journal, vol. 6, pp. 12-15, 2003.
[7] P. Ward and Z. Honggeng, "Impact of Information Technology Integration and Lean/Just-In-Time Practices on Lead-Time Performance," Decision Sciences, vol. 37, pp. 177-203, 2006.
[8] J. K. Liker, The Toyota Way: 14 Management Principles from the World's Greatest Manufacturer. USA: McGraw-Hill Education, 2004.
[9] R. R. Fullerton, C. S. McWatters, and C. Fawson, "An examination of the relationships between JIT and financial performance," Journal of Operations Management, vol. 21, pp. 383-404, 2003.
[10] A. Nahm, M. Vonderembse, and X. Koufteros, "The impact of organizational culture on time-based manufacturing and performance," Decision Sciences, vol. 35, pp. 579-608, 2004.
[11] P. Rondeau, M. vonderembse, and T. Ragu-Nathan, "Exploring work system practices for time-based manufacturers: Their impact on competitive capabilities," journal of Operations Management, vol. 18, pp. 509-529, 2000.
[12] J. M. Farrar, S. M. Abou Rizk, and X. Mao, "Generic Implementation of Lean Concepts in Simulation Model," Lean Construction Journal, vol. 1, pp. 1-23, 2004.
[13] S. M. Wagner and V. Silveira-Camargos, "Decision model for the application of just-in-sequence," International Journal of Production Research, vol. 49, pp. 5713-5736, 2011.
[14] G. Howell, "What is Lean Construction," in Proceedings of the 7th Conference of the International Group for Lean Construction, Berkeley, California, USA, 1999.
[15] W. J. Stevenson, Operations Management, 7 ed. Boston: McGraw-Hill/Irwin, 2002.
[16] J. Freire and L. F. Alarco´n, "Achieving Lean Design Process: Improvement Methodology," Journal of Construction Engineering & Management, vol. 128, p. 248, 2002.
[17] G. Ballard, "The Lean Project Delivery System," Lean Construction Journal, vol. 4, pp. 12-46, 2008.
[18] L. Koskela, Application of the new production philosophy to construction vol. 72: Stanford university (Technical Report No. 72, Center for Integrated Facility Engineering, Department of Civil Engineering). Stanford, CA, 1992.
[19] R. Hwang and H. Katayama, "Integrated procedure of balancing and sequencing for mixed-model assembly lines: a multi-objective evolutionary approach," International Journal of Production Research, vol. 48, pp. 6417-6441, 2010.
[20] M. Hook and L. Stehn, "Lean Principles in Industrialized Housing Production: the need for a cultural change," Lean Construction Journal, vol. 1, pp. 20-33, 2008.
[21] G. Ballard and G. Howell, "Lean Project Management," Building Research & Information, vol. 31, pp. 119-133, 2003.
[22] S. Shingo, "A Study of the Tovota Production System From an Industrial Engineering Viewpoint," 1989.
[23] S. Sernaratne and D. Wijesiri, "Lean Construction as a Strategic Option: Testing its Suitability and Acceptability in Sri Lanka," Lean Construction Journal, vol. 2, pp. 38-44, 2008.

[24] J. T. Roscoe, Fundamental research statistics for the behavioral sciences, 2 ed. New York: Holt, Rinehart and Winston 1975.

[25] R. A. Rahim, F. W. Jalaludin, and K. Tajuddin, "The importance of corporate social responsibility on consumer behaviour in Malaysia," Asian Academy of Management Journal, vol. 16, pp. 119-139, 2011.

[26] O. M. Anifowose, "Cost comparison between built-in security components and some physical characteristics of buildings in Nigeria: Case studies of residential, commercial and institutional buildings in Abuja and Minna," Journal of Building Performance, vol. 2, pp. 18-26, 2011.

[27] R. Diprose, "Physical Safety and Security: A Proposal for Internationally Comparable Indicators of Violence I am grateful for the comments and inputs of Sabina Alkire, Proochista Ariana, Afsan Bhadelia, Alex Butchart, Anna Hiltunen, María Ana Lugo, Andrew Mack, Luca Mancini, Emma Samman, Frances Stewart, Diego Zavaleta and workshop participants. Without their support this paper would not have been possible. A longer, more detailed working paper can be found at www.ophi.org.uk or www.crise.ox.ac.uk," Oxford Development Studies, vol. 35, pp. 431-458, 2007.

WOOD ASH FROM BREAD BAKERY AS PARTIAL REPLACEMENT FOR CEMENT IN CONCRETE

Akeem Ayinde Raheem[1*], Olumide A. Adenuga[2]

[1]Civil Engineering Department, Ladoke Akintola University of Technology, Ogbomoso, Nigeria
[2]Building Department, University of Lagos, Akoka-Lagos, Nigeria

*Corresponding E-mail : raheemayinde@yahoo.com

Abstract

This paper reports the results of experiments evaluating the use of wood ash from bread bakery as partial replacement for ordinary Portland cement in concrete. The chemical composition of the wood ash as well as the workability and compressive strength of the concrete were determined. Wood ash was used to replace 5% - 25% by weight of the cement in concrete. Concrete with no wood ash serves as the control. The mix ratio used was 1:2:4 with water to binder ratio maintained at 0.5. The Compressive strength was determined at curing ages 3, 7, 28, 56, 90 and 120 days. The results showed that wood ash from bread bakery is a Class F fly ash since the sum of (SiO2 +Al2O3 +Fe2O3) is greater than 70%. The compressive strength of wood ash concrete increases with curing period and decreases with increasing wood ash content. There was a sharp decrease in compressive strength beyond 10% wood ash substitution. It was concluded that a maximum of 10% wood ash substitution is adequate for use in structural concrete.

Keywords: *Wood ash, Bread bakery, Pozzolan, Workability, Compressive strength*

1.0 Introduction

Concrete is a versatile construction material that is widely used in virtually all structural works. It is a composite material comprising cement, aggregates, water and admixtures. Wood ash concrete is a type of concrete with wood ash as partial replacement for cement.

Fly ash is defined as a material collected from a practical combustion device after the fuel is consumed [1]. It is an industrial by- product that possesses hydraulic and pozzolanic properties. This type of material imparts technical advantages to the resulting concrete and also enables larger quantities of cement replacement to be achieved [2]. Wood ash is similar to fly ash since it is obtained from a practical combustion device – bread bakery in this case.

2.0 Literature Review

The fly ashes that have been widely studied are those obtained from power plants where coal is used as a fuel [3-8]. This study sought an alternative source of fly ash by investigating the properties of the wood ash obtained from bread bakery. Bread is a staple food in Nigeria that is consumed regularly by a large number of people. Thus, bread bakery is located in every nooks and cranny of the country to cater for the high demand for the product. The bakeries use wood as fuel for their operation thereby resulting in high volume of ash as by-product. The ash particles solidify as microscopic, glassy spheres that are collected from the bakery's exhaust before they can "fly" away, hence the product's name - fly ash [9].

Fly ash with potential for use in concrete is classified into two categories as Class C and Class F [10]. The classification is based on sum of (SiO2 +Al2O3 +Fe2O3). Where the sum is greater than or equal to 70%, the ash is classified as Class F, while it is Class C if the sum is greater than or equal to 50% [11].

This study examines the use of wood ash from bread bakery as partial replacement for ordinary Portland cement in concrete. It involves the determination of the chemical composition of the ash and evaluation of the workability and compressive strength of the concrete.

3.0 Methodology

The methodology adopted in carrying out the study is discussed under two sub headings of Materials and Specimen Preparation for the tests carried out.

3.1 Materials

The wood ash used for this study was collected from Ladoke Akintola University of Technology Ogbomoso, Nigeria, bread bakery located behind Food Science and Engineering Laboratory (Owodunni Building). The wood ash was carefully collected at the exhaust of the bakery, after the fuel had burned out, with the aid of hand scoop. It was sieved to separate the ash from any unwanted material.

The wood ash was ground to fine powdery form and only those retained on a 0.425 μm sieve was used for the experiments. The ash was taken to Larfarge Cement, West Africa Portland Cement Company (WAPCO) Sagamu, Ogun State, Nigeria, for chemical analysis. The Ordinary Portland cement (Dangote, Brand) used was obtained from a local retailer in Ogbomoso. Sharp sand obtained from river bed was used as fine aggregates and granite as coarse aggregates. The fine and coarse aggregates used were also obtained from Ogbomoso, Nigeria.

3.2 Specimen Preparation

Wood ash was used to replace 5% - 25% by weight of the cement in concrete. Concrete with no wood ash serves as the control. The mix ratio used was 1:2:4 (binder, sand and granite) with water to binder ratio maintained at 0.5. The details of the concrete mixtures used in the tests are listed in Table 1.

Table 1: Proportioning of Concrete

Percentage of Wood Ash (%)	Cement (kg)	Wood ash (kg)	Fine Aggregate (kg)	Coarse Aggregate (kg)
0	21.48	-	42.96	85.92
5	20.406	1.074	42.96	85.92
10	19.332	2.148	42.96	85.92
15	18.258	3.222	42.96	85.92
20	17.148	4.296	42.96	85.92
25	16.11	5.370	42.96	85.92

Compressive strength measurements were made on 150mm cube concrete specimens. The specimens were cast in three layers, each layer being tamped with 35 strokes of the tamping rod spread uniformly over the cross section of the mould. The top of each mould was smoothened and leveled and the outside surfaces cleaned. The moulds and their contents were kept in the curing room at temperature of 27 + 50C and relative humidity not less than 90% for 24hours. All specimens were demoulded after 24hours and cured in water at 27 + 50C. The Compressive strength was determined at curing ages 3, 7, 28, 56, 90 and 120 days; using compression machine with maximum capacity of 1500 kN (Model 50- C34AC and Serial no 02094910). The strength value was the average of three specimens.

Slump and compacting factor tests were carried out to check the effect of wood ash on the workability of fresh concrete. The tests were carried out in accordance with the requirements of BS 1881: Part 102 (1983) [12] for slump test and BS 1881: Part 103 (1983) [13] for compacting factor test.

4.0 Results and Discussion

This section discusses the results obtained from the various laboratory tests carried out.

4.1 Chemical Composition

The elemental oxides present in the wood ash sample are shown in Table 2. The results indicated that the wood ash had combined percentages of SiO2, Al2O3 and Fe2O3 of more than 70% (73%). This indicates that it is a good pozzolanic material in accordance with the requirements in ASTM C 618. As reported by [11], wood ash from bread bakery falls under the category of Class F fly ash since the sum of (SiO2 +Al2O3 +Fe2O3) is greater than 70%.

Table 2: Chemical Composition of Wood Ash

Chemical Constituents	Percentage Composition (%)			
	Sample 1	Sample 2	Sample 3	Average
SiO2	61.12	62.75	61.74	61.18
Al2O3	7.05	7.95	7.57	7.52
Fe2O3	3.22	2.91	3.44	3.19
CaO	10.56	12.94	11.80	11.77
MgO	2.34	2.00	3.11	2.48
SO3	1.84	1.80	1.82	1.82
Na2O	1.15	1.00	1.11	1.09
K2O	3.56	4.11	3.75	3.81
CaCO3	6.12	6.22	6.31	6.22
LOI	2.95	3.20	2.99	3.05
LSF	1.20	1.37	1.27	1.28
SR	4.15	4.25	4.56	4.32
AR	7.48	7.56	7.62	7.55
Total SiO2 + Al2O3 + Fe2O3	71.39	73.61	72.75	72.58

4.2 Sieve Analysis

The results of the sieve analysis for fine and coarse aggregates are presented in Figures 1 and 2 respectively. It could be observed from the Figure 1 that the coefficient of uniformity (Cu) and coefficient of curvature (Cc) for fine aggregates are 4.95 and 1.24 respectively. Thus, the sand is classified as being well graded [14]. Similarly, the Cu and Cc for coarse aggregates as obtained from Figure 2 are 0.89 and 1.75 respectively. This shows that the granite is uniformly graded [14]. Therefore, the fine and coarse aggregates are suitable for making good concrete.

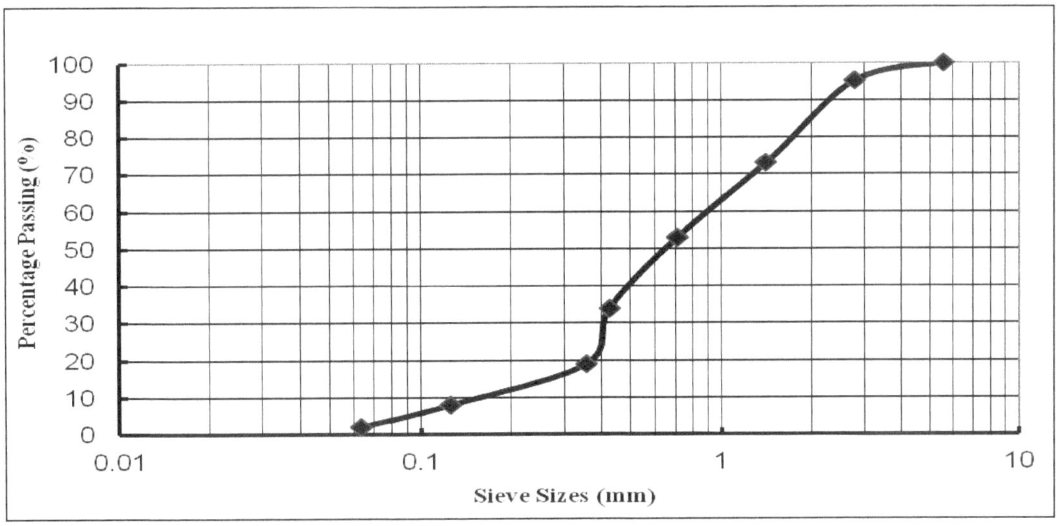

Figure 1: Particle Size Analysis Graph for Sand

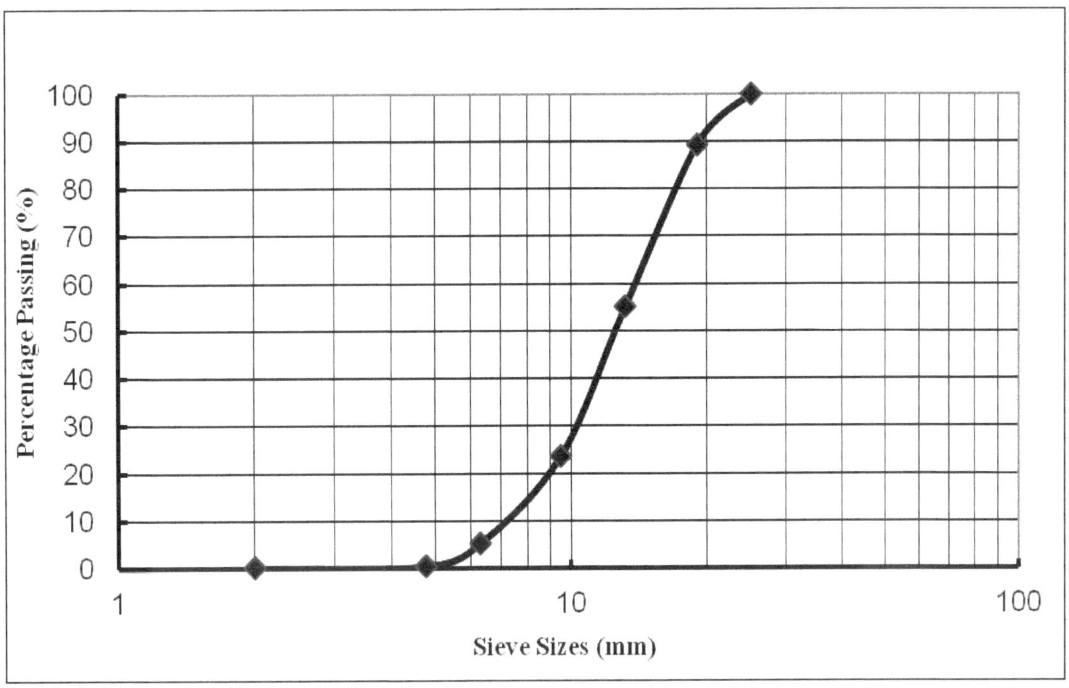

Figure 2: Particle Size Analysis Graph for Granite

4.3 Workability

The results of the slump and compacting factor indicating the workability of the wood ash concrete are shown in Table 3. The table indicates that the slump first decreases for up to 10% wood ash replacement, and then increases with increasing wood ash content. The compacting factor also follows a similar trend. These results indicate that concrete containing wood ash beyond 10% replacement level becomes more workable as the wood ash content increases meaning that less water is required to make the concrete workable. The lower water demand can be attributed to the finer ash particles with filling role in concrete mix.

Table 3: Slump and Compacting Factor of Wood ash Concrete

Percentage of Wood ash replacement (%)	Slump (mm)	Compacting Factor
0	110	0.98
5	85	0.97
10	45	0.96
15	100	0.98
20	140	0.98
25	170	0.98

4.4 Compressive Strength

The compressive strength of concrete cubes tested at 3, 7, 28, 56, 90 and 120 days are presented in Figure 3. The figure indicates that compressive strength generally increases with curing period and decreases with increasing wood ash content. The result at 3 days showed that the compressive strength of concrete with 5% wood ash content is the highest at this level. Similarly, both 5% and 10% wood ash concrete have higher strength than the control at 7 days. These results are in line with the findings of [4] that addition of fly ash increases the early compressive strength of concrete.

At 28 days, the compressive strength ranges from 18.44 Nmm-2 for the control, to 6.67 Nmm-2 for 25% wood ash replacement. The 5% and 10% wood ash replacements have values of 19.10 Nmm-2 and 21.11 Nmm-2 respectively which are greater than that of the control concrete. This shows that pozzolanic action has commenced as evident from the higher percentage of strength increase by the wood ash concrete which can be attributed to the reaction of wood ash with calcium hydroxide liberated during the hydration of cement. This is in line with previous studies on pozzolans [15 - 19].

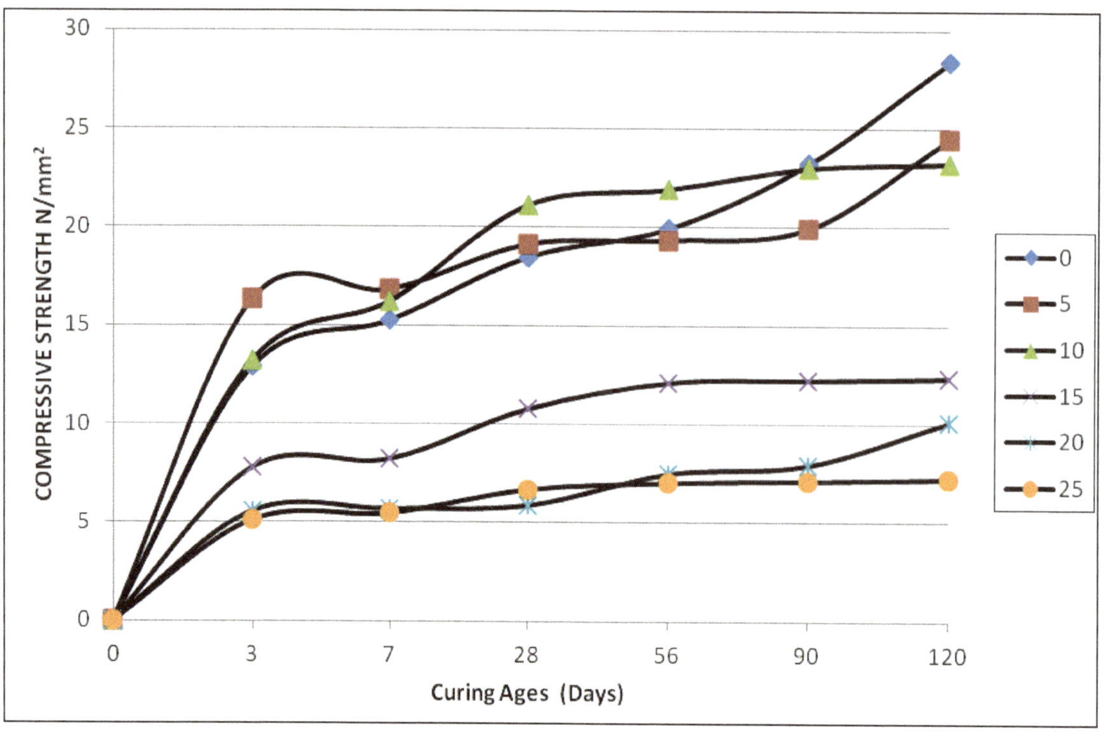

Figure 3: Effect of curing age on the compressive strength of concrete

There was continuous and significant improvement in strength development at 56 and 90 days. This confirms the continuous pozzolanic reaction in wood ash concrete. It is an indication of the role of reactive silica in strength development of pozzolan cement concrete. As the wood ash content increases, more silica is available to react with the lime produced during hydration of cement, thus producing more cementing materials, which contribute to the higher compressive strength [15]. The 10% wood ash replacement still maintains the lead in strength development.

At 120 days, there was continuous increase in compressive strength but the control has the highest value in this case. The 5% wood ash replacement also has a higher value than its 10% counterpart. It could also be observed from Figure 3 that there is a sharp decrease in compressive strength beyond 10% wood ash substitution. This is an indication that the use of wood ash beyond 10% replacement level is not advantageous for structural concrete.

Figure 4 shows the effect of wood ash percentage replacement on the compressive strength of concrete.

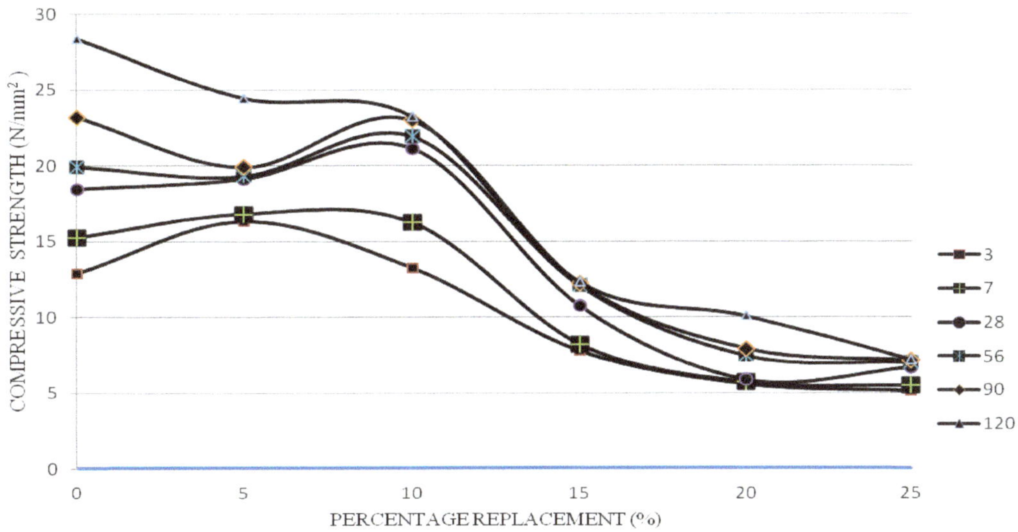

Figure 3: Effect of wood ash percentage replacement on the compressive strength of concrete

As could be observed from the figure, there is a general decrease in compressive strength as the wood ash content increases. However, a notable increase in strength can be observed between 5% and 10% replacement level. Thus, it could be concluded that only up to 10% wood ash substitution is adequate for use in structural concrete. However, since all the concrete cube specimens meet the minimum strength of 6Nmm-2 after 28 days of curing recommended by BS 5224: 1976 [20] for masonry cement, they could be used for general concrete works where strength is of less importance such as in mass concrete, floor screed and mortar.

5.0 Conclusion

From the results of the various tests performed, the following conclusions can be drawn:

1. Wood ash is a suitable material for use as a pozzolan, since it satisfied the requirement for such a material by having a combined (SiO_2 +Al_2O_3 +Fe_2O_3) of more than 70%. It is classified as a Class F fly ash.

2. Concrete becomes more workable as the wood ash content increases. This means that wood ash concrete has lower water demand.

3. The compressive strength generally increases with curing period and decreases with increasing wood ash content. Wood ash concrete has higher strength than the control at early ages.

4. Only up to 10% wood ash substitution is adequate for use in structural concrete.

It is recommended for further studies that other properties of concrete such as tensile strength, flexural strength and durability characteristics be investigated in order to provide an enlarged database on the performance characteristics of wood ash blended cement concrete.

Acknowledgement

The authors acknowledge the management and staff Lafarge Cement, West Africa Portland Cement Company (WAPCO) Sagamu Works, Ogun State, Nigeria; for the opportunity given to perform the various laboratory tests using their facilities.

References

[1] S. Wang, A. Miller, E. Llamazos, F. Fonseca, and L. Baxter, "Biomass fly ash in concrete: Mixture proportioning and mechanical properties," Fuel, vol. 87, pp. 365-371, 2008

[2] S. Wang, and L. Baxter, "Comprehensive study of biomass fly ash in concrete: Strength, microscopy, kinetics and durability," Fuel Processing Technology, vol. 88, pp. 1165-1170, 2007

[3] S. Antiohos, K. Maganari, and S. Tsimas, "Evaluation of blends of high and low calcium fly ashes for use as supplementary cementing materials," Cement & Concrete Composites, vol. 27, pp. 349-356, 2005

[4] R. Siddique, "Performance characteristics of high-volume Class F fly ash concrete," Cement and Concrete Research, vol. 34 (3), pp. 487-493, 2004

[5] C. S. Poon, X. C. Qiao, and Z. S. Lin, "Pozzolanic properties of reject fly ash in blended cement pastes," Cement and Concrete Research, vol. 33, pp. 1857-1865, 2003

[6] N. Bouzoubaa, M. H. Zhang, and V. M. Malhotra, "Mechanical properties and durability of concrete made with high-volume fly ash blended cements using a coarse fly ash," Cement and Concrete Research, vol. 31 (10), pp. 1393-1402, 2001

[7] M. N. Hague, and O. Kayali, "Properties of High-Strength Concrete using a Fine Fly Ash," Cement and Concrete Research, vol. 28 (10), pp. 1445-1452, 1998

[8] T. R. Naik, S. S. Singh, and M. Hossain, "Enhancement in Mechanical Properties of Concrete due to Blended Ash," Cement and Concrete Research, vol. 26 (1), pp. 49-54, 1996

[9] V. N. Dwivedi, N. P. Singh, S. S. Das, and N. B. Singh, "A new pozzolanic material For cement industry: Bamboo leaf ash," International Journal of Physical Science, vol. 1(3), pp. 106-111, 2006

[10] ASTM C 618-9, "Standard Specification for Fly Ash and Raw or Calcined Natural Pozzolan for use as a Mineral Admixture in Portland Cement Concrete," Annual Book of ASTM Standards, Philadelphia, USA, ASTM, 1991

[11] S. Jerath, and N. Hanson, "Effect of fly ash content and aggregate gradation on the durability of concrete pavement," Journals of Material in Civil Engineering, vol. 19 (5), pp. 367-375, 2007

[12] BS 1881: Part 102, "Methods for determination of Slump", London, British Standard Institution, 1983

[13] BS 1881: Part 103, "Methods for determination of Compacting factor," London, British Standard Institution, 1983

[14] G. N. Smith, and G. N Smith Ian, "Elements of Soil Mechanics", Seventh Edition, London, Blackwell Science, 1998

[15] D. A. Adesanya, and A. A.Raheem, "A study of the workability and compressive strength characteristics of corn cob ash blended cement concrete," Construction and Building Materials, vol. 23, pp. 311 – 317, 2009

[16] C. Nuntachai, J. Chai, and K. Kraiwood, "Utilization of Bagasse ash as a Pozzolanic Material," Construction and Building Materials, vol. 23 (11), pp. 3352-3358, 2009

[17] S. T. Lee, H. Y. Moon, and R. N. Swamy, "Sulfate attack and role of silica fume in resisting strength loss," Cement & Concrete Composites, vol. 27, pp. 65-76, 2005

[18] F. V. Riza, "Application of RHA'S Pozzolanic Properties in the Making of CEB," International Journal of Sustainable Construction Engineering & Technology, vol. 2 (2), pp. 32-36, 2011

[19] K. M. A. Hossain, "Chloride induced corrosion of reinforcement in volcanic ash and pumice based blended concrete," Cement & Concrete Composites, vol. 27, pp. 381-390, 2005

[20] BS 5224, "Standard Specification for Masonry Cement," London, British Standard Institution, 1976.

Properties of Compressed Stabilized Earth Blocks (CSEB) For Low-Cost Housing Construction: A Preliminary Investigation

Baba Shehu Waziri[1], Zanna Alhaji Lawan[2], Mustapha, Ma'aji Mala[3]

[1] Department of Civil and Water Resources Engineering, University of Maiduguri, Maiduguri, Nigeria
[2] Works Department University of Maiduguri, Maiduguri, Nigeria
[3] Central Bank of Nigeria, Corporate office, Abuja, Nigeria

*Corresponding E-mail : shehuwaziri@gmail.com

Abstract

Materials used for the construction of walls are normally required to possess adequate strength and erosion resistance. The study investigates the suitability of stabilized laterite soils for the production of compressed earth blocks for low-cost housing construction. Soil samples for the experiment were obtained at two different locations. Sample I was obtained at a borrow pit along Gujba road in Damaturu Yobe state while Sample II was taken at a borrow pit near lake Alau in Borno sate, Nigeria. The results of the study revealed that the specific gravity, bulk density, moisture content and plasticity index of both samples showed satisfactory performance. Different cement stabilization levels of 0%, 2.5%, 5% and 7.5% were used to prepare the specimens for testing. The blocks were moulded using hand operated CINVA-Ram machine. The maximum compressive strength of 2.48N/mm^2 was obtained with stabilization level of 7.5% with sample I at 28 days curing. The strength of the specimens increases with increasing cement content with an average value of 0.35N/mm^2. For higher strength requirements different stabilization options can be considered.

Keywords: *Stabilized earth block, low cost housing, Construction, Nigeria*

1.0 Introduction

The investigation of alternative materials for the construction of low cost housing has been the focus of many studies in many developing countries. The present rate of construction in developing countries according to [1] is generally sufficient to meet the needs of only 10% of the net increase in population per year. This is partly due to the unavailability and the soaring costs of conventional building materials. As this shortage of building is becoming worse, more efforts are being made to develop cheap, serviceable and energy efficient construction materials for the construction of affordable sustainable buildings.

Masonry is one of the most popular materials for housing construction due to its useful properties such as durability, relatively low cost, good sound and heat insulation, acceptable fire resistance, adequate resistance to weathering and attractive appearance [2, 3]. The global interest about the environment according to [4] has increased the use of earth as a building material. The direct use of earth without modification for wall construction in any form has the disadvantage of low performance. The shortcomings principally are low mechanical characteristics, unsatisfactory resistance to weathering and liability to volume change especially in the case of clay. These disadvantages can be improved to make the material compatible with desired application in construction by combined chemical and mechanical action technically known as stabilisation [1, 2, 3, 5, 6]

Compressed Stabilized Earth Blocks (CSEB) offer a number of advantages which includes increased utilization of local material and reducing the cost of transportation as the production is in situ, makes quality housing available to more people, and generates local economy rather than spending for import materials [2, 4]. Other advantages are faster and easier construction method resulting in the lesser requirement of skilled labour, good strength, insulation and thermal properties, less carbon emission and embodied energy in the production phase, create extremely low level of waste and cause no direct environmental pollution during the whole life cycle. Earth bricks have the ability to absorb atmospheric moisture which creates healthy environment inside a building for its occupants. One of the drawbacks of using earth alone as a material for construction as posited by [7] is its durability which is strongly related to its compressive strength. Because most soil in their natural condition lack the strength, dimensional stability and durability required for building construction. These properties can be enhanced through stabilization [8 - 9].

2.0 Literature Review

2.1 Cement Stabilization

Soil stabilization according to [1] is the alteration of any property of a soil to improve its engineering performance. The chief factors affecting stabilization are soil type, cement content, compaction and method of mixing with soil type being the most important [1]. The modification of the properties of soil-water-air system makes the soil compatible with desired applications in construction [2]. One of the main functions of the stabilizing medium is to reduce the swelling properties of the soil through forming a rigid framework with the soil mass, enhancing its strength and durability [10]. Portland cement is the most widely used stabilizer for earth stabilization. Cement has the ability to reduce liquid limit (LL) and increase plasticity index (PI) and hence increases the workability of soil. The addition of chemical stabilizers like cement and lime has twofold effects of acceleration of flocculation and promotion of chemical binding. The chemical binding depends upon the type of stabilizers employed [11]. The study of [7] revealed that soils with Plasticity Index (PI) less than 15% are suitable for cement stabilization. In cement stabilization [12] observed that the content of the cement binder in the mix ranges between 4% and 10% of the soil dry weight. However, [4] posited that if the content of the cement binder is greater than 10% it becomes uneconomical for the production of CSEB.

2.2 Stabilized Pressed Blocks

The materials used for the construction of wall are normally required to possess adequate compressive strength and erosion resistance. Such properties of the soil can be improved by stabilizing it with cementious admixtures such as cement and lime [1]. The strength of stabilised soil can further be improved by the process of compaction which leads to higher densities, thereby higher compressive strength and better resistance to erosion. Exploring the stabilization and compacting techniques, a cheap yet strong and durable material for wall construction is the stabilized earth block. The merit of these blocks is low-cost, use of locally available material, blocks can be made at site with no transportation cost and simplicity in manufacture [1, 2, 3, 6, 13]. As noted by [6] the strength of such blocks increases with density. CSEB brick requires compaction whether it is static, dynamic or vibro-static methods [4]. In preparing the soil for block production, there is need for careful and correct selection of the soil to get the best result and after the mix was put in the mould, it should be given proper compaction. Proper curing should also be made which prevents rapid drying.

A striking contrast between CSEB and conventional bricks is the energy consumed during the production process and carbon emission. CSEB creates 22 kg CO_2/tonne compare to that of concrete blocks (143 kg CO_2/tonne), common fired clay bricks (200 kg CO_2/tonne) and aerated concrete blocks (280-375 kg CO_2/tonne) during production [15]. In average, cement stabilized earth bricks consumed less than 10% of the input energy as used to manufacture similar fired clay and concrete masonry unit [8]. Production of CSEB requires moderate to low skilled worker since the CSEB manufacture is very simple. It only takes 3 stages process which are: soil preparation, mix compression and the curing [2, 4, 6, 13].

3.0 Materials and Methods

The soil samples for the experiment were obtained at two different locations. Sample I was obtained at a borrow pit along Gujba road in Damaturu, Yobe state, Nigeria. Sample II was obtained at a borrow pit near Lake Alau in Borno State, Nigeria. Different tests were conducted to determine the characteristics of the soils (specific gravity, bulk density, loose density and moisture content). Sieve analysis was conducted using the BS set of sieves in accordance with [15]. The atterberg limit test (plastic limit, liquid limit and the plasticity index) were determined in accordance with [16]. Compaction test (standard proctor test) was performed to determine the optimum moisture contents and dry densities of the soil samples. For the determination of the OMC a series of soil mixes were prepared over a range of moisture contents. When the dry density was plotted against the respective moisture content, the two samples produced typical moisture-density curves with clearly identifiable optimums. The specimens were prepared with the desired levels of stabilisations of 0%, 2.5%, 5%, and 7.5% of Ordinary Portland Cement (OPC) in accordance with [16]. The blocks were moulded using the CINVA-Ram machine which is a hand operated press for block making. The CINVA-ram machine delivers a capacity pressure of 2 N/mm^2. The CINVA-Ram was also considered to represent an achievable and economic option [17]. The machine comprises of a rammer and a lever arm which is used to push out the finished blocks of geometry 210mm×100mm×80mm from a four compartment box. The blocks were then cured for 28days.

Finally the compressive strength test for the different levels of cement stabilization at the optimum w/c ratio at different curing ages for the two soil samples was undertaken. The objectives being to investigate the suitability of blocks made with stabilized soil at the optimum w/c ratio for use as masonry units. The crushing strengths of the blocks were tested using the universal testing machine. A total of 64 blocks were prepared and crushed at different curing ages of 7, 14, 21 and 28 days. A metal plate that measures 250 mm×120×10mm was placed on top and bottom of the sample to allow for even distribution of the load. The load was then applied at a constant rate of displacement. The bottom plate advancing at 1.02 mm/min following the failure of the specimen, the maximum load was recorded and crushing strength was evaluated on the basis of equation below:

$$P = \frac{F}{A}$$

Where;

P is the crushing stress (N/mm^2)

F is the failure load (N)

A is the area of specimen (mm^2)

4.0 Results and Discussion

4.1 Characteristic Properties

The results of the different properties of the soil samples are presented in Table 1. The evaluation of the different properties revealed that the S.G of 2.67 and 2.71 of sample I and Sample II are within the range of 2.55 and 4.0 recommended by [18]. The values are greater than the value obtained for earth worm cast by [13]. The bulk densities for the samples I and II were found to 1736 kg/m^3 and 1758kg/m^3 respectively. The natural moisture contents of the two samples were found to be 15.3% and 21.2%. The results of the tests on the properties of the two samples revealed that the soil samples exhibit similar properties.

Table 1: Characteristic Properties of soil Samples

Properties	Sample I	Sample II
Bulk Density(kg/M^3)	1736	1758
Moisture content (%)	15.3	21.2
Liquid Limit (%)	26.9	32.1
Plastic Limit (%)	14.3	17.8
Plasticity Index	12.6	14.4
Specific gravity	2.67	2.71

4.2 Particle size Distribution

The result of the particle size distribution carried out in accordance with [15] for sample I and Sample II are presented in Table 2 and Table 3 respectively.

Table 2: Particle Size distribution for Sample I

S/No	Sieve Size	Wt Retained	%Wt Retained	Cum. % Wt Retained	% Wt passing
1	4.75mm	2.0	0.56	0.56	99.44
2	1.70mm	7.5	2.11	2.67	97.33
3	1.18mm	4.5	1.27	3.97	96.06
4	425µm	14.0	3.94	7.88	92.12
5	300 µm	7.0	1.97	9.85	90.15
6	212 µm	14.0	3.94	13.79	86.21
7	150 µm	1.50	0.42	14.21	85.79
8	75 µm	35.5	10.0	24.21	75.79

Table 3: Particle size distribution for Sample II

S/No	Sieve Size	Wt Retained	%Wt Retained	Cum. % Wt Retained	% Wt passing
1	4.75mm	0	0	0	100
2	1.70mm	4.3	1.21	1.21	98.79
3	1.18mm	6.1	1.72	2.93	97.07
4	425µm	13.4	3.77	6.7	93.30
5	300 µm	8.3	2.34	9.04	90.96
6	212 µm	12.9	3.63	12.67	87.33
7	150 µm	2.0	0.56	13.23	86.77
8	75 µm	31.6	8.9	22.13	77.87

The result indicates that the particles of both samples are greater in the fine category (600 µm - 750µm). Using the [19] method of classification, both samples fall into group A2-6 (silty or clayey gravel and sand).

4.3 Atterberg limit

Atterberg limit test by the casagrande method was performed in accordance with [16] to ascertain plasticity index (PI) of the soil samples. The results are presented in Table 3. The PI of 12.6% and 14.4% for sample I and II does not exceed the maximum value of 35% stipulated by

[16] thus indicating a good laterite soil that is cohesive and hence able to receive proper compaction to enhance the strength and durability characteristics of the laterite. Both samples are suitable for cement stabilization as their PI values does not significantly exceed 15% as stipulated by [4].

4.4 Optimum Moisture content (OMC)

For the determination of the OMC a series of soil mixes were prepared over a range of moisture contents. When the dry density was plotted against the respective moisture content, the two samples produce typical moisture density curves with clearly identifiable optimums as shown in Fig. 3. Subsequent samples were moulded at these optimum values. The OMC of a soil at the time of compaction significantly affects the dry density which in turns affects at least the strength of the sample [17]. The OMC serves as a guide for the preparation and mixing of the block units. Moisture contents affect strength development and durability of the material and have a significant influence on the long term performance of stabilized soil material especially on bonding with mortars at the time of construction. When the brick is dry, water is rapidly sucked out of the mortar preventing good adhesion and proper hydration of the cement and when the brick is very wet the mortar tends to float on the surface without gaining proper adhesion [8 and 20]. Different types of compaction have influence on the optimum water content in the stabilized mixes. As reported by [21] dynamic compaction can reduce the optimum water content from 12% to 10% with about 50% increase in the compressive strength. And optimum water content range between 10 to 13% for static compaction [21].

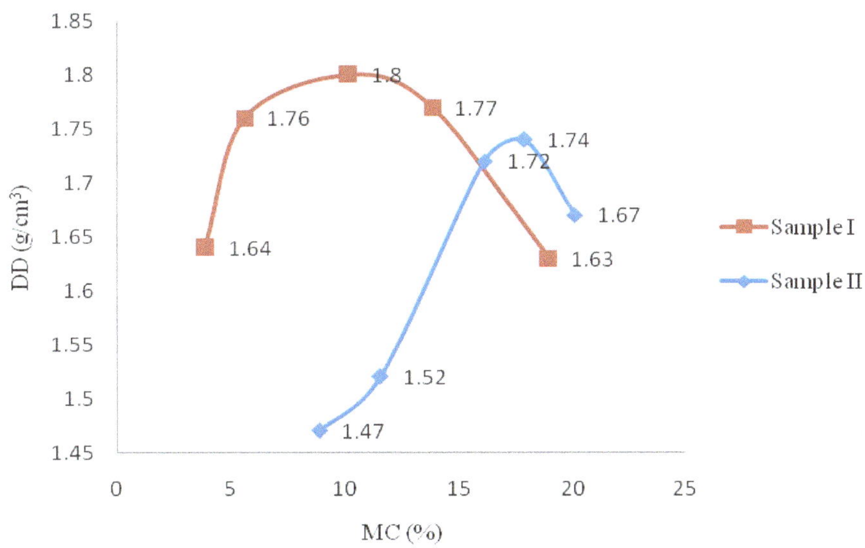

Figure 1: Compaction Curves for Sample I and II

4.5 Compressive Strength

The compressive strength is the most universally accepted value for determining the quality of bricks. Nevertheless, it is intensely related with the soil type and the content of stabilizer. The crushing strengths of the blocks were tested using the universal testing machine. A total of 64 specimens were prepared and crushed at different curing ages of 7, 14, 21 and 28days. Compressive tests were conducted on the blocks at different ages to indicate the rate of strength gain and the strength at a point in time [22].The results for the compressive strength against curing ages for sample I and Sample II are presented in Fig 2 and 3 respectively. The result shows that the highest compressive strength of 2.84N/mm^2 was obtained from sample I with 7.5% cement content at the curing age of 28 days. According to [4] the optimum cement content is in the range of 5%-10%. The strength obtained also compares favourably with the minimum British Standard requirements of 2.8N/mm^2. For the different stabilization levels from 0% -7.5%, the average value of compressive strength decrease by about 0.35N/mm^2.

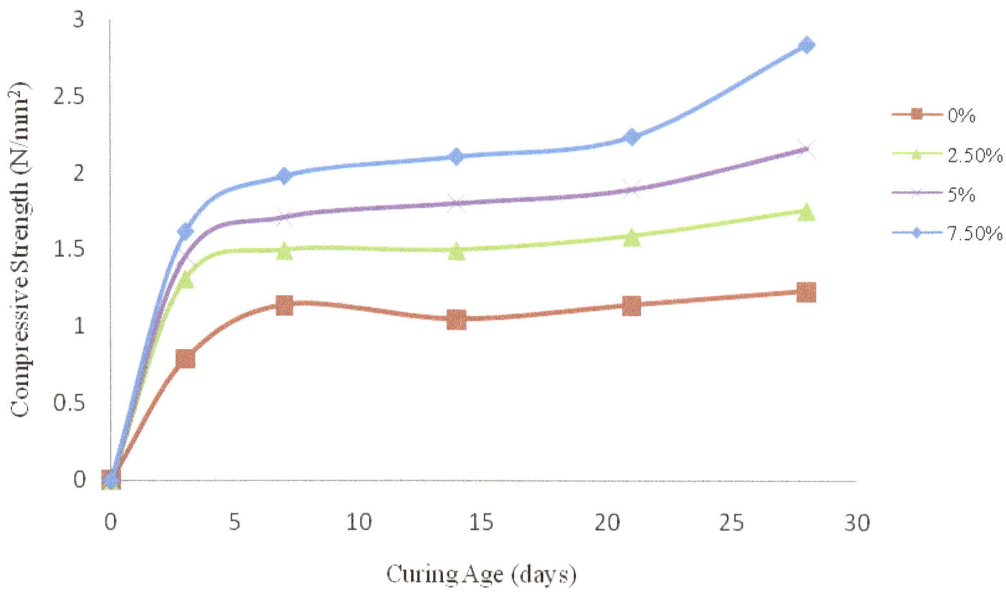

Figure 2: Compressive Strength Against Curing Age for Sample I

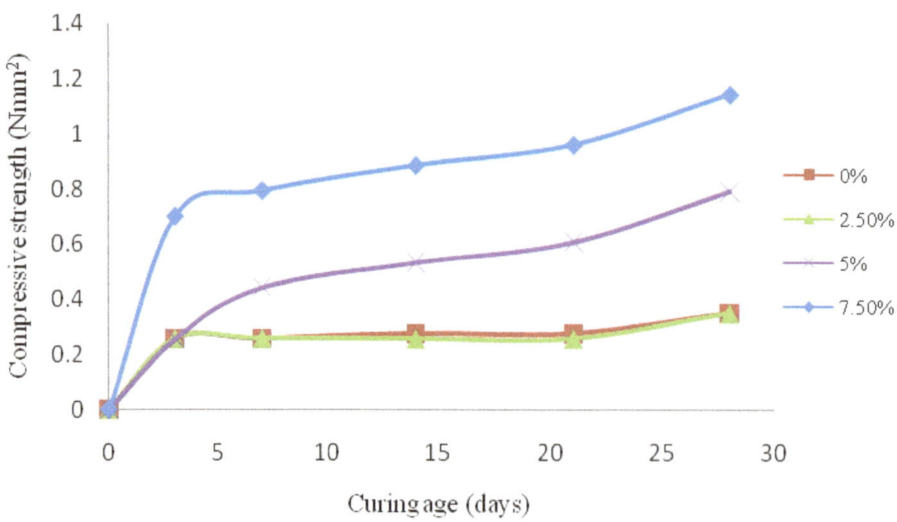

Figure 3: Compressive Strength against Curing Age For Sample II

5.0 Conclusions

From the investigation the following conclusions were drawn.

1. Both the soil samples have significant characteristics that make them suitable for cement stabilization with plasticity index of 12.4% for sample I and 14.6% for Sample II.
2. The strength of both samples increased with increasing stabilisation level from 0% through 7.5% with an average value of 0.35N/mm^2.

3. The highest compressive strength of $2.84N/m^2$ was achieved with 7.5% cement stabilization level of Sample I at the curing age of 28days.
4. Both the samples exhibit similar characteristics and are all suitable for the production of compressed stabilised soil blocks for walling in low cost sustainable buildings.
5. The maximum strength obtained at 28 day strength for both samples are within the minimum strength requirement recommended for the construction of low-rise buildings.

References

[1] S. Krishnaiah, and P. S. Reddy, "Effects of Clay on Soil Cement Block," The 12th International Conference of International Association for Computer Methods and Advances in Geomechanics (IACMAG) October 2008, Goa, India. pp. 4362-4368, 2008

[2] S. Deboucha, and R. Hashim, "Effects of OPC and PFA Cement on Stabilised Peat Bricks," International Journal of the Physical Sciences, vol. 5 (11), pp. 1671-1677, 2010.

[3] S. Deboucha, and R. Hashim, "A Review on Bricks and Stabilized Compressed Earth Blocks," Scientific Research and Essays, vol. 6 (3), pp. 499-506, 2011

[4] V. R. Fetra, I. A. Rahman and A. M. Zaidi, "Preliminary Study of Compressed Stabilised Earth Block (CSEB)," Australian Journal of Basic and Applied Sciences, vol. 5 (9), pp. 6-12, 2011

[5] O. F. Job, "Cement Made with Pulverised Burnt Clay as Partial Replacement," Nigerian Journal of Construction Technology and Management, vol. 1 (1), pp. 29-33, 1998

[6] E. E. J. Kamang, "Effects of Water-Cement Ratio on the Strength Properties of Stabilised Laterite soils," Nigerian Journal of Construction Technology and Management, vol. 1 (1), pp. 58-62, 1998

[7] A. Guettala, H. Houari, B. Mezghiche, and R. Chebili, "Durability of lime stabilized earth blocks," Courrier du Savoir, vol. 2, pp. 61-66, 2002.

[8] P. J. Walker, "Strength, Durability and Shrinkage Characteristics of Cement Stabilized soil blocks," Cement and Concrete Composites, vol. 17 (4), pp. 301-310, 1995

[9] N. Billong, U. C. Melo, F. Louvet, and D. Njopwouo, "Properties of Compressed Lateritic soil stabilized with a Burnt Clay-Lime Binder: Effect of Mixture Components," Construction and Building Materials, vol. 23 (6), pp. 2457-2460, 2008.

[10] A. Y. B. Anifowose, "Stabilisation of Lateritic Soils as a Raw Material for Building blocks," Bulletin of Engineering Geology and the Environment, vol. 58, pp. 151-157, 2000.

[11] M. Janz, and S. E. Johansson, "The Function of Different Building Agents in Deep Stabilisation," Swedish Deep Stabilisation Research Centre. Linkoping, 2002

[12] A. Mesbah, J. C. Morel, P. Walker and K. Ghavami, "Development of a Direct Tensile Test for Compacted Earth Blocks Reinforced with Natural Fibres," Journal of Materials in Civil Engineering, vol.16 (1), pp. 95- 98, 2004

[13] E. E. J. Kamang, "Strength Properties of Compressed Earth Brick with Earthworm Cast as Stabilizers," Journal of Environmental Sciences, vol. 1 (2), pp. 65-70, 1998

[14] T. Morton, "Earth Masonry Design and Construction Guidelines," Berkshire: Construction Research Communications Limited, 2008

[15] B. S. 812, "Methods of sampling and Testing of Mineral aggregates, sands and fillers," British Standard Institution, London, 1975

[16] B. S. 1377, "Soil Classification," British Standard Institution, London 1975.

[17] Y. G. Balami, and Y. D. Izam, "Soil-Bitumen as a Walling Materials," Journal of Environmental Sciences, vol. 1 (2), pp. 29-34, 1998.

[18] R. Maignien, "Review of Research on Laterite," Natural Resource Research IV. UNESCO Paris, France, p. 148, 1966.

[19] AASHTO, "Standard Specification for Transportation Materials and Methods of Sampling and Testing," 14th Edition. American Association of State Highway and Transportation Officials (AASHTO), 1986

[20] J. E. Oti, J. M. Kinuthia, J. Bai, "Stabilised Earth Masonry Technology incorporating Industrial By-Products," Proceedings of the 11th International Conference on non-Cementious Material and Technology, Bath UK, 2009

[21] R. Bahar, M. Benazzoug and S. Kenai, "Performance of Compacted Cement-Stabilised Soil," Cement and Concrete Composites, vol. 26 (7), pp. 811-820, 2004.

[22] O. S. Abejide, "Assessment of the Performance of Soil Soil-CementBlocks with Plant Extract Additive," Construction Focus, vol. 1, pp. 55-63, 2007

10

DEVELOPMENT OF AN ACID RESISTANT CONCRETE: A REVIEW

Anwar Khitab*, Muhammad Tausif Arshad, Faisal Mushtaq Awan, Imran Khan

Department of Civil Engineering, Mirpur University of Science & Technology, Mirpur, Azad Kashmir, Pakistan

*Corresponding E-mail: anwarkhitab@yahoo.com

Abstract

This review paper addresses the measures taken to prevent or minimize the deterioration of concrete, which confronts an acidic environment. Primarily, the mechanism of reaction between alkaline concrete and acid is clearly demonstrated. The mechanism of reaction clearly sets guidelines as to how the chances of this disastrous reaction should be minimized or eliminated at all. The suggested preventive measures are two-fold i.e. the improvement of the basic microstructure of concrete and the provision of barriers against acids. Concrete can be made acid resistant using classical as well as novel techniques like nanotechnology. There exists an immense need that these measures are recognized and implemented by the construction industry to put a stop to huge money losses.

Keywords: *Concrete, Acid attack, mechanism of reaction, preventive measures, nanotechnology*

1.0 Introduction

Concrete is defined as a mixture of cement, sand, gravels, water and sometimes admixtures and additives. Concrete has long been used in civil engineering industry for construction of foundations, footings, retaining walls, slabs, pavements, tunnels, bridges, basins, canals, dams, drains, sewerage lines and many other structures and structural members. Concrete is the backbone of any country's infrastructure. Concrete hardens and gains strength within days. Its relative low cost, ease of application and relative long term service life compared to other materials is the main cause of its popularity. The disadvantage of using concrete is that the micro structure of concrete allows the penetration of water and other destructive species that will cause premature failing of the concrete surface. A permeable concrete will allow infiltration of aggressive agents (chlorides, carbon dioxide and acids) to the steel reinforcement bars causing complete failure of the structure [20].

In general, concrete has a low resistance to chemical attack. The common forms of chemical attack on concrete and the embedded reinforcement in reinforced concrete are chloride attack, sulphate attack, Carbonation due to Carbon dioxide, Alkali-aggregate reactions and acid attack [9, 22].

Chlorides in de-icing salts or in soils, sea water and ground water can enter concrete and destroy the passive oxide film, which normally protects the steel against corrosion. Chloride can enter either through its ingredients (like chloride containing aggregates, and water) or through the environment (such as de-icing salts or seawater). As a result, rust layers build up on steel, whose volume is enough to exert disruptive tensile stresses on surrounding concrete causing it to crack or even spall [8, 23].

Sulphates present in soils, groundwater and seawater react with Tricalcium Aluminate (C_3A) of cement to form expansive compounds. The reaction leads to an increased volume of the products resulting in disintegration of the concrete. Sulphate resisting cement, which has a lower content of C_3A (ASTM Type V cement) is best suited in conditions where an extensive exposure to sulphates is expected [1, 21].

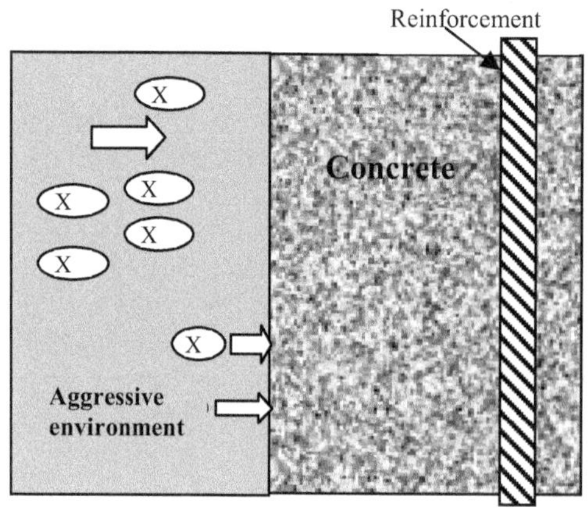

Figure 1: Reinforced concrete exposed to an aggressive environment

The pore solution of concrete is alkaline in nature due to the presence of portlandite or Calcium hydroxide ($Ca(OH)_2$). This alkaline solution protects steel from corrosion. The presence of CO_2 may lead to the destruction of this alkalinity. This process of loss of alkalinity due to the reaction between $Ca(OH)_2$ in concrete and CO_2 from air is known as carbonation [24]. Carbonation leads to the formation of Calcium Carbonate and water, as a result the alkalinity of concrete is lost. The steel thus becomes vulnerable to rusting in the presence of oxygen and moisture.

The reaction between the silica of aggregates (in siliceous aggregates) and the alkali content in cement is called alkali-aggregate reaction. The product of the reaction expands by absorbing water and increases in volume leading to cracking of concrete [1].

2.0 Mechanism of acid attack

As mentioned earlier concrete is alkaline in nature due to the presence of Calcium Hydroxide ($Ca(OH)_2$). If attacked by an acid HX (where X is the negative ion of the acid), the components of the cement matrix break down in accordance with the following famous acid-base reaction [25]:

$$2\,HX + Ca(OH)_2 \rightarrow CaX_2 + 2\,H_2O$$

The decomposition of the concrete depends on the porosity of the cement paste, on the concentration of the acid, the solubility of the acid calcium salts (CaX_2) and on the fluid transport through the concrete [2]. Insoluble calcium salts may precipitate in the voids and can slow down the attack. Acids such as nitric acid, hydrochloric acid and acetic acid are very aggressive as their calcium salts are readily soluble and removed from the attack front. Other acids such as phosphoric acid and humic acid are less harmful as their calcium salt, due to their low solubility, inhibits the attack by blocking the pathways within the concrete such as interconnected cracks, voids and porosity. Sulphuric acid is very damaging to concrete as it combines an acid attack and a sulphate attack [26].

3.0 Special concretes for acid attack prevention

The risk of acid attack on concrete can be minimized by providing due consideration to concrete porosity. Lesser the porosity, lesser will be the chances of acid attack on concrete. In general it is said that all high performance concrete mixtures show a better resistance against acid attack than the reference ordinary type concrete [9]. By High performance concrete is meant a concrete with superior qualities including low permeability and low diffusion [2]. In another way, the concrete resistance to acids can also be provided by giving its surface an acid resistant coating. In Fig. 2, two ordinary concrete cylinders having the same mix design are shown. The cylinder on left was immersed in 5% sulphuric acid for 28 days having a 6% loss of mass at the end of the immersion period.

Figure 1: Concrete cylinder immersed in 5% H_2SO_4 for 28 days (Left)

3.1 Silica concrete

Silica fume (SF) is an industrial byproduct obtained from an electric arc furnace process. During this process, high purity quartz are reduced to silicon at high temperatures producing silica vapors. These vapors oxidize and condense in the low temperature zone to tiny particles known as silica fumes [27]. Fly ash are fine solid particles of ash carried into the air during the combustion of pulverized fuel in power stations. Silica fume and fly ash have proved to be very useful additives for concrete. Using fly ash in combination with micro silica fume, results in an improved resistance against acid attack [14, 15]. Fly ash produces a densely packed mixture of cement paste and aggregates, while silica fume reacts pozzolanically and transforms the calcium hydroxide into Calcium-Silicate-Hydrate (CSH) gel in accordance with the following equation [31]:

$$SiO_2 \text{ (solid)} + Ca^2+ + 2OH- \rightarrow CaO.SiO_2.H_2O \text{ (CSH gel)}$$

The CSH gel is the source of strength in concrete. On one hand it increases the bond between the cement paste and aggregates and on the other hand, it increases the compressive strength and chemical resistance of the concrete. The additional CSH produced by silica fume is more resistant to attack from aggressive chemicals than the weaker CH. Therefore, the combination of silica fume and fly ash results in a denser concrete having lesser quantity of calcium hydroxide, which considerably increases the acid resistance. Silica fume without fly ash produces micro cracks in concrete, which increase the path of acid attack.

3.2 Air entraining concrete

Air entrainment is the process of incorporating minute air bubbles into concrete. Although, air entrainment is mainly practiced to increase freeze-thaw resistance of the concrete, it can also be used as a means of acid resistance. The air entrainment increases the acid resistance because the air voids block micro capillaries and prevent the acid from invading the concrete through these canals. Air-entraining agents are available as additives as well as admixtures. As additives, the agents are interground with cement in fixed proportions, as in cements Type IA, IIA and IIIA [32]. Requirements and specifications of air-entraining agents to be used in concrete are covered in

ASTM C 260 [17]. As admixtures, Vinol resin and fatty acid salts were used as air-entraining agents, which have now been largely replaced by synthetic admixtures: Alkyl sulphates, olefin sulphonates, Diethanolamines etc. [16].

3.3 Blast Furnace concrete

Blast Furnace Slag is formed when iron pellets, coke and limestone/dolomite flux are melted together in a blast furnace. During this process, the lime chemically combines with the aluminates and silicates of the ore and coke ash to form blast furnace slag [18]. Blast furnace slag significantly increases the durability of the concrete. It has been found to be very effective against chloride and sulphate attack; however it does not provide any considerable improvement in acid resistance [19]. The reason for this is probably that the slag does not change the amount of calcium hydroxide in the concrete. In addition to the use of slag as admixture, it is also used as additive by intergrading with the Portland cement clinker. The combined cement is known as slag cement [32].

3.4 High performance concrete

High performance concrete (HPC) is defined as a concrete having properties much superior than an ordinary concrete [28]. Along with other properties like strength and durability, it is supposed to have higher resistance against chemical attack. High-performance concretes are made with carefully selected high-quality ingredients and optimized mix design. The concrete is batched, mixed, placed, compacted and cured to the highest standards. Typically, such concretes will have a low water-cement ratio of 0.25 to 0.4 [1]. Various admixtures are usually used to make these concretes fluid and workable [29]. Owing to lesser porosity, HPCs also offer significant acid resistance. The low porosity is achieved by cement contents in excess of 500 kg/m^3 of concrete, low water to cement ratio, adequate compaction and curing and the incorporation of a super plasticizing admixture [32]. With this combination, the strength as high as 10000 psi or above can be achieved.

4.0 Use of nanotechnology for acid attack prevention

Nanotechnology is a new branch of material sciences that promotes the use of nano particles in different domains. As obvious from its name, nano particles have their sizes in nano meters. A nano particle is defined as the one, which has at least one of its dimensions in nanometers or 10^{-9} m [11]. Materials at nano scale display properties somewhat different from those at micro or macro scale. For example, inert titania is used for pigmentary purposes while a nano titania is a photo-catalyst. Similarly, opaque copper becomes transparent when ground to nano size [12]. Knowing that, attempts were made to apply nanotechnology in different fields including construction engineering. Apart from other beneficial effects, nanotechnology has also found its importance in increasing the acid resistance of concrete as given below.

4.1 Significance against acid attack

In construction materials, the materials density and strengths are the most important properties in which the experts are interested. As mentioned under heading 3.4, the High performance concrete is quite effective against acid attack. However, it is a common observation that materials having high strengths are also associated with higher densities and thus the dead weight of the structure considerably increases. On the other hand, nanotechnology offers quite interestingly certain materials with higher strength while possessing low to medium densities. The examples are carbon nanotubes (CNTs) and carbon nanofibers (CNFs) [7]. As an example, it is believed that the design life of a concrete bridge can be increased from 30 to 45-50 years, using nanotechnology [30].

4.2 Nano cementitious materials

Cemetitious materials in civil engineering include concrete, mortar and cement paste. These materials were developed due to strong bonding characteristics and very rapid hardening and setting of cement. Several new cement composites have been developed during the recent years by combining it with nano-Titania (TiO_2), Carbon nanotubes, nano-silica (SiO_2), nano clay and nano-alumina (Al_2O_3), which have significantly improved the performance of these materials [6,13]. Besides nano composite cementitious materials, efforts are also going on to introduce nano-cement at commercial scale. The present commercial cement particle has its size in micrometers [1]. It is a well-established fact that finer particles (nano) having more specific surface area can fill more effectively the pores in cement matrix, densify its structure, and might lead to superior strength due to faster chemical reactions with water (hydration reactions) [33]. Or in other words nano materials might enhance the chemical resistance of concrete against acid attack.

4.3 Nano coating materials

With the help of nanotechnology, coatings with a molecular structure which simply rejects adhesion by foreign bodies have been created. Many nano-coatings are in the market, which effectively provide effective resistance against acid attack. [34]. Appropriate nanotechnology can be used to create eco-friendly sealers and coatings that deliver pure Performance on concrete. These coatings can be applied by spraying, rolling or brushing [4, 5].

4.4 Nano food additives

New concretes with nano food additives are being developed at National Institute of Standards and Technology (NIST) USA [3]. The technique is named as VERDiCT (Viscosity Enhancers Reducing Diffusion in Concrete Technology). With nano additives in concrete pores, the viscosity of the concrete pore solution can be increased, which will slow down the ingress of external species (including acids).

5.0 Conclusions

This review paper has attempted to collect the state of knowledge and information in the field of concrete, confronting an acidic environment. Emphasis was given to understand the problem and to identify a compatible solution. The solutions include the improvement of internal structure of the concrete or the provision of an external acid resistant membrane. Additionally, the paper also describes the novel nanotechnologies that can be used to achieve acid resistance goals for concrete.

References

[1] A. M. Neville, "Properties of Concrete," ISBN: 0273755803, 9780273755807, Pearson Education, 2011.

[2] C. Shi, Y. L. Mo, "High-Performance Construction Materials Sciences and Applications," World Scientific Publishing Co. Pte. Ltd, Singapore, 2008.

[3] L. Triandafilou, "Strengthening concrete with nanoparticles", FHWA Resource Center, Baltimore, 2009.

[4] http://nanoprotech.com.au/

[5] http://chemicalcontainment.com/nanotech.html

[6] G. Hüsken, M. Hunger, H. Brouwers, "Experimental study, photocatalytic concrete products for air purification," Building and Environment, Vol. 44 (12), pp. 2463-2474, December 2009.

[7] K. Gopalakrishnan, "Nanotechnology in Civil Infrastructure: A Paradigm Shift", ISBN: 3642166563, 9783642166563, Springer, USA, 2011.

[8] E. Poulsen, "Diffusion of Chloride in Concrete - Theory and Application", ISBN 0419253009, 9780419253006, Taylor & Francis, 2005.

[9] R. D. Woodson, "Concrete Structures: Protection, Repair and Rehabilitation", ISBN 1856175499, 9781856175494, Butterworth-Heinemann, 2009.

[10] R. Bassi, S. K. Roy, "Handbook of coatings for concrete", ISBN 1870325826, 9781870325820, Whittles, The University of Michigan, 2007.

[11] A. Khitab, M. Qureshi, M. Nadeem, "Nano Engineered Building Products", International Conference on Engineering and Applied Sciences, 24-27 July, Beijing, China, 2012.

[12] D. M. Sabatini, "Leading Edge Nanotechnology Research Developments", ISBN: 1600219004, 9781600219009, Nova Publishers, 2007.

[13] Z. Li, "Advanced Concrete Technology", ISBN: 047043743X, 9780470437438, John Wiley & Sons, 2011.

[14] V. Sata, C. Jaturapitakkul, K. Kiattikomol, "Influence of pozzolan from various by-product materials on mechanical properties of high-strength concrete", Construct. Build. Mater. Vol. 21(7), pp. 1589–1598, 2007.

[15] ASTM C618 - 12a, "Standard Specification for Coal Fly Ash and Raw or Calcined Natural Pozzolan for Use in Concrete", 2012.

[16] Admixture Technical Sheet (TS5), "Air-entraining admixtures", Cement Admixtures Association, UK, Dec 2012.

[17] ASTM C260 / C260M - 10a, "Standard Specification for Air-Entraining Admixtures for Concrete", 2010.

[18] http://www.nationalslag.org/blastfurnace.htm

[19] R. Snellings, G. Mertens, J. Elsen, "Supplementary cementitious materials", Reviews in Mineralogy and Geochemistry, Vol. 74, pp. 211–278, 2012.

[20] T. Nochaiya, W. Wongkeo, A. Chaipanich, "Utilization of fly ash with silica fume and properties of Portland cement-fly ash-silica fume concrete", Fuel, Vol. 89(3), pp. 768-774, 2010.

[21] P.C. Varghese, "Building materials", PHI Learning, New Delhi, 2010.

[22] J. F. Lamond, J. H. Pielert, "Significance of Tests and Properties of Concrete and Concrete Making Materials", ASTM International, 2006.

[23] X. Shi, N. Xie, K. Fortune, J. Gong, "Durability of steel reinforced concrete in chloride environments: An overview", Construction and Building Materials, Vol. 30, pp. 125-138, 2012.

[24] M. Liang, R. Huang, S. Fang, "Carbonation service life prediction of existing concrete viaduct/bridge using time-dependent reliability analysis", Journal of Marine Science and Technology, Vol. 21(1), pp.94-104, 2013.

[25] J. Kenkel, "Basic Chemistry Concepts and Exercises", CRC Press, 2011.

[26] E. K. Attiogbe, S. H. Rizkalla, "Response of Concrete to Sulfuric Acid Attack", ACI Materials Journal, pp. 481-488, Nov-Dec 1988.

[27] D. Whiting, "Silica fume concrete for bridge decks", Transportation Research Board, 1998.

[28] P. Aitcin, "High Performance Concrete", Taylor & Francis, 2011.

[29] A. Wypych, "Plasticizers Databook", Elsevier, 2013.

[30] www.nanoandme.org/nano-products/paints-and-coatings/

[31] B. W. Langan, K. Weng, M. A. Ward, "Effect of silica fume and fly ash on heat of hydration of Portland cement", Cement and Concrete Research, Vol. 32, pp. 1045-1051, 2002.

[32] A. M. Neville, "Concrete Technology", Pearson Education Ltd, 2012.

[33] L. Raki, J. Beaudoin, R. Alizadeh, J. Makar, and T. Sato, "Cement and Concrete, Nanoscience and Nanotechnology – Review", Materials, Vol. 3, pp.918-942, 2010.

[34] Nanotechnology for Simple, successful concrete repair: A specifiers Guide available at http://www.basf-cc.com.au/sitecollectiondocuments/basf_nanocrete_24xa4_aus_ebook.pdf

11

SUSTAINABLE CONSTRUCTION: WATER USE IN RESIDENTIAL BUILDINGS IN PORTUGAL

Miguel Pires Amado[1], Luis Marques Barroso[2]

[1] Department of Civil Engineering, Universidade Nova de Lisboa, Caparica, Portugal
[2] GEOTPU, Faculdade de Ciencias e Tecnologia, Universidade Nova de Lisboa, Caparica, Portugal

*Corresponding E-mail : ma@fct.unl.pt

Abstract

Water is the most important resource to all life on our planet. The impact that buildings have on the consumption of this resource throughout their life cycle (planning, construction, operation and deconstruction) but mainly during the operation phase, has resulted in very high and unsustainable consumption patterns. Sustainable development, while a long-term goal, requires that the preservation of natural resources becomes a central point of any project strategy. Water is of vital importance to human life, it should be preserved to ensure its availability for a long period of time. However, little concern has been given to the importance of introducing a selection of more efficient solutions to save water in buildings and consequently increase sustainability in the construction sector. In this context it is on the design phase that the main resource saving water measures should be considered. Research on saving water involves an analysis of quantitative and qualitative measures and their adaptation to different contexts. Nowadays there are systems of voluntary certification of sustainable construction, which can help the design phase to achieve sustainability. This paper discusses a set of measures to reduce water consumption and enable a more efficient use of this resource in residential buildings. The measures presented are focused on user awareness campaigns, systems of rainwater and grey water recycling, the use of more efficient devices and reduction of leaks. The measures are analyzed and compared taking into account the consumption patterns for each device both inside and outside a domestic building and its effective reduction in water consumption.

Keywords: *Certification, efficiency, water, residential buildings*

1.0 Introduction

The present article discusses a new dialectical and comprehensive understanding of sustainable construction as an approach to the efficient use of water in buildings. This "efficient water use" must encourage the construction of buildings based on a set of solutions that strive to reduce the natural resource consumption throughout the buildings entire life cycle. The main objective of this article is to analyze different measures that improve water efficiency in residential buildings, based on the calculation of water savings associated with their use.

The construction of new buildings has huge impact on the environment because the construction sector is responsible for the largest physical man-made objects on Earth and the corresponding consumption of a large amount of natural resources.

The large amounts of natural resources used in the construction phase are obviously less of then the amount consumed in the buildings entire life cycle. In this context it is mandatory to seek alternatives and in this context the use of water emerges as one of the most important issues because it is a resource vital for humans and their activities. The vital importance of this resource justifies the continuous study of its use so that, in the long term, humans continue to enjoy the services it provides.

It important to enumerate the main benefits that result from the reuse and recycling of storm waters: reduction of the consumption of potable water, reduced need for wastewater treatment plants (WWTP); efficient operation and maintenance; reduction in underground water consumption; aid in urban flood control; reduction of CO_2 emissions by using more efficient systems [1]. The main benefits of reuse and recycling of gray waters are: reduction of the volume to be treated by the WWTP, reduction of consumption of potable water; enables integrated management of water resources, reduction of the costs of invoicing for users, high social acceptance, reduction of volume of the reservoirs of potable water; allows the creation of alternative supply chains for different uses and activities [1]. The key question this study aims to address is, given all the aforementioned benefits, which systems produce the biggest efficiency gains?

2.0 Literature Review

The concept of Sustainability has evolved over the years, having come to prominence since the late '70s, firstly in an economic perspective and with little environmental concern. Only in the late 80s, this concept became associated to environmental issues, leading to the well-known definition of sustainable development [2]. The need for sustainable construction of new buildings and for sustainable renovation of existing buildings arises from the current reality of construction in the developed world. The quality and efficiency of buildings has a great impact on the environmental, economic and social conditions of populations. The purpose of sustainable construction is to ensure that, through improvement in all dimensions of human activity: economic, social and environmental, the livelihoods of future generations will be unaffected and the quality of life of existing generations maintained or improved [3].

A building can only be considered sustainable when the different dimensions of sustainable development - environmental, economic, social and cultural - are considered from the planning to the implementation phase [3]. The main objective of certification systems is to gather all the available data and information and produce reports. These reports are the base for the complex decision-making processes taking place during the various stages of a building's life cycle in sustainable construction. There are parameters considered at both the scale of the building and to assess the building's interaction with the environment in which it is inserted, evaluating sustainability in a broad fashion [4]. Typically, the parameters that serve to support the assessment of sustainability are related directly or indirectly with the following general objectives:

- Reducing the use of nonrenewable energy and materials;
- Reducing of water consumption;
- Reducing waster production;
- Cutting pollutant emission.

Internationally there are two widely used certification systems to evaluate the "sustainable performance" of buildings, the Building Research Establishment (BREEAM) and the Leadership in Energy and Environmental design (LEED) systems. The BREEAM system appeared in 1988 and was the first certificate system, followed by LEED in 1998. In Portugal the Liderar pelo Ambiente (LIDERA) certification system was introduced in 2005. These three systems of evaluation use similar methods to assess sustainability, differing mainly in the importance that each system gives to a particular category and in its subdivisions. Regarding the assessment of water sustainability the LEED system allocates the biggest weight to it with 10%, compared to the 9% in the BREEAM system and the 8% in the LIDERA system. It also noted the proposal of the BREEAM system that mentions daily per capita water consumption of all those 80 litres as a goal to reach in the United Kingdom [5, 6, 7].

Currently, the estimated total volume of water on Earth is around 1400 million km3, of which only 2.5% (35 million km3) corresponds to freshwater. Most of this water is in glaciers and is not used for consumption. Therefore, the main sources for human consumption are lakes, rivers, ground waters and underground aquifers, which contain an estimated 200,000 km3 of water, less than 1% of all freshwater and only 0.01% of all water on Earth [8].

Although access to drinkable water is growing, [in 1990 about 79% of the population (4.1 billion) had access to drinkable water and in 2000 this number surged to 82% (4.9 billion)], 1.1 billion people still lack access to freshwater and 2.4 billion lack access to sanitation facilities, directly and indirectly causing more than 5 million deaths per year. Today one third of the world population lives in countries with a moderate to high level of water stress, that is where water consumption exceeds 10 per cent of renewable freshwater resources and it is predicted that and within 25 years two thirds of the world population will live under these conditions [8].

According to Barroso [8], Portugal needs 7500 x 106 m3/year of freshwater, the equivalent to 1,880 million Euros. The largest share is consumed by the agricultural sector in irrigation, about 6550 x 106 m3/year (87% of total). The second largest share is in domestic use (mainly from showers, baths and cistern discharges) in the urban water supply with 570 x 106 m3/year (8% of total). Finally, the manufacturing industry uses 385 x 106 m3/year (5% of total) (Figure 1) [8].

From an economic perspective the urban sector is the key sector to make investments. It is a sector that uses 8% of water but has associated costs of 46% (Figure 2) [9]. It is necessary to reduce inefficiency in this sector to bring down costs and losses. For example, the use of high grade filters to produce drinkable water when it will be used in car washes or watering the garden is unnecessary. Another issue afflicting urban water efficiency is losses occurring by adduction when the distances required for distribution are long [9].

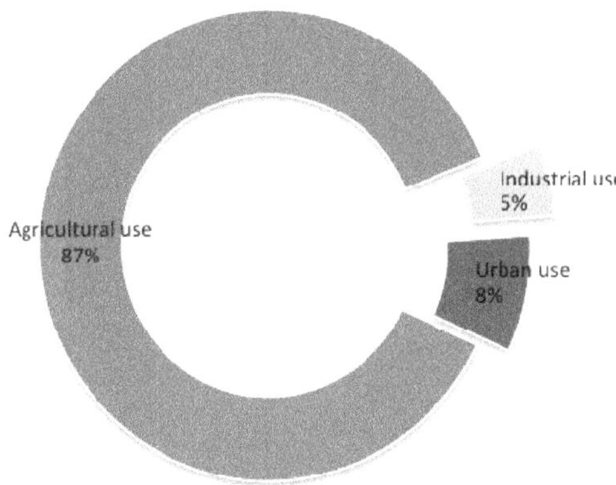

Figure 1: Water use by sector

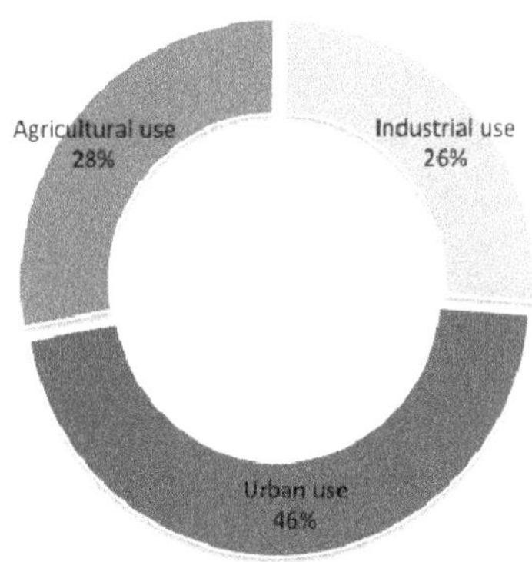

Figure 2: Costs associated with the water

In terms of the associated costs with water use, the urban sector is followed by the agricultural and industrial sectors, both with roughly equivalent costs (28% and 26% respectively). Several measures have been taken internationally to reduce water consumption in these sectors and also to tackle the shortages that may occur in some of the areas where these measures were put to practice. These measures where, among others: grey water treatment, fog retention, and desalination [9, 10]. In Portugal, there are no studies or statistical analysis characterizing domestic consumption patterns, although it can be calculated approximately considering the capacity of water consuming devices (Table 1) [8].

Table 1 – Distribution of daily average consumptions

Use	Consumption (L/capita/day)			
	Multifamily House		Single-family House	
Cistern/Toilet	43	31%	43	27%
Taps	22	16%	22	14%
Bath/Shower	52	37%	52	32%
Washing Machine	13	9%	13	8%
Dishwashing Machine	3	2%	3	2%
Leaks	7	5%	7	4%
External Use	-	-	20	13%

Considering the consumption of a single-family house, it is possible to see that the highest consumption of domestic water in Portugal is associated with baths and showers (32%) followed by cisterns (27%); these devices represent about 59% of all domestic consumption. In the case of multifamily dwellings, where the portion of the external water use is excluded, the consumption of both these devices increases to 68% [11].

According to Pedroso [11] campaigns of awareness and motivation must allude to the reality that surrounds us, creating new ways to reduce unnecessary expenses. Still on the scope of reduction, actions like the use of low-consumption taps; creation of systems for rainwater utilization and the reuse of some waste waters (grey waters) are typically encouraged in campaigns. To reduce and mend past mismanagement, it is important to reduce the cost of consumed water. If limits are created for everyday water use then penalty measures can be applied to incentivize sustainable behaviors. It may also be noted that the additional revenues should be used in promoting the measures that increase water sustainability.

According to Barroso [8], some of the existing water losses in buildings, either cold or hot, occur due to leaks in the devices and appliances used. To prevent this occurrence, campaigns must allude to these causes; examine in detail the origins of these leaks and present preventive measures. It is possible to use more efficient devices to reduce the high levels of water consumption, such as cisterns with smaller discharge volumes and with dual discharge, thermostatic taps for showers, low-flow taps systems, and handles for rapid closure of water showers, low-flow shower heads and velocity amplifiers for discharge. Several of these devices are currently certified by the Associação Nacional para a Qualidade nas Instalações Prediais – (ANQIP) (National Quality in Building Installation Association), the responsible identity for this certification in Portugal.

The use of rainwater raises different opinions on some cases; for instance, the use of rainwater for washing clothes is conditioned by Brazilian standards (for bacteriological reasons) [13], but accepted by German standards. In Portugal there is no specific regulation and this water may be used for washing clothes, flushing toilets, for several external uses such as washing floors and cars, watering of green areas and several industrial uses such as cooling towers, fire systems, networks and HVAC systems [14] with special attention to bacteriological issues.

The volume of rainwater needed for storage must take several factors into account, which should consider the type of dwelling, the type of roof and the location where the dwelling is located [13]. Such factors include:

- Run off coefficient of the roof covering;
- Average annual rainfall;
- Catchment area of the roof surface;
- Hydraulic efficiency of filtering;

The storage of rainwater and its subsequent use for domestic purposes may be, in many cases, one of the best solutions to reduce freshwater consumption and an adequately dimensioned rainwater supply can help to reduce the consumption rates around 43% [9, 11, 13].

The reutilization of these waters will contribute to reduce the need for water treatment and their subsequent distribution. This will both generate economic gains and environmental benefits [11].

The use of gray water – wastewater from showers and washbasins – after appropriate treatment is already a recognized method for saving water in several countries. Although there are some countries which limit the possibility of using these waters citing the risk to public health resulting from splashes in its use that may cause problems. Their contribution to water saving is given by its discharge in toilets, use in irrigation systems and firefighting systems. In general, it needs an appropriate treatment (filtration and disinfection) depending on the quality of the water expected in its future use [9, 12].

3.0 Methodology

In terms of water saving in building sector, the research should carry out in the following three fields: buildings; equipment's and water. In these fields the natural resource - water is present and the different way of water potential use are integrated in the model framework of water efficiency. The methodology used to develop the study uses a simple 6 step structure:

Step 1: Identification of water saving using equipment/devices applicable in residential buildings both single family and multifamily.
Step 2: Characterization of the potential savings from each system.
Step 3: Comparison between current devices and the alternatives.
Step 4: Accounting for the potential use of rainwater
Step 5: Accounting for the potential use of gray water
Step 6: Estimated return of investment period

Considerations are made in the study: the use of rainwater is based on ETA ANQIP normative [15, 16, 17]; the sizing of the reservoirs is calculated for 30 days of demand; the dishwasher is considered to consume about 70% of total consumption; faucets and dishwasher reutilization comes from cisterns enhancing availability. This method enables a correct assessment between consumption needs and water saving potential in different conditions. Comparison between models is realized in terms of savings according to efficiency class [15, 16, 17] and the ecological and economic performance. The last step examines the various solutions in a cost-benefit framework and determines the return on investment periods.

4.0 Results and discussions

Using the described method a case study water savings was specified to three different dwellings with four occupants each one in three different rainfall catchment geographical conditions in Portugal – north, center and south. National data of rainfall will be providing by Portuguese Meteorological Center. The focus on the three different regions pretend to put out the effective capacity of implemented the water saving system in multifamily and single family building. These results take into account various considerations, such as:

- Number of days per year that each inhabitant remains inside his home – 330 days;
- Number of inhabitants in family houses – 4 people;
- Number of residents in multifamily buildings – 3 people;
- Roof catchment area of single-family building – 200m2;
- Roof catchment area of multifamily building – 48m2;
- Different rainfall catchment locations – Lisboa, Faro, Porto

The study considering also the different usage patterns and values of the flow, it is possible to draw a comparison between efficient devices and conventional devices (Table 2 and 3).

Table 2 - Savings multifamily building using the efficient devices

Multifamily Building	Consumption without saving measures			Saving with efficient equipment	Consumption with saving measures			
Devices/Appliances	Consuptiom (l/hab.day)	Consumption (m3/hab.year)	Consumption (m3/agr.year)		Consumpt ion (l/hab.day)	Consumtion (m3/hab.year)	Consumption (m3/agr.year	
Cisterns/Toilets	43	14,2	42,6	57%	18,5	6,1	18,3	
Faucets Dishwasher	15	5	14,9	67%	10,0	3,3	9,9	
Basin tap	7	2,3	6,9	63%	5,0	1,7	5,0	
Showers	52	17,2	51,5	67%	17,2	5,7	17,0	
Washing machine	13	4,3	12,9	44%	7,3	2,4	7,2	
Disheasher	3	1,0	3,0	50%	1,5	0,5	1,5	**Absolut savings**
Leaks	7	2,31	6,9	-	-	-	-	
Total	140	46	138,6	-	59	20	58,8	**58%**

Table 3 - Savings single family building using the efficient devices

Single family Building	Consumption without saving measures			Saving with efficient equipment	Consumption with saving measures			
Devices/Appliances	Consuptiom(l/hab.day)	Consumption (m3/hab.year)	Consumption (m3/agr.year)		Consumption (l/hab.day)	Consumtion (m3/hab.year)	Consumption (m3/agr.year)	
Cisterns/Toilets	43	14,2	56,8	57%	18,5	5,7	24,4	
Faucets Dishwasher	15	5	19,8	67%	10,0	3,3	13,2	
Basin tap	7	2,3	9,2	63%	5,0	1,7	6,6	
Showers	52	17,2	68,6	67%	17,2	5,7	22,7	
Washing machine	13	4,3	17,2	44%	7,3	2,4	9,6	
Disheasher	3	1,0	4,0	50%	1,5	0,5	2,0	
Exteriors	20	6,6	26,4	-	20,0	6,6	26,4	**Absolut savings**
Leaks	7	2,3	9,2	-	-	-	-	
Total	160	52,8	211,2	-	79	26	105	**51%**

The comparison of the Table 2 and 3 can be seen in Table 4, where the class of water efficiency according to ANQIP can also be found [15, 16, 17].

Table 4 – Savings associated with efficient equipment and appliances

Devices/Appliances	Certification according to ANQIP	Savings (%)
Cisterns/Toilets	A	57
Showers	A+	67
Basin Taps	A	62,5
Faucets Dishwasher	A	67
Washing Machine	A	44
Dishwasher	A	50

The evaluation of the potential for rainwater harvesting was assessed using the procedure of ANQIP [17]. To study and establish a comparative analysis of rainwater harvesting in buildings made use of the same context referred before.

The total amount of available rainwater is calculated using the following formula

$$V_a = C \times P \times A \times \eta_f \text{ .... Ref[15]}$$

Where:

Va - Annual volume of rainwater usable (liters)
C - Coefficient of run off of the roof
P - Cumulative average annual rainfall (mm)
A - Catchment area
ηf - Hydraulic efficiency of filtering

The value of hydraulic efficiency of filtration is determined at: $\eta f = 0.9$ and the values for run off of various roof types are displayed in Table 5.

Table 5 – Values to run off coefficient of the roof, Source: Barroso,(2010)

run off coefficient of the roof (C)		
Type A	Impermeable roof	0,8
Type B	Gravel roof	0,6
Type C	Extensive green roof	0,5
Type D	Intensive green roof	0,3

Thus, for the case of a single-family housing, each measure provides savings that vary between 40% and 76.3% – from the use of grey water to the use rainwater together with efficient devices, respectively (Table 6). Although the savings of 76.3% would result in reducing the flow from 160 to 37.9 L/capita/day, which may seem very ambitious, such reduction is possible, because this consumption covers all the needs of water for one person.

Table 6 – Comparison of measures in single family building

Measure	Normal consumption (m3/agr.year)	Reduction (m3/agr.year)	Final Consumption (m3/agr.year)	Consumption (L/capita/day)	Savings (%)
Gray Waters	211.2	83.2	128	117	40
Rainwater	211.2	91.2	120	90.8	43.2
Efficient Appliances	211.2	106.2	105	79	50
Rainwater + Efficient Appliances	211.2	161.2	50	37.9	76.3

For the case of a multifamily building, the use of grey water and rainwater is quite similar, with both being the measures with lower efficiency, nevertheless providing savings ranging between 31% and 31.5%. If it is a system of rainwater harvesting in conjunction with efficient devices, it becomes an ideal choice because it allows reducing consumption by about 73.7%, which translates into a consumption of 36.7 L/capita/day (Table 7).

Table 7 – Comparison of measures in multifamily building

Measure	Normal consumption (m3/agr.year)	Reduction (m3/agr.year)	Final Consumption (m3/agr.year)	Consumption (L/capita/day)	Savings (%)
Gray waters	138.6	42.5	96.1	97.1	31
Rainwater	138.6	43.7	94.9	95.9	31.5
Efficient appliances	138.6	79.8	58.8	59	58
Rainwater + efficient appliances	138.6	102.2	36.4	36.7	73.7

The potential of the measures analyzed and presented translates into economic gains and environmental benefits. These gains are strengthened in technical terms by an economic analysis of the proposed measures. In this sense, the analysis of the various proposed devices and water-saving systems, equated with the objective of estimating the cost-benefit and to validate the period of return on investments, seems to be advantageous.

The payback period of the various measures presented is estimated, considering the total costs of benefits and the total investment. In Table 8 is possible to see the return of the various saving measures in the three different geographical locations in Portugal under consideration, for the case of private houses and in Table 9 for the case of multifamily houses (zones defined according to the level of rainfall and water prices).

Table 8 – Estimated return of investment period on single family buildings (years)

Measure	Location		
	Lisboa	Faro	Porto
Gray water	> 18	> 15	> 14
Rainwater	> 35	> 34	> 32
Efficient appliances	> 7	> 6	> 6
Efficient appliances + Rainwater	> 30	> 27	> 24
Effic. appliances (without wash machine and dishwasher)	< 2	< 2	< 2

Table 9 – Estimated return of investment period in multifamily buildings (years)

Measure	Location		
	Lisboa	Faro	Porto
Efficient appliances	> 12	> 10	> 9
Efficient appliances (without wash machine and dishwasher)	< 3	< 2	< 2

It is evident that the return period of the investment is quite high. Only the implementation of efficient appliances constitutes a measure that leads to a shorter return period. However, the social value of water resources must not be forgotten; a situation that can lead to consider what should be the way forward in the long term, i.e. a decision-making framework for sustainability.

In this cost/benefit analysis, where the suggested measures are economically analyzed, there is an important factor of difficult quantification, given the unpredictability and uncertainty of the availability of water resources in the future. Within 25 years two thirds of the world population will live in water stress conditions and we can't predict what changes will occur in the water supply industry due to this factor of scarcity, and how this situation will be reflected in the value of water. The rate of water availability may not increase at today's commonly used annual rate, and could see its value fairly inflated due to resource scarcity. Moreover, the availability of water from other sources, like the sea, through the use of other technologies has a cost of production, associated with the high-energy consumption they require, which also contributes to the increasing of the cost of water.

Therefore it is important to diffuse knowledge and establish an operating procedure applicable to a project that will push the implementation, throughout the building's life cycle, of measures to contribute to the saving of water resources. This will lead to the economic efficiency of the process, which ultimately translates into a more sustainable project.

5.0 Conclusions

Given the fact that the scarcity of water is expected to occur in foreseeable future, it is essential that we develop a set of measures aiming to increase the efficiency of water use, a situation that has been analyzed in this study, through the various measures of action proposed, whose performance tests results were presented above.

Although the values of 37.9 or 36.7 L/capita/day seem extremely low when compared to current consumption, they are closer to the minimum value of 50 L/capita/day, which includes all the basic needs of a human being in their day-to-day life. Thus, it is considered that a final proposal should suggest the following procedures:

- For existing buildings, in a first phase, the consumption dropping to 80 L/capita/day, which is a decrease of 50% for single-family house and a decrease of 42% for multifamily house, could be attained through changes in the devices, which can be replaced by others with greater efficiency and without changes in the water distribution networks. In the next phase it will be possible to study the implementation of systems for rainwater utilization, which is the measure that, coupled with more efficient devices, would translate into lower levels of freshwater consumption.
- In new buildings, a consumption reduction to 40 L/capita/day could be achieved through the implementation of a rainwater collecting system combined with more efficient devices. This is justified with the measures developed in this paper and translates into a growth of a much wider social consciousness.

Given all these factors, it becomes increasingly clearer that water is an endangered resource and all the measures to reduce its consumption or increase the level of efficiency of its use must be taken.

In terms of saving water resources, any of the measures considered will be of great advantage. It is possible to conclude that all the presented measures are valid and feasible to implement taking into account the results presented above, which point to a clear reduction of water resources for the same activities that today are used in unsustainable manners.

Acknowledgement

Researchers express the acknowledgement to GEOTPU FCT UNL, for the important support provide to the development of this research.

References

[1] M. Neves, and E. Bertolo, "Aproveitamento e Reutilização da Água para Usos Domésticos", Jornadas de Hidráulica, Recursos Hídricos e Ambiente, FEUP (in portuguese), 2006.

[2] M. P. Amado, "Planeamento Urbano Sustentável", Editora Caleidoscópio, Casal de Cambra, Lisboa, ISBN: 9789728801748, (in portuguese), 2005.

[3] M. P. Amado, "The Sustainable Building Process", In CD: Ron Wakefield (eds): RMIT University, Australia, ISBN: 978-1-921166-68, pp. 65-69, 2007.

[4] L. Bragança, "Sustentabilidade de soluções construtivas", Universidade do Minho, Braga (in portuguese), 2006.

[5] LEED, "LEED for New Construction and Major Renovations Rating System", U.S. Green Building Council, LEED, USA, 2009.

[6] BREEAM, "Code for Sustainable Homes", Technical guide, BREEAM, London, 2009.

[7] LIDERA, "Lidera Sistema de Avaliação da Sustentabilidade", Versão para Ambientes Construídos (V2.00b) Lidera, IST UTL, Lisboa (in portuguese), 2009.

[8] L. Barroso, "Construção Sustentável – Soluções Comparativas para o uso eficiente da água nos edifícios de habitação", Dissertação de Mestrado em Engenharia Civil, FCT UNL (in portuguese), 2010.

[9] S. Muthukumaran, K. Baskaran, and N. Sexton, "Quantification of potable water savings by residential water conservation and reuse – A case study", Resources, Conservation and Recycling, vol. 55: pp. 945-952, 2011.

[10] L. Domènech, and D. Saurí, "A comparative appraisal of the use of rainwater harversting in single and multifamily buildings of the metropolitan area of Barcelona (Spain): Social experience, drinking water savings and economic costs," Journal of Cleaner Production, vol. 19, pp. 598-608, 2011.

[11] V. M. E. Pedroso, "Medidas para um uso mais eficiente da água nos edifícios, Informações Cientificas Técnicas", LNEC, Lisboa (in portuguese), 2009.

[12] M. C. Almeida, P. Vieira, and R. Ribeiro, "Uso Eficiente da água no sector urbano", Série Guias Técnicos, Ed: IRAR, INAG e LNEC, Lisboa (in portuguese), 2006.

[13] E. Ghisi, "Potential for potable water savings by using rainwater : An analysis over 62 cities in southern Brazil", Building and Environment Journal, vol. 42 (2), pp. 204-210, 2006.

[14] A. Arpke, and N. Hutzler, "Operational Life-Cycle Assessment and Life-Cycle Cost Analysis for Water Use in Multioccupant Buildings", Journal of Architectural Engineering, vol. 11 (3), pp. 99-109, 2005.

[15] ANQIP, "ETA 0806 – Especificações para atribuição de rótulos de eficiência hídrica ANQIP a duches e sistemas de duche", ANQIP, Aveiro (in portuguese), 2009.

[16] ANQIP, "ETA 0808 – Especificações para atribuição de rótulos de eficiência hídrica ANQIP a torneiras e fluxómetros", ANQIP, Aveiro (in portuguese), 2010.

[17] ANQIP, "ETA 0906 – Certificação de sistemas prediais de reutilização e reciclagem de águas cinzentas", ANQIP, Aveiro (in portuguese), 2011.

12

COMMUNITY PARTICIPATION IN LOCAL GOVERNMENTS: PUBLIC CONSULTING AND TRANSPARENCY IN GAZA STRIP, PALESTINE

Adnan Enshassi, Ahmed S. Kullab

Department of Civil Engineering, Islamic University IUG, Gaza, Palestine

*Corresponding E-mail: aenshassi@gmail.com

Abstract

The objectives of this study are to identify and examine the community participation strategies concerning consultation and information disclosure and to identify and evaluate major barriers to community participation development. A questionnaire survey was developed to elicit the perceptions of the municipality's officials (top management and/or mayors of the selected sample) of Gaza Strip municipalities regarding community participation practice. The results indicated that there is a weak transparency (information disclosure) and community consultation in the municipalities of Gaza Strip. The findings revealed that, the major barriers of information disclosure approach are due to legal challenges, public awareness, lack of community members' skills and knowledge and social factors. In addition, the major barriers of community consultation are mainly due to lack of community members' skills and knowledge, lack of social capital and trust of local people in their leaders, political, economic challenges, community culture and municipality council awareness. It is recommended to expand the scope of public participation and developing strategies that maximize citizen input in community development activities in local governments of Gaza Strip. The findings can assist in identifying new directions for enhancing public participation in Gaza Strip local governments.

Keywords: *Community Participation; Public policy; Municipalities; Local government; Palestine*

1.0 Introduction

The beginning of the local governments in Palestine has witnessed a course of four historical periods: the Ottomans, the British, the Jordanians, and the Palestinian Authority. During the last ruling period, the Ministry of Local Government (MoLG) was formed by a decision from the Palestinian leadership in Tunisia on 25 February 1994. The MoLG has established a local government structure and ensure that local government consists of local councils in the form of municipalities, town councils, and joint services councils that work for the benefit of the residents. Each local committee has an area of jurisdiction within the boundaries of the state. The MoLG adopted the following four basic objectives [1]:

1. Advancing the concepts of local government and decentralized management; and to create local government institutions that support national objective of building Palestinian local communities that embrace democratic elections.
2. Raising the quality of services in the Palestinian rural community to bridge the gap between rural and urban areas.
3. Developing the abilities of local councils.
4. Reviewing the performance of local councils established before the inception of the Palestinian national authority in order to arrive at a local government vision concurrent with the Palestinian agenda.

The scope of community participation in local government is considered very limited. This phenomenon is found in all levels of Palestinian institutions. It is manifested in an uniform lack of development vision, failure to promote a philosophy of involving the community in bearing the responsibility for the development and reaping its rewards, and by the prevalence of overly centralized, ineffective work coordination methods [2].The objectives of this study are to identify and evaluate the community participation strategies concerning consultation and information disclosure and to identify and assess major barriers to community participation development.

2.0 Literature Review

The debate about the community participation aspect in local governments and its effects on developing local government are getting more important all over the world. There are numerous research works considered the community involvement in local governments. Silverman [3] examined the socio-economic characteristics and public participation strategies of municipalities in USA and Canada. The Study aimed to survey the mechanisms used by municipalities to stimulate community participation and to argue that contrasts available between the socio-economic make-up of central cities in the USA and Canada and to explain the divergent techniques used for community participation. The results indicated that Canadian municipalities adopt a boarder range of public participation techniques related to: voluntarism and public engagement, neighborhood and strategic planning, and e-government. In contrast, the USA municipalities are more likely to promote community participation through mechanisms such as annual community meetings and referendum on public issues.

Another more recent study was conducted by Monfardini [4] purposed to analyze the effect of new public management-based reforms on public accountability in two countries, Italy and Sweden, explaining what strategies can be used to enhance accountability toward citizens. In which, two case studies have been analyzed in order to compare how public accountability has been enhanced in different cultural and geographical contexts. Monfardini [4] argued that the enhancement of public accountability, in both countries, is at stake both in the political debate and the public policies, but the strategies implemented resulted are path dependent especially in the choice between more disclosure and more citizen participation.

Other researches were built to clarify the benefits that could be achieved from development of community partnerships at the local level. In Ponta Delgada (Archipelago of the Azores), for the first time in Portugal, the relevant stakeholders were involved in the development process of sustainable mobility plan. The case study proved that, on a participatory basis, sustainable transport planning provides more satisfactory and efficient solutions. Gil et al[5]found that the active and participatory management plans are mandatory in order to fully and truly address the mobility issues. Sebina and Rosenzweig [6]presented the process and results of a local-level South African action research project on introducing foresight methods into a local government planning process. They found positive outcomes in the King Sabata Dalindyebo (KSD) foresight process included a high level of stakeholder engagement and senior management buy-in, supported by positive evaluations by diverse participants. The learning was subsequently incorporated into long-term development plans and proposals such as the ten-year development plan for the region.

Participatory governance mechanisms have been widely promoted in developing countries. They are claimed to bring about several public policy benefits, including increased accountability, higher government responsiveness, and better public services [7]. In this context, Njoh [8] reported that the citizen participation is a viable cost-saving strategy in the development process especially in resource scarce settings. Ngowi and Mselle [9] concluded that all communities need a variety of facilities in order to function well. While some of these facilities need outside input to implement, the majority may be planned, implemented and managed by the beneficiaries in one way or another. Therefore, it is confirmed that community involvement is key to fostering significant and sustainable changes in under-developed communities [10]as well as public inputs will enhance the public's confidence in governments [11, 12] .

Some researchers are intended to identify the barriers and obstacles hindering people involvement in local governments. In this context, Adamson [13] examined the community first program, an area-based regeneration policy, in Wales (UK) to explore the barriers to community empowerment. He considered three related research projects provide data to inform the discussion of community empowerment and to consider the implications of delivery of the policy for theorizing the relationship between the citizen and the state as mediated through regeneration partnerships. The findings identified major barriers to the achievement of community empowerment including issues of community capacity, institutional capacity, organizational cultures and regulatory frameworks. Waheduzzaman[14] tried to find out the barriers to people's participation for good governance in developing countries. The specific objective was to explore the attitude of stakeholders responsible for ensuring people's engagement with local development programs. His study revealed that one of the major barriers to people's participation in local government affairs is the traditional attitude of related stakeholders. This study thus unlocked the practical knowledge about making people's participation effective in implementing development programs, specifically in the developing countries.

In Palestine, the community participation approach was mainly studied in an ordinary way. Rare studies have focused in participation issues in local governments. However, this study is an extension of some past researches on public participation in the policy processes of municipalities in Palestine. For instance, this research builds upon Sabri and Jaber [15] of managerial performance of Palestinian local authorities. They surveyed the Palestinian municipalities examining whether the contribution of residents in the local government aspects have been increased or not during the rule of the elected councils and mayors of Palestinian municipalities.

Similarly, this study offers extensions to Khalifa [16] research on planning with citizen participation in the Palestinian novelty municipalities. He tried to examine the current environment of community participation in planning in the Palestinian Novelty Municipalities and to Pinpoint the adopted strategies of participatory planning as practiced in the Palestinian novelty municipalities and analyze the effectiveness of such strategies. He concluded that there exists a gap of understanding between the community and the municipalities regarding this issue, since the participatory approach is new and both the municipality and the community are entrenched in the old paradigm. Each party blames the other, as the municipality is concerned that participation will diminish its role while the community thinks of participation as waste of time. Ammar et al [17] investigated the effect of residents' participation in management and maintenance works on residents' satisfaction in new multi-story housing projects in Gaza, Palestine. The result of the analysis showed a significant relation between the level of the residents' participation and their satisfaction. From the previous literature and other research works, it is concluded that community participation in local governments can be classified into five levels as following:

- Information disclosure: (some researchers mentioned the synonym 'Transparency') This level stated that the local government should provide the public with balanced and objective information on municipal services and procedures, plans, challenges, available resources and opportunities for development, and achievements [18-20].
- Public consultation: In which the public's input on matters affecting them is sought, in other word, this level ensures soliciting feedback from the public on the local government plans, budget, performance, municipal service procedures, and activities [21, 22] .
- Participation in the planning process: such as the public's participation in developing strategic development plans and identifying the community's needs, and the participation of specific groups (such as youth) in the designing of a community project [6, 23, 24].
- Participation in the decision making process: meaning that public should be involved in exploring alternatives and deciding on the best appropriate option, which takes into consideration interest and priorities of the local community [25-27].
- Participation in municipal budgeting and financial contributions: participation in the form of financial and in-kind contributions is mostly reliant on initiatives from citizens or institutions. These contributions include: voluntary work, implementation of programs/projects authorized by the local government, and financial contributions [28].

Figure 1: Community participation strategies

Table 1: List of barriers for each level of community participation approach

List of barriers	Group\Level	Reference								
		Ngowi and Mselle[9]	Kastlungeret *al* [29]	Adamson [13]	Monfardini [4]	Sabri and Jaber[15]	Waheduzzaman [14]	Wallis and Dollery[30]	Cumbers and McMaster [31]	PytlikZilliget *al* [11]
None existence of obligatory community membership.	Community Consultation		√	√						
Weakness of statutory agencies capacity.		√		√			√			
Lack of community members' skills and knowledge.				√		√				
The local governments' organizational structures.			√	√	√			√	√	√
Financially expensive strategies and programs.		√								√
Lack of incentives and promotions to the public.		√				√				√
Political and economic issues.					√		√			√
Lack of awareness of the value of people's participation.		√		√			√			√
Administrative centralization.				√		√	√	√	√	√
Lack of social capital.			√		√	√	√	√		√
Local governments' organizational structures.	Information Disclosure		√	√	√	√		√	√	√
Lack of community members' skills and knowledge.		√		√	√	√	√			√
Political and economic challenges					√	√			√	√
Financially expensive strategies		√			√					
Awareness of local governments representatives					√	√	√			
Community culture		√		√	√		√			√
Social factors		√			√					√
Difficulty to ensure authenticity and quality of information.					√					
Legal challenges			√		√	√		√		√

This paper considers the first two levels (Information disclosure and public consultation), trying to examine the existing strategies conducted to ensure those two levels and to find out the barriers hindering its implementation. Based on the reviewed literature, a list of barriers hindering the first two levels of community participation was considered as shown in Table 1 above. Figure 1 above illustrates the community participation strategy.

3.0 Methodology

A questionnaire survey was developed to elicit the perceptions of the municipality's officials (top management and/or mayors of the selected sample) of Gaza Strip municipalities regarding community participation practice. The questionnaires seek also to elicit the respondents' opinion regarding major barriers that face municipalities' administration and hindering community participation development. A pilot study was carried out in order to test the validity and reliability of the questionnaire. The pilot study included selected sample of top managerial staff as well as mayors of Gaza Strip municipalities. The comments of the respondents have been taken into consideration; the final research instrument has been formulated in Arabic.

The total number of municipalities in Gaza Strip is 25 municipalities. The number of respondents was 21 which yield 85 percent of the total target sample (table 2).The officials who filled out the questionnaire were either the mayor, or the general manager of the municipality. The questionnaire had been filled out directly through interviews or sent by e-mail. For municipalities that were contacted by e-mail, two letters of reminder were sent after one week in order to assure the response, while others were contacted again by telephone to return the filled questionnaire.

Table 2: List of Gaza Strip municipalities and respondents of the questionnaire

No.	Municipalities of Gaza Strip	Respondents	No.	Municipalities of Gaza Strip	Respondents
1.	Gaza Municipality	√	14.	Rafah Municipality	√
2.	Al-Zahra Municipality	√	15.	Almosadder Municipality	
3.	Wadi-Gaza Municipality		16.	Nuseirat Municipality	√
4.	Beithanoun Municipality	√	17.	Alburaj Municipality	√
5.	Bietlahia Municipality		18.	Alzawaida Municipality	√
6.	Jabalia-Alnazla Municipality	√	19.	Wade-Alsalqa Municipality	√
7.	Deir Al-Balah Municipality	√	20.	OmalnaserMunicipality	√
8.	Almghazi Municipality	√	21.	Absan-Alkabera Municipality	√
9.	Khanyounis Municipality	√	22.	Absan-Aljadida	
10.	Banisuhaila Municipality	√	23.	Khuzaa Municipality	√
11.	Alqrara Municipality	√	24.	AlnaserMunicipality	√
12.	Alfukhare Municipality	√	25.	Almoghraqa Municipality	√
13.	Alshoka Municipality	√			

Validity can be defined as an instrument as a determination of the extent to which the instrument actually reflects the abstract constructs being examined, it refers to the degree to which an instrument measures what it is supposed to be measuring [32]. High validity is the absence of systematic errors in the measuring instrument. When the instrument is valid; it truly reflects the concept it is supposed to measure. Achieving good validity required the care in the research design and sample selection. Internal consistency of the questionnaire was measured by a scouting sample, which consisted of 11questionnaires, through measuring the correlation coefficients between each paragraph in one field and the whole field. Table 3 shows the correlation coefficient and p-value for each field items. As shown in the table, the P- Values are less than 0.05 or

0.01.The correlation coefficients of this field were significant at α = 0.01 or α = 0.05.Therefore, it could be said that the paragraphs of this field were consistent and valid to be the measure what it was set for.

Table 3: Correlation coefficient and p-value for testing questionnaire validity

No.	Statements	Pearson coefficient	p-value	Sig. level
Information disclosure barriers				
1	The present local governments' organizational structures are not ready for information disclosure to the public.	0.719	0.000	**
2	Public awareness is considered a barrier to information disclosure.	0.608	0.001	**
3	Lack of community members' skills and knowledge are hindering the local government to disclose information.	0.889	0.000	**
4	Local governments are afraid of information disclosure approach due to the political and economic challenges.	0.736	0.000	**
5	Strategies to disclose information are financially expensive to be implemented.	0.666	0.000	**
6	In your opinion, information disclosure is not important for the public and it has a negative effect on the community development.	0.680	0.000	**
7	The community culture is considered a reason for hindering information disclosure approach.	0.692	0.000	**
8	Social factors restrict some information to be published to the public. (For example: it May affect privacy)	0.555	0.004	**
9	Difficulty to ensure authenticity and quality of information is considered an obstacle to information disclosure.	0.530	0.006	**
10	In your opinion, information disclosure will lead to legal challenges.	0.674	0.000	**
Community consultation barriers				
11	None existence of obligatory community membership is an obstacle to community consultation.	0.822	0.000	**
12	Weakness of statutory agencies capacity to work in community-sensitive ways is considered an obstacle.	0.607	0.001	**
13	Lack of community members' skills and knowledge are hindering their involvement in local government consultation. (community culture)	0.575	0.003	**
14	The present local governments' organizational structures are not ready for community consultation approach.	0.513	0.009	**
15	Strategies and programs improving community consultation are financially expensive to be implemented.	0.815	0.000	**
16	Community consultation is weak due to lack of incentives and promotions to the public.	0.461	0.021	*
17	Community consultation is negatively affected by the recent political and economic issues.	0.763	0.000	**
18	The reason behind community consultation weakness is that none of the local government's representatives and local people are aware of the value of people's participation.	0.646	0.000	**
19	One of the factors that hindering public consultation in local governments is the administrative centralization.	0.545	0.005	**
20	Lack of social capital, mainly trust of local people in their leaders, is hindering true people's consultation.	0.427	0.033	*

** Correlation coefficient is significant at the α = 0.01
* Correlation coefficient is significant at the α = 0.05

Reliability of an instrument is the degree of consistency with which it measures the attribute it is supposed to be measuring [33]. The test is repeated to the same sample of people on two occasions and then compares the scores obtained by computing a reliability coefficient. For the most purposes, reliability coefficient above 0.7 is considered satisfactory. Period of two weeks to a month is recommended between two tests. But due to complicated conditions that the municipalities' mayors are always too busy, so that it was too difficult to ask them to responds to our questionnaire twice within a short period. Therefore, in order to overcome the distribution of the questionnaire twice, the reliability measurement can be achieved by using Cronbach's Alpha coefficient and Half Split Method.

Half split method depends on finding Pearson correlation coefficient between the means of odd rank questions and even rank questions of each field of the questionnaire. Then, correcting the Pearson correlation coefficients can be done by using Spearman Brown correlation coefficient of correction. The corrected correlation coefficient (consistency coefficient) is computed according to the following equation: Consistency coefficient = 2r/(r+1), where r is the Pearson correlation coefficient. The normal range of corrected correlation coefficient 2r/(r+1) is between 0.0 and + 1.0 [33]. As shown in Table 4, the general reliability for all items equal 0.892, and the significant (α) is less than 0.05, so all the corrected correlation coefficients are significance at α = 0.05. It can be said that according to the Half Split method, the dispute causes group are reliable.

Table 4: Split-Half Coefficient method

Section	Person-Correlation	Spearman-Brown Coefficient	P-value
Barriers Hindering Community Participation	0.805	0.892	0.000

Cronbach's coefficient alpha method is used to measure the reliability of the questionnaire between each field and the mean of the whole fields of the questionnaire. The normal range of Cronbach's coefficient alpha value between 0.0 and + 1.0.The higher values reflect a higher degree of internal consistency. The general reliability for all items equal 0.920. This range is considered high; the result ensures the reliability of the questionnaire [Table 5].

Table 5: Reliability Cronbach's Alpha

Section	No. of Items	Cronbach's Alpha
Barriers Hindering Community Participation	20	0.920

4.0 Results and discussion

4.1 Community participation strategies used in local governments

4.1.1 Information disclosure (Transparency)

The results shown in Table 6 indicate that all municipalities in the Gaza Strip did not publish the internal performance assessment reports of their employees. Eighty six percent of municipalities did not let the public know about its annual budget documents and discuss the final statements of budget with residents. Eighty one percent of the municipalities did not have official councils meetings that are open for public attendance. Just 33 percent of municipalities have an updating website to publish their activities. Sixty two percent of municipalities have a strategic development plans and follow-up assessment for these plans are published to public. Sixty seven percent of the municipalities appoint their employees according to an advertisement and using competition among applicants and 81 percent of municipalities guide their residents for their services procedures.

Table 6: The existed information disclosure strategies

Information disclosure (Transparency)	Yes (%)	No (%)
Are the internal performance assessment reports published to public?	0	100
Are the major annual budget documents published to public?	14	86
Are the final statements of budget discussed by residents?	14	86
Are official sessions of the council opened to public?	19	81
Is there updating website to publish the local government activities?	33	67
Are the local strategic development annual implementation plans, as well as the results of follow-up assessment for these plans published to public?	62	38
Are the employees appointed according to advertisement using fair competition?	76	24
Are the municipal services procedures, especially, the procedures for the permits and basic services published?	81	19

The findings showed that there is a real need to enhance the concept of open councils meeting to public and local citizens, to publish a summary of the internal performance assessment to public. This issue may implicate encouragement to municipalities' employees, to discuss the final statements of budget with residents, to publish the major budget documents annually, to call citizens for attending the official sessions of municipality council. It is important for every municipality to have an updating website to present its activities to public, this issue is fetal to enhance confidence and support the two-way commitment between local council and the community. This concept are internationally recognized and recommended by various studies such as [3, 18, 20, 29].

4.1.2 Community consultation

The results in Table 7indicated that the available strategy to insure community consultation is the existence of a system for receiving citizens' complaints. The second most available strategy is the hearing sessions conducting periodically. Seventy one percent of municipalities concerned involving their community in selecting projects priorities and the process of projects development. Sixty seven percent of them have an activated advisory and sector committees. On the other side, around two thirds of municipalities did not have an active internet communication with its citizens, 48 percent of municipalities did not consult a feedback from the public on the quality of services and fairness of its distribution, and 43 percent of them did not periodically conduct town hall meetings between the municipality council and citizens. This result agreed in part with Sabri and Jaber [15] study of managerial performance of Palestinian municipalities.

Table 7: The existed community consultation strategies

Community consultation	Yes (%)	No (%)
Is there an internet communication between the municipality and citizens?	38	62
Is the feedback on the quality of services and fairness in distribution consulted from the public?	52	48
Are there periodical town hall meetings between municipality council and citizens?	57	43
Are an advisory and sector committees formed and activate?	67	33
Is the local community involved in projects development process?	71	29
Are the projects priorities are selected through participation of citizens?	71	29
Are there hearing sessions held periodically?	76	24
Is there a system for receiving citizens' complaints?	90	10

Community consultation concept can be implemented by activating the internet communication between municipalities and citizens since it's the easiest and most comfortable way for citizens to communicate the local council. Feedback consulted from the public on the quality of services and fairness in distribution has a great impact on community consultation and satisfaction, and there is a real need for conducting town hall meetings that calling citizens and municipality council for attendance.

4.2 Major barriers of community participation development

4.2.1 Information disclosure

Table 8 illustrated that the major barriers of information disclosure approach that facing the municipalities of Gaza Strip are mainly due to legal challenges, public awareness, lack of community members' skills and knowledge and social factors. Other effective barriers are due to community culture, municipality organizational structure and the difficulty to ensure authenticity and quality of information. Some barriers that have a small degree of effect are due to the political and economic challenges, municipality council's awareness, and financially expensive strategies to be implemented.

Table 8: Major barriers of information disclosure that facing municipalities

Barriers to community participation	**Mean (5)**	**Rank**
Legal challenges.	3.84	1
Public awareness	3.68	2
Lack of community members' skills and knowledge	3.56	3
Social factors (For example: it May affect privacy)	3.48	4
Community culture	3.08	5
Local governments' organizational structures	3.04	6
Difficulty to ensure authenticity and quality of information	2.88	7
Political and economic challenges.	2.64	8
Municipality council's awareness.	2.32	9
Financially expensive strategies to be implemented.	2.24	10
All barriers	3.076	

According to the results, legal challenges are the most barrier of information disclosure concept in Gaza municipalities, common reason of this are that municipality plans and orientations may assault the rights of citizens and this make the municipality council afraid of disclosing information to public. This challenge can be dissolved by discussing the municipality plans and orientations with the public before adopting it. The next significant barriers of information disclosure concept are the public awareness, lack of community members' skills and knowledge, social factors and community culture. This can be enhanced by conducting awareness meetings and publications in order to make the community aware of the value of his contribution in community development.

4.2.2 Community consultation

The major barriers of community consultation approach facing the municipalities of Gaza Strip as presented in Table 9 are mainly due to lack of community members' skills and knowledge, lack of social capital, mainly trust of local people in their leaders, political, economic challenges, community culture and municipality council awareness. Other less effective barriers are due to administrative centralization, non-existence of obligatory community membership and the weakness of statutory agencies capacity to work in community-sensitive ways. Factors such as lack of incentives and promotions to the public, local governments' organizational structures, financially expensive strategies and programs to be implemented have an insignificant effect on community consultation development.

Table 9: Major barriers of community consultation that facing municipalities of Gaza Strip

Barriers	**Mean (5)**	**Rank**
Lack of community members' skills and knowledge.	4.16	1
Lack of social capital, mainly trust of local people in their leaders.	3.68	2
Political and economic issues.	3.34	3
Community culture and municipality council awareness.	3.04	4
Administrative centralization.	2.88	5
None existence of obligatory community membership.	2.8	6
Weakness of statutory agencies capacity to work in community-sensitive ways.	2.68	7
Lack of incentives and promotions to the public.	2.44	8
Local governments' organizational structures.	2.04	9
Financially expensive strategies and programs to be implemented.	1.72	10
All barriers	2.87	

The results indicated that the most barrier of community consultation approach is the lack of community members' skills and knowledge and as mentioned in the barriers of information disclosure concept. This barrier can be dissolved by conducting awareness meetings and publications and by conducting training courses that feeds the community members with the required skills to make them ready for community participation. There is a great need to enhance the social capital and community trust on municipality council. This can be achieved by using the open-government concept in managing the local government. Another important issue is the political and economic challenges that the citizens of Gaza Strip are facing, therefore the municipality council is required to consider this issue in the laws and legislations managing the local government.

5.0 Conclusion and recommendations

This study concluded that there is a weak transparency (information disclosure) situation and community consultation in the Gaza Strip municipalities. This fact is true due to the lack of trust and confidentiality between citizens and municipality councils since as stated in the results. The internal performance assessment reports and major annual budget documents are not published to public, the absence of community in the final statements of budget discussion. Local municipalities did not use internet for creating a two way communications with the citizens, and the weak consideration of public feedback on the quality of services and fairness in distribution. It may be concluded that the municipalities of Gaza Strip face two groups of community participation barriers.

The first group of barriers hindering information disclosure approach was: the major barriers are due to legal challenges, public awareness, lack of community members' skills and knowledge and social factors. Other barriers were due to community culture, municipality organizational structure and the difficulty to ensure authenticity and quality of information. Less effective barriers are due to the political and economic challenges, municipality council's awareness, and financially expensive strategies to be implemented.

The second group of barriers hindering community consultation approach was: the major barriers are mainly due to lack of community members' skills and knowledge, lack of social capital and trust of local people in their leaders, political, economic challenges, community culture and municipality council awareness. Other less effective barriers are due to administrative centralization, non-existence of obligatory community membership and the weakness of statutory agencies capacity to work in community-sensitive ways. Factors such as lack of incentives and promotions to the public, local governments' organizational structures, financially expensive strategies and programs to be implemented have an insignificant effect on community consultation development.

Based on the findings of this study it is recommended:

- Municipality councils meetings should be open to the public; this implicates community confidence and trust to the local councils.
- Each municipality should establish an updated website through which they can inform, interact with the public, and present its services and activities.
- Municipalities should discuss its plans and orientations with the public before adopting it in order to avoid legal challenges created between citizens and local governments. This in turn will stimulate the information disclosure approach and enhance confidentiality between citizens and municipality councils.
- The official budget should be published inside the municipality and to the public.
- It is recommended to apply training courses, awareness meetings and publications for citizens to encourage their participation in the municipal activities since the lack of public knowledge and skills is a common barrier for both information disclosure and community consultation approaches. Such courses should feed the community by the right and effective methods to express their opinions and ideas about municipality services and activities.
- Establishing a special division in each municipality that can facilitate the public participation process, in order to guide local government representatives and provide them with knowledge on how to promote public participation.
- It is essential for each municipality to ensure that technical and financial resources needed for community participation processes are in place.

References

[1] Human Development Report, "Community Participation in Local Government and Development",Chapter3, Development Studies Programme, Birzeit University, Birzeit, Palestine, 2002.

[2] Human Development Report, " Human Development in Palestine: A Participatory Approach", Development Studies Programme, Birzeit University, Birzeit, Palestine, 1998.

[3] R. M. Silverman, "Central City Socio-Economic Characteristics and Public Participation Strategies: A Comparative Analysis of the Niagara Falls Region's Municipalities in the USA and Canada", International Journal of Sociology and Social Policy, vol. 26, pp. 138-153, 2006.

[4] P. Monfardini, "Accountability in the New Public Sector: A Comparative Case Study", International Journal of Public Sector Management, vol. 23, pp. 632-646, 2010.

[5] A. Gil, H. Calado, and J. Bentz, "Public Participation in Municipal Transport Planning Processes – the Case of the Sustainable Mobility Plan of Ponta Delgada, Azores, Portugal", Journal of Transport Geography, vol.19, pp. 1309-1319, 2011.

[6] G. Sebina and L. Rosenzweig,"A Case Study on Localizing Foresight in South Africa: Using Foresight in the Context of Local Government Participatory Planning", Foresight, vol. 14,pp. 26-40, 2012.

[7] J. Speer, "Participatory Governance Reform: A Good Strategy for Increasing Government Responsiveness and Improving Public Services?", World Development, vol. 40, pp. 2379-2398, 2012.

[8] A. Njoh, "Municipal Councils, International NGOs and Citizen Participation in Public Infrastructure Development in Rural Settlements in Cameroon", Habitat International, vol. 35, pp. 101-110, 2011.

[9] A. Ngowi and P. Mselle, "Community Participation in Facility Management", Facilities,vol. 16, pp. 314-318, 1998.

[10] M. Pardasani, "Tsunami Reconstruction and Redevelopment in the Maldives: A Case Study of Community Participation and Social Action", Disaster Prevention and Management, vol. 15, pp. 79-91, 2006.

[11] L. PytlikZillig, A. Tomkins, M. Herian, J. Hamm, and T. Abdel-Monem, "Public Input Methods Impacting Confidence in Government", Transforming Government: People, Process and Policy, vol. 6, pp. 92-111, 2012.

[12] T. Cohen, "Can Participatory Emissions Budgeting Help Local Authorities to Tackle Climate Change?", Environmental Development, vol. 2, pp. 18-35, 2012.

[13] D. Adamson, "Community Empowerment: Identifying the Barriers to "Purposeful" Citizen Participation", International Journal of Sociology and Social Policy, vol. 30, pp. 114-126, 2010.

[14] Waheduzzaman, "Value of People's Participation for Good Governance in Developing Countries", Transforming Government: People, Process and Policy, vol. 4, pp. 386-402, 2010.

[15] R. N. Sabri and Y. R. Jaber, "Managerial Performance of Palestinian Local Authorities", Transforming Government: People, Process and Policy, vol. 1, pp. 350-363, 2007.

[16] S. Khalifa, "In Search for a Model: Planning with Community Participation in the Palestinian Novelty Municipalities", Master Thesis, Al-Najah National University, Nablus, Palestine, 2011.

[17] S. Ammar, K.Hj Ali, and N. Yusof, "Effects of Residents' Participation in Management Works on Satisfaction in Multi-Story Housing", Procedia - Social and Behavioral Sciences, vol. 62, pp. 837-843, 2012.

[18] C. Nyman, F. Nilssonand B. Rapp, "Accountability in Local Government: a Principal-Agent Perspective", Journal of Human Resource Costing & Accounting, vol. 9, pp. 123-137, 2005.

[19] J. Hochtl, P. Parycek and M. Sachs, "E-participation Readinessof Austrian Municipalities", Transforming Government: People, Process and Policy, vol. 5, pp. 32-44, 2011.

[20] C. Armstrong, "Providing a Clearer View: An Examination of Transparency on Local Government Websites", Government Information Quarterly, vol. 28, pp. 11-16, 2011.

[21] D. Scottand P.Vitartas, "The Role of Involvement and Attachment in Satisfaction with Local Government Services", International Journal of Public Sector Management, vol. 21, pp. 45-57, 2008.

[22] G. Lee and Y. Kwak, "An Open Government Maturity Model for Social Media-Based Public Engagement", Government Information Quarterly, vol. 29, pp. 492-503, 2012.

[23] B. Checkoway, T. Allison and C. Montoya, "Youth Participation in Public Policy at the Municipal Level", Children and Youth Services Review, vol. 27, pp. 1149-1162, 2005.

[24] G. Hassan, A. El-Hefnawi and M. El-Refaie, "Efficiency of Participation in Planning", Alexandria Engineering Journal, vol. 50, pp. 203-212, 2011.

[25] P. S. Scherrer, "Directors' Responsibilities and Participation in the Strategic Decision Making Process", Corporate Governance, vol. 3, pp. 86-90, 2003.

[26] B.S cott-Ladd, and V.Marshall, "Participation in Decision Making: A Matter of Context?", The Leadership & Organization Development Journal, vol. 25, pp. 646-662, 2004.

[27] C. Pita, R. Chuenpagdee, G. J.Pierce, "Participatory Issues in Fisheries Governance in Europe", Management of Environmental Quality: An International Journal, vol. 23, pp. 347-361, 2012.

[28] K. Razaki, "Municipal Budgeting: Positives, Pitfalls, and Politics", ASBBS Annual Conference, vol. 19, pp 775-786, 2012.

[29] B. Kastlunger, E. Lozza, E. Kirchlerand A. Schabmann, "Powerful Authorities and Trusting Citizens: The Slippery Slope Framework and Tax Compliance in Italy", Journal of Economic Psychology, vol. 34, pp. 36-45, 2013.

[30] J. Wallis and B. Dollery, "The Impact of Alternative Styles of Policy Leadership on the Direction of Local Government Reform", International Journal of Social Economics, vol. 32, pp. 291-306, 2005.

[31] A. Cumbers and R. McMaster, "Rethinking Public Ownership and Participation", On The Horizon, vol. 20, pp. 172-181, 2012.

[32] M. J. Miller, Western International University , RES 600: Graduate Research Methods Lecture notes, 1998.

[33] S. G. Naoum, "Dissertation Research and Writing for Construction Students". 2nd ed. UK: Elsevier Ltd, 2007.

13

AN ANALYSIS OF COST MANAGEMENT STRATEGIES EMPLOYED BY BUILDING CONTRACTORS ON PROJECTS IN ZIMBABWE

Benviolent Chigara[1], Tirivavi Moyo[2], Fungai Hamilton Mudzengerere[3]

[1] Department of Landscape Architecture and Urban Design, National University of Science and Technology, Bulawayo, Zimbabwe
[2] Department of Quantity Surveying, National University of Science and Technology, Bulawayo, Zimbabwe
[3] Faculty of The Built Environment, National University of Science and Technology, Bulawayo, Zimbabwe

*Corresponding E-mail : benviolent@gmail.com

Abstract

The construction industry is a major consumer of capital resources by virtue of the magnitude, complexity and nature of the works it executes. Yet improving cost performance still remains a chronic challenge facing the construction industry worldwide. Apart from advances made by contractors to devise mechanisms to manage project cost, studies across the world reveal that construction projects are hardly completed within budget. The Zimbabwean construction industry has not been spared either. However, this background is considered unhealthy for an industry whose capital consumption is heavy and concomitant cost overruns spell disastrous financial consequences on the investor. This study explored cost management strategies employed by contractors on building projects and also investigated the challenges that contractors encounter in managing project cost. Exploratory research design was employed making use of both questionnaires and interviews to collect data from selected contractors registered under category A, B and C of the Construction Industry Federation of Zimbabwe (CIFOZ). The study observed that in the majority of cases, contractors' efforts to manage projects costs are centred on management of project resources. Additionally, monthly cost reports, variance management, project budgets and cash flow forecasting are among the other strategies. However, in spite of the growing trend to automate cost control systems, there is evidence to suggest that the traditional paper based systems of storing and managing information is widely used in the industry. Contrary to having cost management systems in place, contractors admit experiencing cost overruns on their projects. These cost overruns are attributable to organizational related challenges, shortage of skilled personnel, and unavailability of some building materials on the local market, variations and deficiency in cost control systems.

Keywords: *Construction, Cost management strategy, Cost overrun, Zimbabwe*

1.0 Introduction

The construction industry is a key industry to economic growth of any nation providing shelter for economic and social activities including on-site and off-site infrastructure to facilitate the smooth functioning of these activities. The industry does not only touch on the lives of virtually everyone on a daily basis; it occupies a fundamental national position in many national economies - the bellwether of economic growth [1] As a result of the sizeable nature of projects executed; the industry requires substantive injection of capital and any loss through failure or abandonment has a crippling effect on the capabilities of the investors and financiers [2]. Notwithstanding the complex nature of the work undertaken by the construction industry, cost and time need to be effectively monitored and controlled if the anticipated profit margin will be realized for the contractor and for the project to be completed within the budget cost of the client [3]. To this end, financiers and executors (contractors) of construction works are bound to be cost conscious if their business objectives are to be realized from the project. In Zimbabwe, despite containment of inflation (from around 231 million percent in December 2008 to less than 3% by the first quarter of 2013), the prevailing liquidity crunch has the potential to influence the stance taken by both clients and contractors in respect of management of project cost. According to [4] contractors who do not stay current on costs may suffer 'disastrous' consequences on their already slender profit margin. To safeguard their primary objectives (of survival, growth and profitability), contracting organisations should put in place an effective cost management system within the structure of their organisations [5]

Irrespective of the economics generated through effective cost management, most projects are delivered over-budget. Nine out of ten projects faced cost overrun in the range of 50 to 100% [6]. The problems of project cost overrun are considered to be more severe in developing countries where the overrun sometimes exceed 100% of the anticipated cost of project [7]. In a previous study based on selected building projects conducted in Zimbabwe after the introduction of multi-currency regime, it was noted these projects (at various stages of construction) had overspend their budgets by an average overrun of 34% [8]. Prior to this period, Joina centre, a commercial building in the heart of Harare, had its budget revised several times. In Nigeria, the rate at which construction projects fail or are abandoned is retrogressive [2] and the minimum average percentage escalation cost of projects was 14% [9]. Such margins on cost overrun are huge when compared against arguments presented by [10] where cost overruns of between 10-15% are considered substantial and has potential to offset the client's overall budget. However, the problem of cost overrun is not peculiar to developing countries. In Norway, the challenge of project overruns is so prevalent that [11] summed that public building projects will always overrun their estimated costs; it's just a question of by how much. And yet a project cost which is out of control affect viability of a construction project and often adds to investment pressure, increases construction cost, reduces investment profit, affects investment decision-making and, above all, it causes disputes that turn down the overall project progress [12, 13]. With such a background, the need for effective cost management systems in contracting organisations cannot be overemphasized. In fact, cost control should be seen as an important management tool that is crucial for the survival of a construction company [14]. It is an essential objective of any firm and as such a cost control system is considered the nerve centre of a well-run contracting organization [15, 16]. However, apart from the importance of cost management to the construction industry, it has not received the attention that it has in other industries, for example, production engineering [3].

This research, which is part of a broader study on cost management in the Zimbabwean construction industry, sought to investigate the cost management strategies employed by contractors on building projects in Zimbabwe and examine the factors that militate against contractors' cost management efforts. Despite the contractors' ability to pass on project costs to clients, they remain the immediate victims of the consequences of poor cost management. And for that reason, the study focussed on contractor's cost management. Although a number of manuscripts are available worldwide discussing construction cost management, there is notably lack of such literature in Zimbabwe. This research will, among other things, contribute to the development of this body of knowledge on construction cost management in Zimbabwe. In addition, the research findings will provide a platform upon which remedies may be instituted for the betterment of the industry.

2.0 Review of Related Work

Cost management is an approach used to realize decisions made for planning, controlling and developing competitive strategies, and it is noteworthy to say that making a balance between this factor and other dimensions of competition such as quality and time is required [17]. It is the control of costs through the formal process of budget development, monitoring, and adjustment to achieve the maximum amount of work at a specified level of quality where unknowns and uncertainty may cause costs to increase beyond acceptable levels [18]. The Project Management Institute [19] considers cost management as primarily concerned with the cost of resources needed to complete project activities and would be achieved through a process involving estimating, budgeting, and cost control. However, cost management is much more than simply maintaining records of expenditure and issuing cost reports - it means understanding how and why costs occur and promptly taking the necessary response in light of all the relevant information [4]. Thus the key question to stay competitive in construction enterprise is cost management and control [4]. In a study in Thailand, [20] concluded that contractors perceive cost control as a prerequisite for profit maximization. However, to deliver a project within budget requires careful planning and taking appropriate action especially during project execution. Thus during project execution, procedures for project control and record keeping become indispensable tools to managers and other participants in the construction process [21]. Two ready-made types of contractor cost management systems: the big picture monthly review of the project and a more detailed analysis of the sections within the project in order to identify those sections of works that are underperforming [4].

Over the years, contractors developed several cost management strategies to assist them to record all financial transactions that occur as well as giving managers an indication of the progress and problems associated with a project [21]. Some of the strategies, tools and techniques used by contractors to manage cost, include but not limited to: cost-value reconciliation, earned value analysis, variance analysis [4, 19, 22], cash flow forecasting [15, 4, 19], the plan and schedule, combination estimate and schedule [19], project status report, profit and loss at valuation dates or overall [22] and monthly cost reports [21]. These tools may be used singularly or in combination. However, apart from earned value, the other cost control techniques suffer from lack of real time reporting. Earned value management has gained more acceptance and use in developed countries. For example, in the USA, it is a requirement that contractors and sub-contractors working on defence projects employ earned value analysis [23]. On the other hand, traditional cost control processes, common with most developing countries, are criticized because they simply identify variances by monitoring actual performance against cost estimates. For example, cost projection is not usually taken into consideration [24]. In addition to the above stated strategies, cost control software such as Microsoft Project [19, 22] and Asta Power Project, Project Costing System (PCS), Construction Industry Software (COINS), WinQS, Microsoft Excel, and other bespoke systems [25] are used for cost control. Whilst there is no doubt that effective formal control systems are necessary for effective project control, every day on-going control is even more important, and informal activities are a vital part of this part [26]. The choice of the system depends on, among other factors, economic considerations. In that regard, for a cost control system to be cost effective, cost of control should never exceed the value of control, must operate in a timely manner and should be as simple as is possible to operate [27].

However, apart from the proven importance that cost carries with regards to project performance and the advances that have been made by the construction industry to manage cost, most projects are delivered above agreed budgets. A cost overrun is measured as actual out-turn costs minus estimated as a percentage of estimated costs [28]. Lower than expected productivity, higher than expected wage costs, higher than expected material costs, among other factors, cause cost overrun on projects [21]. With these factors in mind, [29] suggest that cost control will be achieved by selecting the right man for the right job, the right equipment for the right job and the right quality of materials, in the right quantity from the right source, at the right price and delivered at the right time. A critical requirement of cost management procedure is timely availability of cost information with very little delay between fieldwork and management review of performance [1]. This would give the project manager enough space to take corrective action while opportunity exists for that. However, the existing cost control systems suffer from lack of real time reporting [21]. The cost of this approach is a cost management crisis where contractors are made aware that a project has over-spent when it is too late to take corrective action. Apart from providing reasonable ground upon which an informed decision can be made, a cost management system offers no panacea to cost management challenges. In fact what is needed is the managerial decision that something should be done differently and effecting such decision [28].

3.0 Methodology

The research methodology is discussed under research design and data collection methods.

3.1 Research Design

The research was largely qualitative, employing exploratory design. Exploratory design is particularly suitable in areas where there is little experience to serve as a guide [30]. It is an initial research which analyses the data and explores the possibility of obtaining as many relationships as possible between different variables [31]. This methodology was found befitting for this research due to unavailability of published literature on this subject in Zimbabwe. The methodology, therefore, provided a conduit for diagnosing the situation, discovery of ideas and insights.

Although the study is meant to reflect on the whole country, however, due to the dominance of Harare and Bulawayo, the majority of the targeted companies were taken out of these two cities. Harare is the capital city of Zimbabwe while Bulawayo is the second largest city. Based on 2013 statistics of construction companies registered with the Construction Industry Federation of Zimbabwe (CIFOZ), 80.64% of contractors in Category A, B and C are in Harare, while 7.5% are from Bulawayo. Category A contractors enjoy unrestricted caps on tendered jobs, whilst category B and C contractors are registered for work of a maximum value of US$ 6,000,000 and US$ 3,000,000 respectively. Collectively, Harare and Bulawayo account for 88% of CIFOZ registered contractors in these categories. Contractors that participated in the study were first stratified into three selected categories and convenience sampling was then used to select individual companies in each category. The research had a target of 30 contractors (representing 32.25% of the population). A sample size of 30 has been corroborated by [32, 33] and Griffin and Hauser cited by [34]. The Economist [32] advice of a sample size of a minimum of 30 for statistical analysis and that this provides a useful rule of thumb. The magnitude of projects undertaken by contractors in category A to C is usually medium-to-large and such ventures involve substantial capital injection hence the interest to manage costs on such projects is great. A cost overrun on such projects implies serious financial consequences on both the contractor and the client.

3.2 Data Collection Methods

The study used semi-structured questionnaires and interviews to solicit information from construction industry practitioners within the construction firms, who included among others: project quantity surveyors, project managers, contracts managers and directors. The questionnaire had both closed-ended questions and open ended questions. Questions relating to strategies employed by contractors to manage costs and challenges encountered were left open-ended. The basis for this approach was two-fold: to check contractors' understanding of cost control and to allow them to brain storm and reflect on the strategies they use on a day to day basis. In addition to questionnaires, follow-up interviews were also conducted with selected five experienced respondents to further triangulate data obtained from the questionnaire survey and to explore the experiences of the respondents in relation to the strategies employed, their effectiveness and constraints encountered to achieve cost efficiency. Data gathered was analysed using Statistical Package for Social Scientists (SPSS version 16). Data from open ended questions was pre-coded before entry into SPSS. The study relied on literature for pre-analysis coding of responses from questionnaires. In addition to primary data, the study also reviewed literature from textbooks, journals, conference presentations and selected websites on cost management.

4.1 Results and Discussions

This section presents results, analysis and inferences drawn from the study. These will be discussed under the following headings.

4.1 Demographics of Respondents

Figure 1 below shows the distribution of respondents.

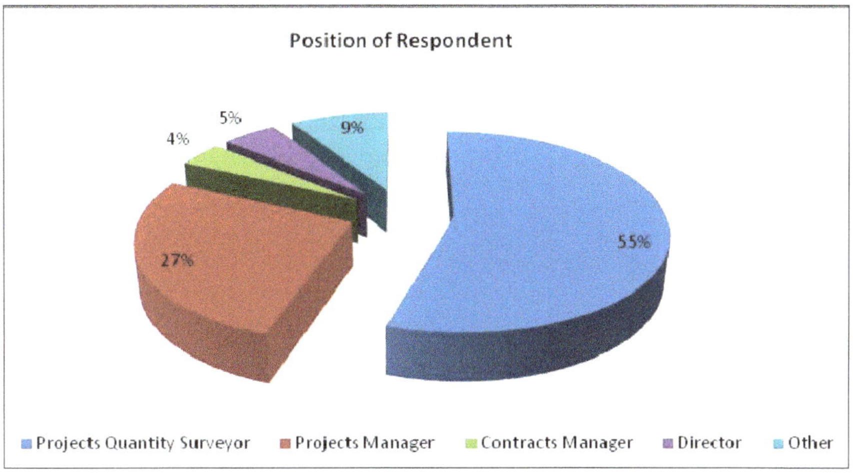

Figure 1: Position of Respondent in a Contracting Company

Out of the targeted 30 contractors, 22 responded giving a response rate of 73%, that is, 23.65% of the population. The respondents, who had a cumulative construction industry work experience of 166 years (average of 7.22 years), were distributed in the following order: project quantity surveyors (55%), project managers (27%), contracts managers, directors and others (14% collectively). The study drew 72.4%, 18.2% and 9.1% of its respondents from category A, B and C contractors respectively. Despite loss of experienced personnel in 2008 following the macro-economic meltdown in the country, the demographics of the respondents who participated in the study shows that they had relevant technical experience to provide valid responses to questions included in the questionnaire and interviews.

4.2 Strategies for cost management

The strategies employed by contractors to manage cost are summarized in Figure 2 as follows: cost reports (31.8%); cost estimating and budgeting (36.4%), variance analysis (54.5%); resources management (59.1%), cost value reconciliation (CVR) (31.8%) and cash flow analysis and work programmes (18.2%), and other factors, for example, project meetings (13.6%). Cost estimating and budgeting, variance analysis and cost value reconciliation are mostly used by contractors in category A and B, while resources management and variance analysis are used by contractors in all the three categories. However, contractors in category C did not show much use of cost reports, cost estimates and budgets, and cost value reconciliation. Although there is a growing trend to automate cost control systems (as shown by an increase in use of such software as Microsoft Project, Microsoft Excel and other in-house systems), however, overreliance on the mechanistic Microsoft Excel and paper-based systems pose challenges especially where there is need for real-time reporting. These findings are not unique to Zimbabwe. Research in Uganda indicates that work programmes, reports, project budget and inspection of works, among others, are the techniques used to manage project cost [35]. In Nigeria, [5] noted that project budget was the most frequently used tool followed by monitoring of labour and material cost respectively. However, what differs between these countries is the weighting attached for each strategy. For instance, in Uganda work programmes are weighted high, while this same factor has a very low frequency in Zimbabwe. Figure 2 below summarises contractor cost management strategies. The identified strategies are discussed in the subsequent sections.

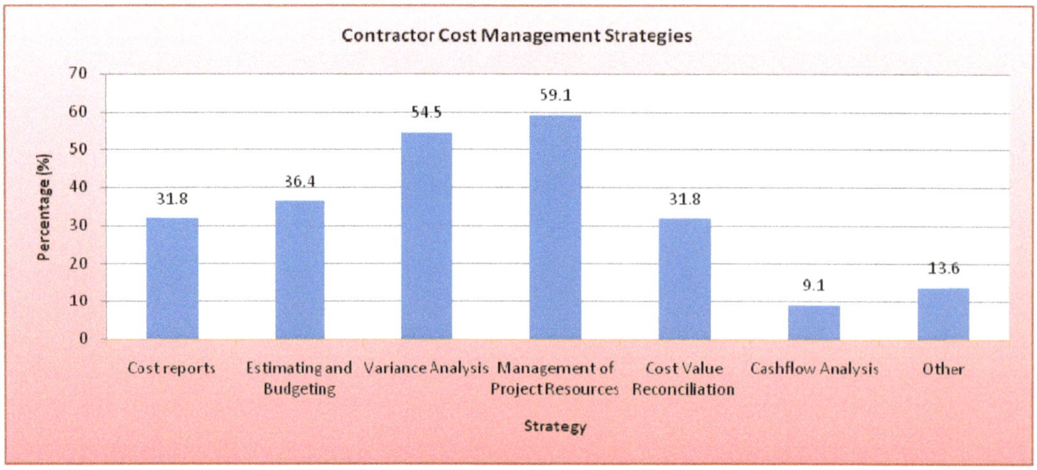

Figure 2: Cost Management Strategies on Building Projects

4.2.1 Resources management related strategy

Cognisant of the impact of resource utilisation on project cost, 59.1% of contractors employ a resource-centric strategy to cost management. This resources management strategy is divided into materials management (including but not limited to: bulk-buying of materials, secured storage of materials, monitoring site materials movements, material reconciliation); labour management (labour engagement strategy, supervision of site labour, monitoring labour outputs, performance based labour rewarding, employment of site security personnel); and plant and equipment management (utilisation schedules, timing of plant availability). Most contractors buy key materials like bricks, cement, just to mention some, directly from manufacturers where they also receive trade discounts. On site, material registers, plant registers and labour time cards are used to manage movement of resources and minimize idleness and misuse of resources. In addition, the project manager through the site foremen oversees site resource utilisation and/or loss through theft, vandalism and wastage. These efforts are meant to control variances in resource utilisation (quantity and efficiency) which may impact negatively on project cost. Earlier studies by [21, 19, 8] reveal some close relationship between resources consumption and project cost. Project cost is identified as a result of resource consumption and therefore the success of a project depends upon performance management of resources when controlling costs. These results are also confirmed by [5] in Nigeria and [35] in Uganda although at varied ratings. According to [36] the need to control costs is important in order to eliminate the unnecessary wastage of resources.

4.2.2 Variance Management (variance analysis and cost value reconciliation)

Under this strategy contractors employ collectively or independently variance analysis and cost value reconciliation. Variance analysis and cost value reconciliation (CVR) are employed by 54.5% and 31.8% of the contractors respectively. Whilst variance analysis is used by all the three categories of contractors, CVR is more prevalent with category A and B contractors. Variance analysis compares budgeted against actual spending on a project. On the other hand CVR is not limited to comparison of planned against actual expenditure but also focus on value generation. CVR helps contractors to assess the magnitude by which their profit is changing as a project progresses. Variance analysis forms an essential part of CVR reporting procedure. However, it does not portray the true cost of performance of a project. It is because of these limitations that contractors are migrating to CVR especially when projects get bigger and more complex. CVR allows forecasts to be made of the project's cost and profitability to completion, at any given date of reconciliation. Collectively variance management allows the contractor to note the existence of a variance in project costs (labour, material or overhead costs) and thus offering an opportunity for the contractor to assess the possible drivers of cost variance(s) so as to institute appropriate remedial action. Irrespective of the arguments on the importance attached to timeliness and accuracy of variance reporting and its interpretation in cost management [37], this study noted that such assessments are done on a monthly basis – thus corresponding to the 'big picture monthly review' [4]. Although the use of such tools as S-curves was noted, however, the extent of use remains limited.

4.2.3 Cost estimating and budgeting

Cost estimates and budgets are also used by 36.4% of contractors. Contractors prepare cost estimates and produce budgets based on appropriate contract documents which include but not limited to prepared bills of quantities (BoQs) and drawings. Such budgets become the benchmark upon which project cost performance will be monitored and controlled. The use of cost estimates and budgets concur with findings by such researchers like [1, 19, 38, 21, and 40] whose presentations suggest that for one to successfully control project costs one would need to know what goes into making up the total project budget. Thus the estimate and budget form the entry point to cost management. In Thailand, a study by [20] indicated that from a contractors' perspective an accurate estimate is not only vital for winning a contract with the chance of winning profit but also serving as an effective plan for cost control.

4.2.4 Cost Reports

Cost reports provide progress information that is used to assess project performance against set targets. These are used by 31.8% of the contractors. The study observed that 59.1% of the contractors produce cost reports on a monthly basis, while 18.2% report weekly, 4.5% daily and 18.1% any other times. Whilst there is no rule cast-in-stone on reporting intervals, what is clear is that the reports should be prepared often enough so that excessive project cost can be detected while there is enough time to take corrective action [1]. The research noted an inherent relationship between project cost reports and cost value reconciliations. This relationship is evidenced by the two being used by an equal number of contractors under study and at similar period(s), that is, monthly. This consequently suggests that the two strategies have been jointly used and are therefore complimentary to each other.

4.2.5 Cash-flow analysis and work programmes

Cash flow forecasting and work programmes are used by 9.1% of the contractors respectively. Cash flow forecasting involves breaking of the project budget against the construction programme of works to produce a project cash flow. Though not equivalent to earned value analysis, cash flow forecasting tries to relate the project's cash flow against the time as reflected in the schedule of activities. In that regard, contractors use programmes concurrently with project cash flow to manage and control project costs. This approach allows the contractor to relate project cost performance against project programme or schedule. The frequency for cash flow analysis obtaining in Zimbabwe is at variance with arguments put forward by Hwee as cited by [5]. The later argue that that cash flow forecasting is the most important factor that can affect project profitability once the construction project is in progress. Apart from arguments presented in earlier studies that project cost is a function of project schedule [Carr, 1993 in [39], this study observed very low frequency for work programmes.

4.2.6 Other factors

In addition to the above identified factors, project meetings are also used by 13.6% of the contractors to mange project cost. Project meetings provide a platform to discuss project progress and cost performance and also provide an opportunity to identify areas that are not performing well and proffer alternative solutions.

4.3 Statistical Inference of Cost Management Strategies

A chi-square test was performed to analyze the significance of the above strategies. Table 1below shows the outcome of the test.

Table 1: Chi-square Test on Cost Management Strategies

Test Statistics

	Cost Reports	Cost Estimating & Budgeting	Variance Analysis	Resources Management	Work Programmes	Cost Value Reconciliation	Cashflow Forecasting	Others (meetings)
Chi-Square	2.909[a]	1.636[a]	.182[a]	.727[a]	14.727[a]	2.909[a]	14.727[a]	11.636[a]
df	1	1	1	1	1	1	1	1
Asymp. Sig.	.088	.201	.670	.394	.000	.088	.000	.001

a. 0 cells (.0%) have expected frequencies less than 5. The minimum expected cell frequency is 11.0.

Thus Chi-Square 95% at 1 degree of freedom (df) as shown in Table 1 is 3.841 and this eliminates all the strategies with test statistic greater than 3.841 which are work programmes and cash flow forecasting and others such as meetings. The outcome of this statistical inference shows that cost reports, cost estimating and budgeting, variance analysis and resources management are the significant strategies in this study. These results, to some extent, correlate well with findings in a study on cost control practice in Nigeria where establishment of working budget, monitoring labour cost, monitoring material cost were found to be highly significant methods of managing cost. However, these findings contrast with observations by Hwee in [5] and Carr in [39] with respect to work programmes and cash flow forecasting in cost control.

4.4 Problems Encountered by Contractors in Managing Project Cost

The factors that affect contractor's efforts to manage costs are categorised as: labour related challenges (64%); organisational problems (50%); material related challenges (36%); variations (23%); cost control systems related challenges (18%); plant related challenges (14%); funding challenges (14%); inaccurate estimate (9%) and others (5%). These factors are illustrated in Figure 3.

The challenges to cost management presented in Figure 3 above can be grouped into: exogenous and endogenous factors. Exogenous factors include: resources related challenges (labour, materials and plant), funding challenges, and variations; while endogenous factors include: organisational, deficient cost control systems and incorrect estimates.

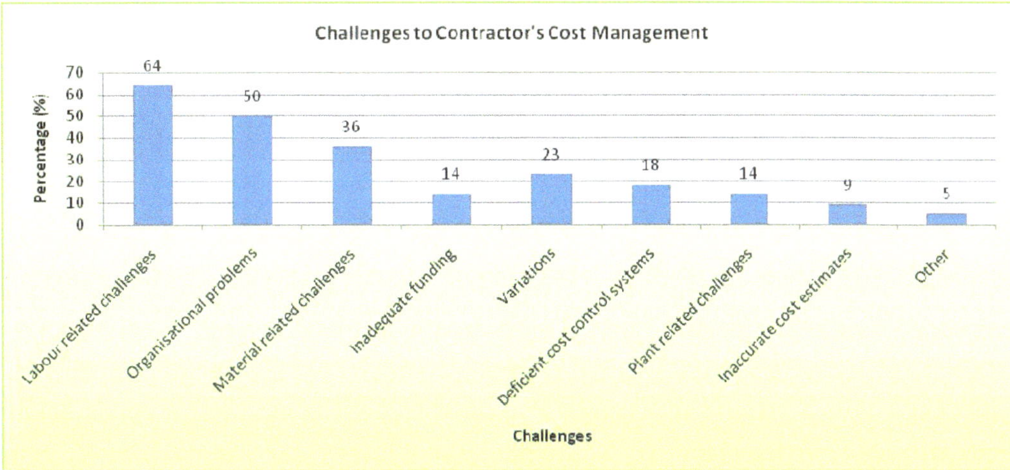

Figure 3: Challenges to Cost Management

4.4.1 Exogenous factors (Externally Induced)

Lack of experienced staff, material price fluctuations, unavailability of certain materials on the local market and inefficient or obsolete plant were identified as the resources management related challenges that impact on contractors' cost management efforts. However, labour-related challenges had the highest frequency. Lack of experienced personnel reflects on the period of mass exodus of skilled labour from Zimbabwe to other countries from the year 2000 to 2008. Most skilled and experienced artisans and professionals left the country for greener pastures in neighbouring countries like South Africa, Botswana and Namibia, and other distant countries like United Kingdom, United Arab Emirates, and New Zealand. A majority of them are still 'locked' up in these countries. Brain drain of such magnitude adversely affected construction output and performance. The spill over effects of inexperienced cost personnel and shortage of skilled labour manifest in deficient cost control systems, inefficient cost control procedures and poor workmanship. Another labour related challenge related to reduced labour productivity. The impact of labour related challenges on cost management practice was also noted in Nigeria, where engagement of inexperienced staff had a high rating factor [41].

On the other hand, unavailability of certain materials on the local market and fluctuations in material prices also presented challenges to cost management. Apart from the gains made through the introduction of multi-currency regime in the country, the key materials manufacturing industries, for example, cement and steel manufacturing just to mention some are yet to be fully capacitated to meet demand for these materials. In 2012 the President of CIFOZ indicated that challenges related to critical shortage of such inputs as cement and bricks, together with under-capitalisation of major construction companies and liquidity constraints all served to limit the sector's capacity to emerge as one of the top contributors of the country's Gross Domestic Product (GDP) [42]. Under such circumstances contractors are held in some waiting period for the supply of certain materials or alternatively resort to importation of some materials with concomitant disadvantages of delaying the project and pushing up project cost through payment of idle labour and increased price of materials. This challenge has not been limited to Zimbabwe. In Uganda, [35] suggest that lack of materials and equipment, among other factors, presented difficulties to contractors' efforts to control project costs. In the Indian construction sector is besieged with project procurement and execution problems ranging from non-availability of materials and equipment as per schedule, delay in payments of the completed work due to paucity of funds [36]. In Nigeria, price fluctuations, inaccurate estimates, design changes and shortage of materials impacts are attributed to project cost overrun [5]. In addition to resources related challenges, variations also affect contractor's efforts to manage project cost. Variations disturb flow of the programme and impact on cost through time. Whilst the 'lower' quartile rating accorded to variations in this study confirms similar findings in Nigeria by [9], however, these findings are at variance with studies in Botswana where variations are considered a strong factor in influencing cost overrun on public sector projects [7]. On the other hand, late honouring of claims by clients, though not a significant factor as shown by a chi-square test in Table 2 is another cost management challenge identified by contractors. In South Africa delayed interim payments and settlement of final accounts is also reported as a common problem [36].

4.4.2 Endogenous factors (internally induced)

Within contracting organisations, organisational deficiencies (poor methods of communication, poor planning, contract administration and site organisation), level of development of cost control systems, inaccurate cost estimates and policies on procurement of resources impact on cost management efforts. For example, inefficient communications systems, management interference in technical aspects of the project, failure to recruit personnel and plant of right 'calibre' negatively affects efforts directed towards cost management. Yet against this background scholarly writings indicate that contractor's management capability has significant impact on cost and time performance of building projects. On the other hand, inaccurate estimates of project cost also pose a challenge to cost management. Due to inexperience and other related factors, contractors sometimes under-price tenders thereby increasing their chances of winning tenders or inexperienced estimators inaccurately-estimate the value of works to be executed. However, inaccurate estimates disrupt construction and budget programmes. A study by [28] observed that cost estimates are often inaccurate. Similar findings were also noted in South Africa where the Construction Industry Development board (cidb) suggest that undue focus on lowest price to the detriment of best value impacts on the performance of contractors who invest their innovation in the winning of the contract, fully prepared for the claims that will be necessary to remain afloat [36]. The study shows that 59% of the contractors produce cost reports on a monthly basis – a time period that is considered 'standard' in the Zimbabwean construction industry – however, such time period may be considered long enough to advice the contractor that the project underperformed a month ago.

4.5 Significance Test of the Challenges to Cost Management

A Chi-square test was performed to show the significance of the challenges identified for cost management. This is shown in Table 2 below.

Table 2: Chi-square Test of the Challenges to Cost Management

Test Statistics

	Labour related challenges	Organisationa l problems	Materials related challenges	Project funding challenges	Variations	Deficient cost control systems	Incorrect cost estimates	Contractor Incompetence	Other
Chi-Square	.727[a]	.182[a]	1.636[a]	11.636[a]	6.545[a]	4.545[a]	11.636[a]	12.250	4.000
df	1	1	1	1	1	1	1	1	1
Asymp. Sig.	.394	.670	.201	.001	.011	.033	.001	.000	.046

a. 0 cells (.0%) have expected frequencies less than 5. The minimum expected cell frequency is 11.0.

Thus Chi-Square 95% at 1 degree of freedom is (df) 3.841 and this eliminates the problems with a chi-square greater than 3.841. Therefore, labour related challenges, organizational challenges and material related challenges are the significant factors affecting contractor's efforts to manage project cost. The other factors, which include: project funding challenges, variations, incorrect cost estimates and contractor incompetence, were found not significant.

5.0 Conclusions and Recommendations

The study observed that in the majority of cases, contractors' efforts to manage projects costs are centred on management of project resources. This resource-centric strategy premised on materials, plant and labour resources management is founded on the principle that resources consumption stimulates building project costs. Statistical analysis performed through chi-square test at 95% shows that: cost reports, cost estimating and budgeting, variance analysis, cost value reconciliation and resources management are the significant factors employed by contractors to manage costs on building projects in Zimbabwe. However, despite the existence of these strategies, most contractors admit that their projects would, in most cases, end with a cost overrun. This may be inferentially explained by a host of challenges affecting cost management practice. Whereas several challenges were observed with respect to cost management, it is labour related challenges, materials related challenges and organisational problems that impact significantly on cost management practice. The study also observed that the cost management process in Zimbabwe, like in other developing countries in Africa and elsewhere is still dominated by the traditional paper based system or 'prematurely' developed automated systems of managing project cost information. In a nutshell, the problem in Zimbabwe is not about having the strategies to manage costs but relates to 'flawed' implementation. The findings of this research suggest that the strategies instituted to manage project cost and, the challenges to such practice are not significantly different from the ones in other developing countries including but not limited South Africa, Botswana, Nigeria and Uganda. To improve on the effectiveness of contractors' cost management strategies in Zimbabwe, the research recommends that contractors pay particular attention to problem areas that significantly affect cost management (labour, material and organisational related challenges), engagement of experienced cost management personnel or upgrading of staff's professional ingenuity through conventional training, refresher courses or participation in seminars; and regular review of cost management systems and strategies. The industry can also take advantage of advancement in ICTS to automate their cost information management systems. This study has provided baseline information on the strategies used to manage project cost and related challenges; however, further research is needed to assess the effectiveness of these strategies to manage project cost.

References

[1] R.H. Clough, G.A. Sears and S.K. Sears, "Construction Project Management," John Wily and Sons, USA, 2000

[2] C.C. Nkwachukwu, E. Ibeawachi and M.N. Okoli, "Project Management Factor Indexes: A Constraint to Project Implementation Success in the Construction Sector of a Developing Economy," European Journal of Scientific Research. Vol. 43 No. 3, pp 392-405, 2010

[3] R. Pilcher, "Project Cost Control in Construction," 2nd ed., Oxford Blackwell Science, 1992.

[4] K. Potts, "Construction Cost Management. Learning from Case Studies," Taylor and Francis, London, 2008

[5] A.O. Sanni and O.D. Durodola, "Assessment of contractors' cost control practices in Metropolitan Nigeria," in S. Laryea, S.A. Agyepong, R. Leiringer and W. Hughes, eds. Procs 4th West African Built Environment Research (WABER) Conference, 24-26 July 2012, Abuja, Nigeria, pp. 125-132

[6] I.A. Rahman, A.H. Memon and A.T.A. Karim, "Significant factors causing cost overruns in large construction projects in Malaysia," Journal of Applied Sciences 13 pp. 286-293, 2013

[7] A.H. Momon, I.S. Rahman and A.A.A. Azis, "Preliminary study on causative factors leading to construction cost overrun," International Journal of Sustainable Construction Engineering and Technology Vol 2, Issue 1, June 2011

[8] B. Chigara and E. Mangore, "An analysis of the implications of resources management on building projects, " International Journal of Marketing and Technology, vol. 2, Issue 9, September 2012

[9] A. Omoregie, D. Radford. Infrastructure delays and cost escalation: Causes and effects in Nigeria," Proceedings of the sixth international postgraduate research, Delft University of Technology and NTO, 3-7 April, Netherlands, 2006

[10] D.K. Chimwaso, "An evaluation of cost performance of public projects: Case of Botswana," Department of Architecture and Building Services, Gaborone, 2001.

[11] P.T. Eikeland, "Cost Control in Building Projects," Kolbotn, Norway, 1997.
[12] H. Li, "Study on Construction Cost of Construction Projects," Asian Social Science, Vol. 5, No. 8, August 2009
[13] M.F. Khamidi, W.A. Khan and A. Idrus, "The cost management of construction projects through earned value analysis," International conference on economics and finance research IPDER, vol. 4, 2011
[14] C. Charoenngam and E. Sriprasert. "Assessment of cost control systems: a case study of Thai construction organisations," Engineering, Construction and Architectural Management," vol. 8, pp. 368-380, 2001
[15] A.O. Sanni and M. Hashim, "Assessing the challenges of cost control practices in Nigeria construction industry," Interdisciplinary Journal of Contemporary Research Business, Vol 4, No. 9, January 2013.
[16] G. Forster, "Construction Site Studies: Production, Administration and Personnel," United Kingdom, Longman, 1996.
[17] G. Reiss, "Project Management Demystified: Today's Tools and Techniques," London, E & F N Spon, 1992.
[18] A. Reizaian, "Time-cost-quality-risk of construction and development projects or investment," Middle-East Journal of Scientific Research 10 (2), 2011
[19] R.D. Staley and G. Jones, "Fort Mackinac Restoration: Cost management project delivery," Association of Preservation Technology International (APT) Bulletin, vol. 38, No. 2/3, 2007
[20] C. Charoenngam and E. Sriprasert. "Assessment of cost control systems: a case study of Thai construction organisations," Engineering, Construction and Architectural Management," vol. 8, pp. 368-380, 2001
[21] C. Hendrickson and T. Au, "Project Management for Construction: Fundamental Concepts for Owners, Engineers, Architects and Builder," Prentice Hall Inc, New Jersey, 2000
[22] Y.A. Olawale and M. Sun "Cost and time control of construction projects: inhibiting factors and mitigating measures in practice," Construction Management Economics, 28:509-526, 2010
[23] G. Reiss, "Project Management Demystified: Today's Tools and Techniques," London, E & F N Spon, 1992.
[24] A.P. Kern and C.T. Formoso, "Guidelines for improving cost management in fast, complex and uncertain construction projects," Proceedings of the 12th Annual Conference for Lean Construction (IGLC-12) Elsinore, August 2004
[25] P.P. Dharwadker, "Construction Management," New Delhi: Oxford and IBH Publishing, 1989
[26] A. Stretton, A. "Notes on Effective Project Cost Control," PM World Today, Vol. XI, Issue X, October 2009.
[27] J.R. Meredith and S.M. Mantel, "Project Management. A Managerial Approach," John Wiley and Sons Inc., 2003
[28] C.C. Cantarelli, B. Flyvbjerg, E.T.E Molin and B. van Wee, "Cost Overrun in Large Scale Transport Infrastructure Projects: Explanations and Their Theoretical Embeddedness," EJTIR Issue 10(1), p. 5-18, March 2010.
[29] F.C. Harris and R. McCaffer, "Modern Construction Management," Blackwell Scientific Publishing, 2002.
[30] C.R. Kothari, "Research Methodology: Methods and Techniques, 2nd ed.," New Age Techno Press. New Dehli, 2004
[31] R. Panneerslvam, "Research Methodology," New Dehli, Prentice Hall, 2004.
[32] The Economist, "The Economists numbers guide," Hamish Hamilton, London, 1993.
[33] Sample Size. www.saifulbahri.com/Medical_education/Medical_Education_Notes_Sem_2/SAMPLE_SIZE.pdf
[34] P. DePaulo, "Sample size for qualitative research" www.quirks.com/articles/a2000/20001202.aspx
[35] G. Otim, F. Nakacwa and M. Kyakula, "Cost control techniques used on building projects in Uganda," Second International Conference in Engineering and Technology, January 2011.
[36] F.A. Emuze, "Performance Improvement in South African Construction," PhD Thesis. Faculty of Engineering, The Built Environment and Information Technology, Nelson Mandela Metropolitan University, South Africa, 2011
[37] B. Cooke and P. Williams, "Construction planning, programming and control, 3rd ed.," Wiley-Blackwell, 2003.
[38] M. Selver, "Capital project cost control. Canada," Project Management Association of Canada, 2009.
[39] W.J. Bender, "A risk-based cost control methodology for construction complex structures with the mobile offshore base as a case study," PhD dissertation submitted to the Faculty of the Graduate School of the University of Maryland, College Park.

[40] Kwakye,"Construction Project Administration in Practice," Pearson Longman, England, 1997.
[41] A.O. Sanni and M. Hashim, "Assessing the challenges of cost control practices in Nigeria construction industry," Interdisciplinary Journal of Contemporary Research Business, Vol 4, No. 9, January 2013.
[42] P. Chiyangwa, " Construction Sector slowly emerging from the woods," Construction Journal, Vol.1, Harare, 2012

14

A WEB-BASED COLLABORATIVE SYSTEM FRAMEWORK FOR GREEN BUILDING CERTIFICATION

Jack C.P. Cheng[1], Vignesh Venkataraman[2]

Department of Civil and Environmental Engineering, The Hong Kong University of Science and Technology, Hong Kong

*Corresponding E-mail : vvaa@ust.hk

Abstract

The built environment is moving towards sustainable development and the number of green buildings increased worldwide in recent years. Green buildings are environmentally, socially and economically desirable; however the certification of green buildings is often expensive and labor-intensive. The document preparation and review process for green building certification is iterative in nature and requires the collaboration of many project participants, certification organizations and third party engineering consultants. This paper aims at developing a system framework that can assist the green building certification process. Different certification standards have been studied and their scope and credit calculation were compared to understand the requirements of the system framework. Based on the study, the required features were designed as follows, (1) role-based access control, (2) document manager and related document discovery, (3) workflow manager, (4) credit manager, and (5) knowledge manager. The features of the system provide a collaborative environment and reduce repetitive works commonly occurring in manual processes. The features of the system framework are discussed and illustrated considering the Hong Kong BEAM Plus standard. This paper also describes how the system framework enhances collaboration in the certification process.

Keywords: *Collaborative system, Document management, Green building certification, Hong Kong Building Environmental Assessment Method (HK-BEAM), Information Retrieval*

1.0 Introduction

What is a green building? The term 'green building' has various definitions. The United States Environmental Protection Agency (U.S. EPA) defines green buildings as "the practice of increasing the efficiency with which buildings and their sites use energy, water and materials, and reducing building impacts on human health and the environment, through better siting, design, construction, operation, maintenance, and removal- the complete building life cycle"[1].American Society for Testing and Materials (ASTM) defines a green building as, "a building that provides the specified building performance requirements while minimizing disturbance to and improving the functioning of local, regional, and global ecosystems both during and after its construction and specified service life," and a building that "optimizes efficiencies in resource management and operational performance, and minimizes risks to human health and the environment"[2].The International Council for Research and Innovation in Building and Construction (CIB) defines 'sustainable construction' as "a holistic process aiming to restore and maintain harmony between the natural and built environments, and create settlements that affirm human dignity and encourage economic equity"[3]. Regardless of these various perceptions, green buildings have been growing rapidly worldwide. As of 2011, for example, there were over 10,000 green buildings certified by the LEED (Leadership in Energy and Environmental Design) standard in the United States alone.

Construction has been accused of causing environmental problems ranging from excessive consumption of global resources both in terms of construction and building operation to the pollution of the surrounding environment [4]. Although green buildings are a favorable solution, the potential increase in cost to construct and operate green buildings has often put developers and tenants on halt while going green. In fact, green buildings may cost more upfront, but save from lower operating costs over building lifetime. According to Katsand Capital (2009), a minimal upfront investment of about two percent of construction costs typically yields life cycle savings of over ten times the initial investment [5].Rating systems and labeling programs are crucial to promote green buildings. However, promotion of green buildings is hindered due to insufficient fiscal incentive, lack of information about eco-label green materials and green building practices, and limited collaboration among the building team and stakeholders.

This project aims to develop a web-based collaborative system framework that can facilitate the green building assessment process, which is labor intensive and multi-participant in nature. This paper is organized as follows: Section 2 reviews different green building assessment standards. Section 3 discusses the developed system framework, its functionalities and potential benefits. Section 4 presents an example scenario considering the Hong Kong green building standard, namely BEAM Plus. Section 5 concludes by outlining the need, benefits and future work.

2.0 Research Background

2.1 Web-based Collaboration Systems

The rapid development of Internet has allowed construction organizations to utilize web-based technologies for improved inter-organizational communication and workflow. A web-based system uses a centralized information integration approach through a shared web server or a central database behind the web server [6]. The Architecture, Engineering and Construction (AEC) industry has utilized web-based systems for Project Management, Supply Chain Management, Knowledge Management and Engineering Design purposes. For example, Nitithamyong and Skibniewski (2004) summarized the different commercially available web-based Project Management systems and their factors of success and failure in the construction industry [7]. Chen and Tien (2007) developed a prototype system, namely ROCCAD, for real-time online collaborative CAD drawings [8]. Zhang and El- Diraby (2011) present a web-based communication system for coordinating information and knowledge flow in the AEC industry [9]. Cheng et al (2010) prototyped a service oriented web-based system for supply chain collaboration and monitoring [10]. There are plenty of examples and literature that support the use of web-based technologies for communication, collaboration and knowledge transfer in the AEC industry. Since the green building certification process requires frequent communication among participants, information storage, retrieval, transfer and reuse, web-based collaborative concepts are used for developing the system framework.

2.2 Information Systems for Green Building Certification

Green building certification processes are information intensive and iterative, involving different collaborators and heavy documentation. LEED Online is a web-based tool that employs a series of active PDF forms to automate filing documentation and communication between project teams and reviewers. Besides LEED Online, there are other few commercial software tools such as IES Tap and Lorax PRO that provide assistance to the green building certification process, but none has been very prominent and successful. A probable reason is that these tools cannot fully fulfill the needs for the green building certification process. Therefore, this study aims to summarize the system requirements and framework design for successful implementation of web-based technologies for green building certification process and presents a web-based collaborative system framework that we have developed according to these requirements.

2.3 System Requirements for Green Building Certification

The certification process is labor intensive and requires efficient collaboration. Li et al. identified that coordination between project participants was found to be a critical factor in determining green building project success [11]. Credit calculation is an integral part of the certification process with the total credit score to earn the desired grade level. Since different certification standards have different scoring procedures, a credit manager to manage all the scoring is a mandatory requirement for a green building certification system framework. Document creation, review and submission are the key activities in the certification process and should be supported in a green building certification system. In addition, different project participants need to work together with collective responsibilities for successful execution. Managing individual roles and responsibilities could be very challenging especially with a big project team and when executing multiple projects at the same time. Therefore, security and confidentiality are also required when performing and managing green building certification. After studying various green building certification standards and their processes, the requirements of a web-based green building certification collaborative system have been summarized. Figure 1 illustrates these system requirements and how they can be fulfilled in various system functionalities. To satisfy these requirements, our proposed system framework includes the functionality of role-based access control for supporting security and confidentiality, document management for managing heavy documentation, workflow management for managing individual responsibilities and tracking responsible collaborators to execute project activities, credit management for planning and strategizing final grade, and document tagging for easy identification for retrieval of documents. The system also requires knowledge management functionality to document best practices and project knowledge for efficient execution of future projects. Based on the nature of the certification process, the required functionalities of the system will be discussed in detail in Section 4.

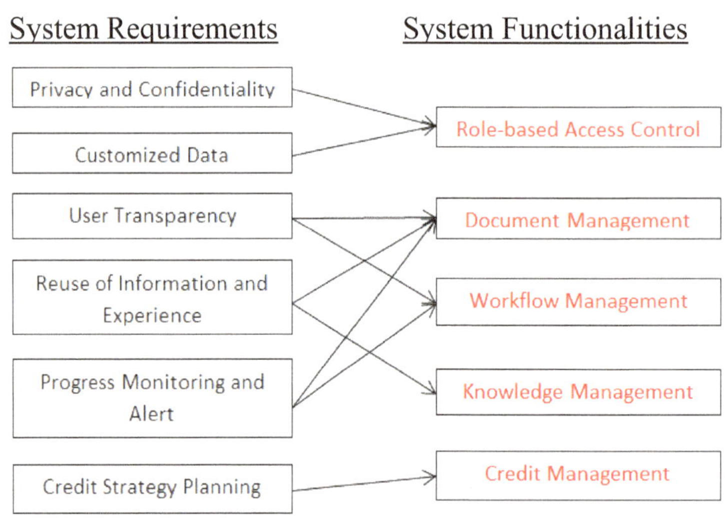

Figure 1: System Requirements and Suggested Functionalities for a Web-based Collaborative Green Building Certification System

3.0 Green Building Certification Standards

3.1 Worldwide Green Building Certification Standards

Green building certification standards provide the guidelines to measure, improve, certify, benchmark, and label whole-life environmental sustainability and the performance of buildings. It is one of the driving factors that promote sustainability in developed and developing nations.

Different countries developed their own certification standards for getting their buildings certified. Certification standards that are widely used and have a great global impact have been studied and compared. The selected standards in this study were (1) LEED from North America- the most widely used green building assessment standard in the world serving the North America and Asia, (2) BREEAM (BRE Environmental Assessment Method) from the United Kingdom- the oldest assessment standard in the world serving the United Kingdom and other parts of Europe, (3) Green Star from Australia- a prominent assessment standard in Oceania and the African regions, and (4) BEAM Plus from Hong Kong- a recently developed assessment standard in the fastest developing infrastructure regions of Hong Kong and mainland China. Apart from these, green building certification standards developed in countries like Japan, Singapore, Taiwan and Germany are also evolving fast and getting popular rapidly. Evolving from new constructions scheme, LEED, BREEAM and Green Star currently have different rating systems for new constructions, existing buildings, schools, industrial buildings, healthcare, courts, prisons, commercial interiors, cores and shells, and so on [12]. LEED, BREEAM, GreenStar, BEAM Plus were studied to understand the credit distribution, coverage and calculations with respective to their latest versions to better design a system framework compatible with different green building certification standards.

Table 1 shows the background of the different certification standards selected in this study. Their place of origin, year of establishment, controlling organization, and total number of projects certified are tabulated in Table 1 for reference. As shown in Table 1, although LEED was introduced after the BREEAM, LEED has the most certified projects thus indicating its popularity.

Table 1: Background of Different Green Building Certification Standards

	BEAM PLUS	LEED	GREEN STAR	BREEAM
Established Year	2009	2000	2003	1990
Country	Hong Kong	USA	Australia	UK
Organization	HKGBC	USGBC	GBCA	BRE
Certified Projects	>500	>10,000	>600	>5000

3.2 Comparison of Scope among Standards

Credits in each green building certification standard are divided into many categories for a comprehensive assessment of different types of projects. Categories from different standards include Site Aspects, Material Aspects, Energy Use, Water Use, Indoor Environmental Quality, Transport, Waste, Pollution and Innovation. Each category constitutes different numbers of credits, and the total number varies with different standards. The difference in the standards is the result of considering the local parameters such as building types, climate, and building codes. Table 2 compares the credit distributions in BEAM Plus (New Buildings 2009 and Existing Buildings 2009), LEED (New Construction 2009 and Existing Building 2009), Green Star (Office 2008 and Residential 2009), and BREEAM (Office 2008 and Residential 2008).As shown in Table 2, BEAM Plus and LEED have very similar categories whereas Green Star and BREEAM have nearly the same categories.

3.3 Comparison of Credit Calculation Method among Standards

The final grade award is determined after calculation of earned credits. Each standard has different credit calculation procedures and criteria for the final grade award. The credit calculation method in LEED is the most simple and straightforward, with the final grade award determined only based on the total points achieved. Green Star assigns weighting to each of its categories.

The ratio of the total weighted points earned to the total available points determines the final grade awarded for Green Star. In addition to following Green Star's procedure, BEAM Plus and BREEAM have additional criteria in specific categories for determining the final grade award. The additional requirements of BEAM Plus will be discussed in the later sections. Table 3 shows the criteria for earning the different grades available in BEAM Plus, LEED, Green Star and BREEAM, respectively. For example, a total of 75 points will earn the highest grade in Green Star whereas it takes 80 points in LEED to earn the highest grades. Table 4 illustrates the different weightings assigned to each individual category.

Table 2: Comparison of Scope among Different Green Building Certification Standards

Category	BEAM PLUS		LEED		GREEN STAR		BREEAM	
	New Building	Existing Building	New Construction	Existing Building	Office	Residential	Office	Residenti al
Site/Land Use & Ecology	25	19	27	26	8	11	10	10
Materials& Resources	23	13	15	10	25	31	12	12
Energy	44	40	38	35	29	26	30	30
Water	10	9	11	14	12	12	9	9
IEQ/Health & Well being	35	33	17	15	27	20	10	10
Management	-	-	-	-	12	18	22	22
Transport	-	-	-	-	11	14	9	9
Pollution/ Emission	-	-	-	-	19	19	13	13
Waste	-	-	-	-	-	-	7	7
Innovation	6	6	6	6	5	5	10	10
Regional Priority	-	-	4	4	-	-	-	-

Table 3: Grade Award Requirements among Different Standards[12]

	BEAM Plus (max. 127)	**LEED (max. 110)**	**Green Star (max. 190)**	**BREEAM (max. 122)**
Outstanding/ Platinum/ 6 Star	75% (Platinum)	80 Points (Platinum)	75 Points (6 Star)	85% (Outstanding)
Exceptional/ Gold/ 5 Star	65% (Gold)	60-79 Points (Gold)	60-74 Points (5 Star)	70% (Exceptional)
Very Good/ Silver/ 4 Star	55% (Silver)	50-59 Points (Silver)	45-59 Points (4 Star)	55% (Very Good)
Good/ Certified/ Bronze	40% (Bronze)	40-49 Points (Certified)	-	45% (Good)
Pass	-	-	-	30% (Pass)

Source: Vignesh Venkataraman 2012

From table 4, it can be noticed that LEED does not weigh any of its categories. In BEAM Plus, LEED, Green Star and BREEAM, the credits earned in the innovation category are included as additional credits and used for calculating the final score to encourage clients to adopt innovative green building strategies and energy conservation technologies. BEAM Plus even sets minimum criteria in the innovation category for their Silver, Gold and Platinum grades

Table 4: Comparison of Category Weightage among Different Standards

Category	BEAM PLUS	LEED	GREEN STAR	BREEAM
Site Aspects	25%	-	-	-
Land Use and Ecology	-	-	7%	10%
Materials& Resources	25%	-	10%	12.5%
Energy	35%	-	25%	19%
Water	12%	-	15%	6%
IEQ/Health & Well being	20%	-	20%	15%
Management	-	-	8%	12%
Transport	-	-	10%	8%
Pollution/ Emission	-	-	5%	10%
Waste	12%	-	15%	6%
Innovation (Additional)	-	-	-	10%

3.4 Summary

As green building certification standards are fast evolving, multiple certifications for a building are becoming a common practice. To facilitate multiple certifications in our system we studied popular green building standards to understand their scope and requirements. The LEED, BREEAM, GreenStar and BEAM Plus standards were considered for comparison. The credit categorizations, credit calculation methods and final grade award criteria were studied and compared. Many green building certification standards consider weightage to each of their credit categories for determining the final grade award. Although the differences in these standards are justifiable, it creates complication for stakeholders, including property investors, who purchase buildings in different countries and apply for certification [13]. Due to the complex nature of credits and calculation methodology among different standards, a web system assisting the certification process should be developed for improving the overall efficiency and catering multiple needs in the assessment process. Such a system can also facilitate multi-participant information sharing and collaboration which are crucial for green building certification processes.

4.0 The Proposed Web-based Collaborative System Framework for Green Building Certification

The green building certification process involves stakeholders from different organizations collaborating at the same time to execute project activities. Credits can be achieved by project stakeholders submitting relevant documents to the certification body, which verifies the documents submitted to award credits and determine final grade based on the calculation methodology specified in the guidelines. This makes document management and credit calculation the most important features of the certification process. Individuals from various organizations work together in the process in the long term and their involvement demands privacy and high security. Based on these requirements, the system framework is designed with five major components, (1) role-based access control for a secure and safe usage (2) document management for storing, organizing and retrieving documents, (3) workflow management to track individual work progress, (4) credit management to help calculate the final grade and grade planning, and (5) knowledge management to document best practices. As the Internet has emerged as the most cost effective means for inter-organizational integration, our proposed system framework is web-based and accessible on the Internet. The proposed system framework is shown in Figure 2 and each of these components will be discussed in the following sections.

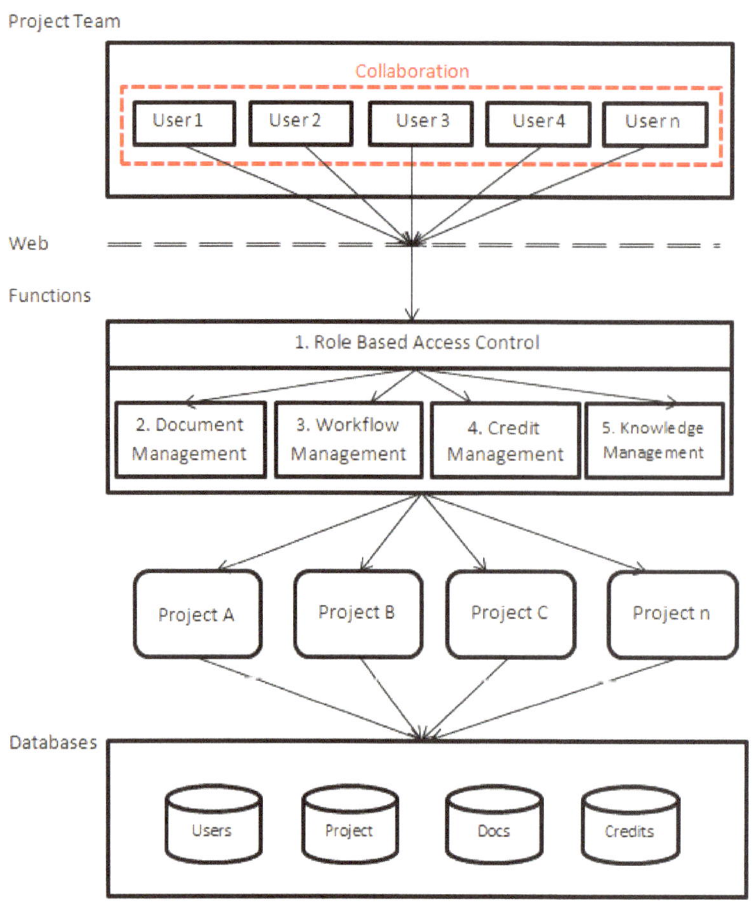

Figure 2: The Conceptual Web-based Collaborative System Framework

4.1 Role-based Access Control

Various types of supporting documents need to be prepared, reviewed and submitted during the green building certification process. The supporting documents can be grouped into the following types.

- Templates and forms, which are generally application forms and cover letters, available from their respective websites,
- Architectural drawings, including building plans, building elevations, and floor plans,
- Engineering drawings, including technical details and sectional views of building components,
- Photos, which provide visual proof for showing installed components such as equipment and car parks,
- Analysis/calculation reports, in the aspect of energy consumption, cost budget, and other areas,
- Materials procurement bills, which provide proof for showing purchased sustainable materials/equipment,
- Utility bills (for certification of existing buildings only), which consists of electricity and water bills, and
- Compliance certification letters, which are letters from suppliers stating the quality and sustainability of the equipment and products.

These documents are submitted by the applicant (the owner/client of the building) to the certification body, but are often created and prepared by different parties. Hence, our system is open to various project participants and adopts a role-based access control. Six types of accessibility can be assigned to each role for each type of supporting document – (1) Authorize, (2) Create, (3) Modify, (4) Review, (5) View, and (6) Not Accessible. Table 5 shows the accessibility of different roles on different types of supporting documents in our proposed system framework. The client has the highest authority in this hierarchy followed by the green building consultant. The other participants are responsible for creating, modifying and reviewing specific documents whenever necessary. For example, architects can create and modify architectural drawings, can review engineering drawings, but cannot access utility bills like electricity and water bills. This feature of the system framework supports the security and confidentiality of the users.

Table 5: Accessibility Roles Assignment Matrix for the Collaborative System

Types of Supporting Documents / Roles	Templates and Forms	Architectural Drawings	Engineering Drawings	Photos	Analysis/ Calculations/ Reports	Bills (materials) (procurement)	Bills (water, electricity, etc.) for existing	Compliance Certification Letters
Client (Applicant)	A,R,V	A,R,V	A,R,V	V	A,R,V	A,R,V	R,V	R,V
Client's Consultant	C,M,V	R,V	R,V	R,V	R,V	R,V	R,V	R,V
Architects	V	C,M,V	R,V	V	V	V	N/A	V
Engineering Designers	V	R,V	C,M,V	V	C,M,V	V	N/A	C,M,V
Contractors	V	V	R,V	C,M,V	N/A	R,V	N/A	N/A
Relevant Subcontractors	V	V	V	C,M,V	C,M,V	R,V	N/A	N/A
Suppliers	N/A	N/A	V	N/A	N/A	C,M,V	N/A	C,M,V
Certification Body	A,R,V	R,V	R,V	R,V	R,V	R,V	R,V	R,V
Government Organizations (not in the system)	N/A	N/A	N/A	N/A	N/A	N/A	C	C,M,V

4.2 Document Management

Document management is one of the primitive features of the certification process. Certification process involves submitting huge amount of documents and some credits require submission of multiple documents for credit award. It requires a proper organization of the documents. The document manager can support uploading and retrieval of supporting documents corresponding to their credit category, and help to monitor the status of the documents. Based on the recent status of the documents, users can communicate in requesting and committing different actions to a document thus enhancing information transparency and inter-organizational collaboration.

4.2.1 Related Document Discovery

A well-managed document repository can support the reuse of information and experience. Various participants in different projects will lead to a massive amount of documents. Managing such huge amount of documents for sorting and retrieval will be painstaking. Relatedness analysis is a useful technique to arrange large quantity of unorganized documents into organized groups for easy identification and retrieval.

Each document is associated with multiple tags which represent the features of the document. The features can be the credits which the document is related to, type of the document (e.g. photos and engineering drawings), type of the project that document is submitted for, and some manual tags like keywords. A feature matrix can be created for relatedness analysis, as shown in Figure 3. A value of 1 is assigned if a tag belongs to a document and a value of 0 is assigned otherwise [12]. For example, as illustrated in Figure 2, the document d1 is a drawing submitted for the credit SA02.

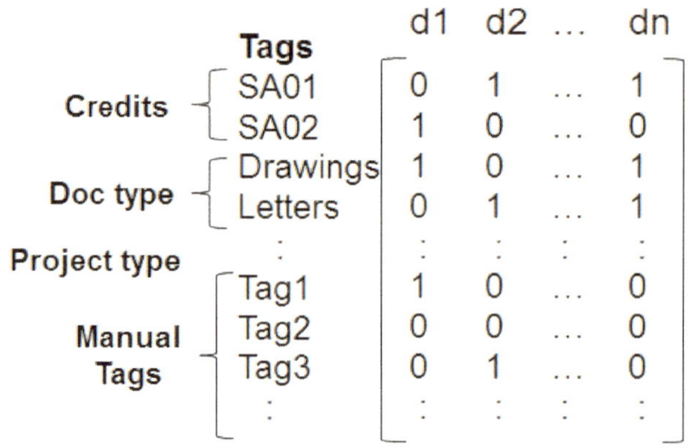

Figure 3: Matrix Representation of Document Tagging

When documents are represented as term column vectors, the similarity of two documents corresponds to the correlation between the vectors, which can be calculated as the cosine of the angle between the vectors, namely cosine similarity. Cosine similarity is one of the most popular similarity measure applied to text documents, such as in numerous information retrieval applications and clustering. The cosine similarity between documents d1 and d2 can be calculated as

$$\mathrm{SIM}(\mathrm{d1}, \mathrm{d2}) = \frac{\overrightarrow{d1}.\overrightarrow{d2}}{|\overrightarrow{d1}| X |\overrightarrow{d2}|}$$

Let us consider the above Equation.1, where $\overrightarrow{d1}$ and $\overrightarrow{d2}$ are the column vectors over a set d= [d1, d2]. Each row represents a term vector with its tag, which is either 0 or 1, where 0 and 1 represent the presence and absence of the tag, respectively. As a result, the cosine similarity is non-negative and bounded between [0, 1]. Since the absence of a tag is valued as 0, the similarity of two documents without any similar tags will be 0. For two documents with all tags identical, the cosine similarity value is 1. In other cases, the similarity values range from 0.1 to 0.9.

4.3 Workflow Management

The collaborative nature of the certification process demands people from different organizations working together. The system provides a platform for organized workflow by assigning individual responsibilities to project team members and provide an easy way to track people concerning specific responsibility. For example, if a specific project work task has not been completed or requires modifications, the system can assist the user by identifying the responsible person and communicate to complete the required changes. Users can also keep track of their work progress and manage their workflow coherently. Furthermore, preparation and submission time for the documents related to the affected work task can be changed accordingly. The change will be announced to the corresponding project participants according the role-based accessibility matrix as mentioned in Section 4.2.

4.4 Credit Management

Different certification standards have different credit calculation procedures for grade awards. Many certification standards also have special criteria for final grade awards. Credit calculation and grade planning can be cumbersome if a team handles different certification standards or applies for multiple certifications. The proposed system framework features a credit dashboard that can handle the credit calculations and plan the final grade. This allows users to have a better approach towards their targeted grade. Grade planning and reviewing can be done more easily with the credit dashboard.

4.5 Knowledge Management and Best Practices

Green building certification process is complex and relatively new to many people in the construction industry. Inexperience and lack of proper documentation have led to inefficient practices and make the certification process complex and time consuming. As the industry is project-based and similar challenges can be met in the future, documenting best practices will create a knowledge database that will help project participants to follow efficient practices and save time in future projects.

4.6 Collaboration Supported by the System

The iterative and collaborative preparation processes for green building certification are supported in our proposed system framework, as shown in Figure 4. In the figure, the left side describes the iterations that appear in the certification preparation process while the corresponding right side identifies the respective participants/roles involved. As shown in Table 5, the roles supported in our systems include the client, the green building consultant, architects, engineering designers, contractors, subcontractors, and suppliers, each having different access rights in the system. After the client and the consultant agree on and set the target grade, the responsible team members prepare the required supporting documents for earning credits. Once the prepared documents have been uploaded, they are reviewed, modified, and authorized for final submission. If there are any necessary changes, the system helps to identify the respective participants and allows them to communicate, thus enhancing collaboration. If the target grade is not achieved, as summarized in the credit dashboard, the team can review their project plan again and the process is repeated [14].

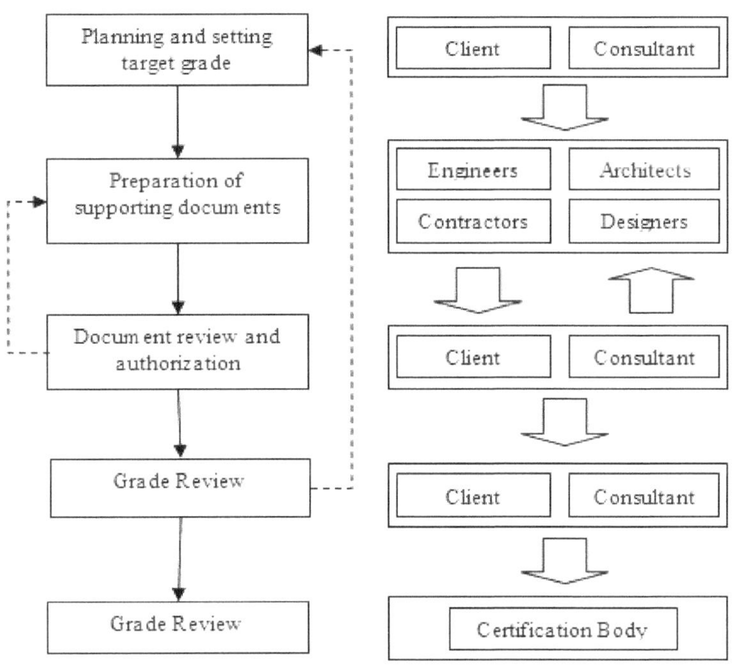

Figure 4: Iterations and Collaborations Supported by the System

5.0 Example Scenario

In the last two decades, the construction industry in the Asian region has grown dramatically, and is poised to continue such growth [15]. The demands have led to new emerging concepts of engineering design and facilities management. For example, the Hong Kong government has been taking an active part in driving green building initiatives [16].The growing economy and government initiatives to go green saw a huge rise in the number of green buildings and certifications. In 1996, the green building certification standard HK-BEAM (Hong Kong Building Environmental Assessment Method) was established by the Hong Kong Green Building Council in Hong Kong. The standard was renamed to BEAM Plus in 2009. As of 2011, the BEAM standard has certified over 500 projects in Hong Kong, Beijing, Shanghai and Shenzhen. On a per capita basis, BEAM has assessed more buildings than any other similar scheme in use worldwide [14]. The BEAM standard was developed carefully and shares similarity in scope with the LEED standard and credit calculation with the BREEAM standard. It had a broad coverage and increasing demand.

We exploited this potential and developed a web-based collaborative system for facilitating the Hong Kong BEAMPlus certification process and for illustrative purpose. The functions of the system are explained and illustrated in the following sections. The document manager will illustrate the management of documents required for certification and people responsible for specific project works. The related document discovery illustrates the system functionality of knowledge management by successfully retrieving and reusing information. The credit dashboard will assist project participants for efficient grade planning and credit management. Confidentiality will also be supported through role based access to these functionalities

5.1 System Configuration

The system was built for document storage and management that serves as a repository of information and documents for data storage, retrieval and submission. The system consists of a database and server support. The scripting languages used for creating the web interface are PHP and HTML. Open source software was used in the system design and implementation. The open source tools used in this system are

- Apache- The Apache HTTP Server project for developing and maintaining an open-source HTTP server for modern operating systems including UNIX and Windows.
- MySQL- The world's most widely used open source relational database management system that runs as a server providing multi-user access to a number of databases.
- PHP- Hypertext Preprocessor is a general-purpose server-side scripting language originally designed for web development for producing dynamic web pages.
- HTML (Hyper-Text Markup Language) is the main markup language for displaying web pages and other information that can be displayed in a web browser

5.2 Hong Kong BEAM Plus Certification Process

The BEAM Plus certification process between an applicant (client/owner) and the certification organizations (Hong Kong Green Building Council and BEAM Society) is illustrated in Figure 5. The applicant submits the BEAM Plus application with supporting documents, which are then reviewed and assessed by the BEAM Society. Before and during the certification process, the applicant also needs to communicate and collaborate with various project participants and gather materials from them for preparing the supporting documents. Submission and review of the supporting documents is an iterative process at the preparation and the submission stages.

5.3 Document Manager for BEAM Plus

The system requires users to log in at the homepage. With a successful login, users can view, upload, revise, and authorize the supporting documents, based on the role-based accessibility granted to the users. The supporting documents are grouped according to the corresponding BEAM sections. On the document summary page, as shown in Figure 6, users can monitor the latest status (e.g. "submitted", "reviewed", "requested", "approved") of each supporting document under each section. The last modified date and user of each document are also displayed for transparency. On the summary page, users cannot see documents that are not accessible ("N/A") for the users. Documents that are requested for submission are displayed in red. With such transparent and personalized document management in our system, document collection, preparation and review can be facilitated to save time for BEAM certification.

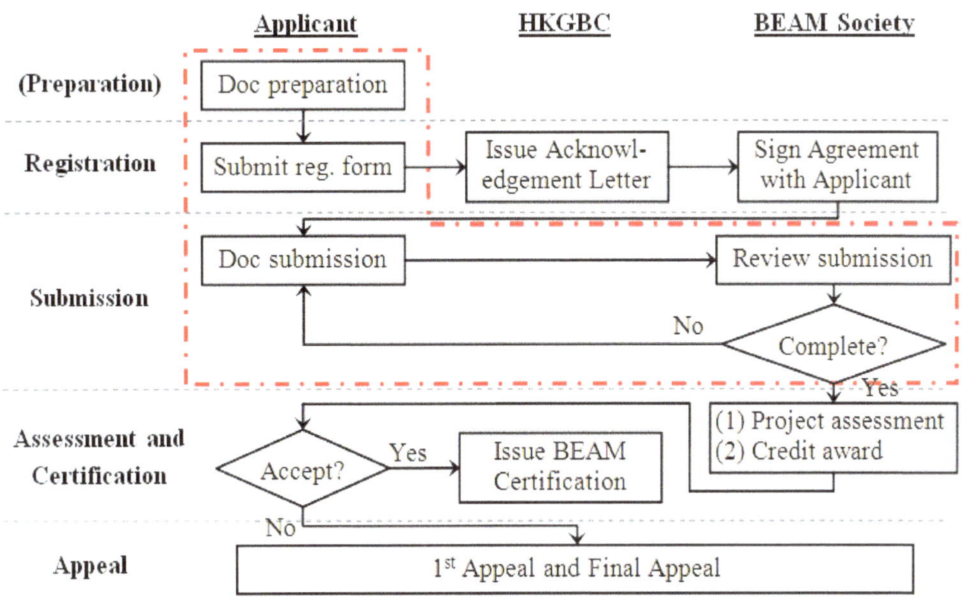

Figure 5: The Certification Process for the Hong Kong BEAM Standard

SITE ASPECTS (SA)					
Section Number	Name	Credits Obtainable	Credits Obtained	Credits Targeted	Supporting Files
SA P1	MINIMUM LANDSCAPE AREA	Required	--	Required	--
SA 1	CONTAMINATED LAND	1B	--	0	--
SA 2	LOCAL TRANSPORT	2+1	3	2	1. Letter from Franklin Consulting Limited confirming the car park space >Submitted, by Keith Lo (9 Feb 2012) 2. Design drawing showing the location and size of car park >Submitted, by Keith Lo (12 Feb 2012) 3. Design plan for shuttle service vehicles parking >Submitted, by Leo Ho (12 Feb 2012) 4. Architectural drawing showing pedestrain access to nearby MTR subway station >Submitted, by Keith Ho (18 Feb 2012)
SA 3	NEIGHBOURHOOD AMENITIES	1+1+1	--	2	--
SA 4	SITE DESIGN APPRAISAL	1+1B	--	1	1. BEAM NB Template > Submitted, by Anderson Kingston (8 Mar 2012) 2. Site design appraisal report from Franklin Consulting Limited >Requested, by Peggy Tsui (25 Mar 2012)
SA 5	ECOLOGICAL IMPACT	1B	--	0	--
SA 6	CULTURAL HERITAGE	1	--	1	1. Site layout plan >Submitted, by Keith Lo (23 Mar 2012) 2. Site survey results report by Jasper Morison Company >Reviewed, by Terry Cheng
SA 7	LANDSCAPING AND PLANTERS	1+2	--	3	--

Figure 6: Document ManagerPage in the Demonstrative System

5.3.1 Related Document Discovery for BEAM Plus

To find related documents, 182 documents were collected in total and arranged in order of their categories. Collected documents comprise of photos, building drawings, bills, reports, etc. Each document was analyzed and provided with multiple tags. Every document should contain tags of the document's credit category, type of document and document information. Project information should be added if dealing with multiple projects. Manual tags can also be provided for better identification of the document. Manual tags can vary depending on the type and content of the document and does not follow any standard protocol. After the inclusion of all tags in the documents, the feature matrix was created with the documents on one side and the tags on the other side. A value of 1 is assigned if a tag belongs to a document and a value of 0 is assigned otherwise. As an example, Table 6 lists three documents chosen for illustration.

Table 6: Illustrative Documents Used for Similarity Calculation

Doc. No	Title	Tag1	Tag2	Tag3
1	Car Parking Drawings	SA2	Drawing	Car Park
2	Demolition Waste Reduction Report	MA10	Report	Waste
3	Construction Waste Reduction Report	MA11	Report	Waste

Case1: Consider document 1 and document 2 for relatedness analysis. As shown in Table 6, document 1 has the tagsSA2, Drawing and Car park whereas document 2 has the tags MA10, Report and Waste. Using Equation 1 presented in Section 4.2.1, the cosine similarity value is given as

$$\mathrm{SIM}(\mathrm{d1},\mathrm{d2}) = \frac{0}{\sqrt{3}\sqrt{3}} = 0$$

Case2: Consider document 2 and document 3 for relatedness analysis. As shown in Table 6, document 2 and document 3 share the Report and Waste tags. Using Equation 1, the cosine similarity is given as

$$\mathrm{SIM}(\mathrm{d2},\mathrm{d3}) = \frac{2}{\sqrt{3}\sqrt{3}} = 0.67$$

Document 1 is a drawing of car parking facilities for the building. It is required for obtaining Credit SA 2 (2-3 credits) in the Site Aspects category. Documents 2 and 3 are demolition and construction waste reduction reports. These documents are required for obtaining credits MA 10 (2credits) and MA 11 (2 credits) in the Materials Aspects category. As shown in the results, document 1 and document 2 have totally different natures and different credit obligations, and therefore have the lowest (zero) similarity value. On the contrary, documents 2 and 3 have similar natures and belong to similar categories but different credit obligations, and therefore have 67 percent similarity. The above results indicate the potential of the similarity analysis algorithm in the searching, sorting, organization and submission of huge amount of documents, particularly across multiple projects. Document submission is the core in this certification process, and the proposed algorithm will be a desirable addition

5.4 Credit Dashboard for BEAM Plus

On the document manager page as shown in Figure 6, the BEAM credits obtainable, the credits obtained, and the target credits for each individual section are shown clearly. However, clients are sometimes more concerned about the final assessment grade that their buildings can achieve. The assessment grade award criteria for BEAM are shown in Table 7. The grade is determined by the percentage (%) of the applicable credits gained under each performance category and its weighting factor, which is shown in Figure 8. Given the importance of the Site Aspects (SA), Energy Use (EU) and Indoor Environmental Quality (IEQ), the minimum percentages (%) of credits for the three categories are also required for the determination of the final assessment grade. A minimum number of credits in the Innovation category are also needed for obtaining Silver Grade, Gold Grade, and Platinum Grade.

Figure 8 shows the credit dashboard on our system that summarizes the credits obtained and reports the assessment grade achieved at the moment. In the example, the overall score is 70.5%, which is higher than the overall percentage requirement for the Gold grade. However, the percentage for the category Energy Use is only 59.5%, which is lower than the EU percentage requirement for the Gold Grade. Therefore, the assessment grade achieved at the moment is only Silver. Using this dashboard, clients can observe the latest fulfillment situation of each assessment criterion and probably strategically put additional efforts on specific BEAM sections or categories to achieve a higher BEAM Assessment Grade.

Table 7: BEAM Assessment Grade Award Criteria

Grade	Overall	S_A	E_U	I_{EQ}	I_A	
Platinum	$\geq 75\%$	$\geq 70\%$	$\geq 70\%$	$\geq 70\%$	3 credits	(Excellent)
Gold	$\geq 65\%$	$\geq 60\%$	$\geq 60\%$	$\geq 60\%$	2 credits	(Very Good)
Silver	$\geq 55\%$	$\geq 50\%$	$\geq 50\%$	$\geq 50\%$	1 credit	(Good)
Bronze	$\geq 40\%$	$\geq 40\%$	$\geq 40\%$	$\geq 40\%$	-	(Above Average)

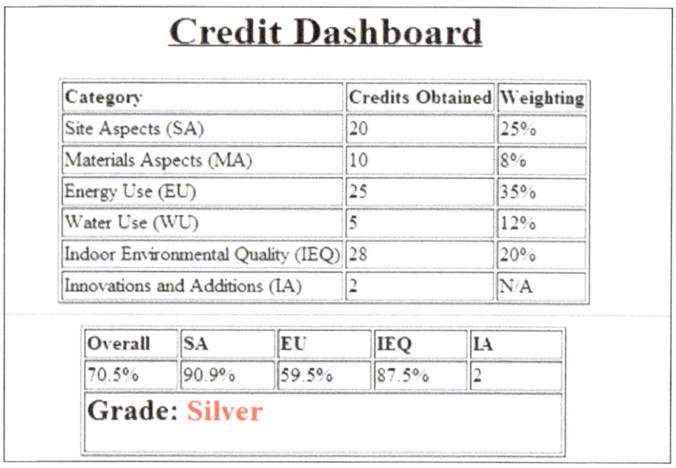

Credit Dashboard

Category	Credits Obtained	Weighting
Site Aspects (SA)	20	25%
Materials Aspects (MA)	10	8%
Energy Use (EU)	25	35%
Water Use (WU)	5	12%
Indoor Environmental Quality (IEQ)	28	20%
Innovations and Additions (IA)	2	N/A

Overall	SA	EU	IEQ	IA
70.5%	90.9%	59.5%	87.5%	2

Grade: Silver

Figure 8: Credit Dashboard in the Demonstrative System

6.0 Conclusions and Future Work

This paper describes the challenges of green building certification and the importance of a web-based collaborative system in assisting the green building certification process. The entire process is iterative and involves the preparation and submission of numerous documents by stakeholders representing different industries and organizations. Manually managing a huge amount of information might leave useful information uncoordinated, and it will also lead to miscommunication and confusion of information stored. The growing demand for building certifications offers no help towards the existing methods but instead, demands efficient ways. We studied different green building certification standards worldwide and summarized the essential system requirements for an efficient green building certification process, which are (1) role-based access control, (2) document management, (3) credit management, (4) workflow management, and (5) knowledge management. According to the system requirements, a web-based system framework was developed for assisting different green building certification processes. The developed system framework supports document management and certification planning with role-based access control. The proposed system framework not only supports the sharing of supporting documents for green building certification, but also facilitates collaborative document management and reviews the decision-making through its various features. This study also successfully proposed an algorithm for discovering similar documents thus eliminating information overload and enhancing information retrieval. Therefore, the proposed system framework could reduce document review iterations, save time, cut cost and increase work efficiency. The system framework can serve as a repository for project information and provide users universal access to the information over the web. The proposed system framework was also illustrated and validated using the Hong Kong BEAM green building standard. The results of this study demonstrated that a web-based collaborative environment can assist the existing manual green building certification process by increasing project team collaboration and organizing project related information more efficiently.

In the future, additional functions would be added to the system for increased efficiency. Forum discussions would be introduced for useful information exchange, and the algorithm for computing document similarity would be improved for more accurate results. Report generators and document suggestions for different credits will make the system automated. The compatibility of the proposed system framework can be extended to certification standards like LEED, Green Star and BREEAM, thus making multiple certifications for buildings effortless.

References

[1] U.S. Environmental Protection Agency (EPA). Available: http://www.epa.gov/greenbuilding/pubs/about.htm
[2] American Society for Testing and Materials (ASTM). "E2114-08 Standard Terminology for Sustainability Relative to the Performance of Buildings".
[3] International Council for Research and Innovation in Building and Construction (CIB). Agenda 21 for Sustainable Construction in Developing Countries.
[4] G. K. C. Ding, "Sustainable construction—The role of environmental assessment tools," Journal of Environmental Management, vol. 86, pp. 451-464, 2008.
[5] G. Kats and E. Capital, Green building costs and financial benefits: Massachusetts Technology Collaborative Westborough, MA, 2003.
[6] B. Becerik, "A review on past, present and future of web based project management & collaboration tools and their adoption by the US AEC industry," International Journal of IT in Architecture Engineering and Construction, vol. 2, pp. 233-248, 2004.
[7] P. Nitithamyong and M. J. Skibniewski, "Web-based construction project management systems: how to make them successful?," Automation in construction, vol. 13, pp. 491-506, 2004.
[8] H.-M. Chen and H.-C. Tien, "Application of peer-to-peer network for real-time online collaborative computer-aided design," Journal of Computing in Civil Engineering, vol. 21, pp. 112-121, 2007.
[9] J. Zhang and T. El-Diraby, "Social Semantic Approach to Support Communication in AEC," Journal of Computing in Civil Engineering, vol. 26, pp. 90-104, 2011.
[10] J. C. Cheng, K. H. Law, H. Bjornsson, A. Jones, and R. Sriram, "A service oriented framework for construction supply chain integration," Automation in construction, vol. 19, pp. 245-260, 2010.
[11] Y. Y. Li, P.-H. Chen, D. A. S. Chew, C. C. Teo, and R. G. Ding, "Critical project management factors of AEC firms for delivering green building projects in Singapore," Journal of construction Engineering and Management, vol. 137, pp. 1153-1163, 2011.
[12] Venkataraman, "Development of a Mobile Platform For Green Building Certification," Masters Thesis, Department of Civil and Environmental Engineering, Hong Kong University of Science and Technology, Hong Kong, 2012.
[13] R. C. Retzlaff, "Green building assessment systems: a framework and comparison for planners," Journal of the American Planning Association, vol. 74, pp. 505-519, 2008.
[14] J. Cheng and V. Venkataraman, "Collaborative System for HK-BEAM Green Building Certification," Cooperative Design, Visualization, and Engineering, pp. 211-218, 2012.
[15] J. Raftery, B. Pasadilla, Y. Chiang, E. C. Hui, and B.-S. Tang, "Globalization and construction industry development: implications of recent developments in the construction sector in Asia," Construction Management & Economics, vol. 16, pp. 729-737, 1998.
[16] E. H. W. Chan, Q. K. Qian, and P. T. I. Lam, "The market for green building in developed Asian cities—the perspectives of building designers," Energy Policy, vol. 37, pp. 3061-3070, 2009.

15

A REVIEW OF LEAN CONCEPT AND ITS APPLICATION TO SUSTAINABLE CONSTRUCTION IN THE UK

Oyedolapo Ekundayo Ogunbiyi*, Adebayo Akanbi oladapo and Jack Steven Goulding

University of Central Lancashire, United Kingdom

*Corresponding E-mail : oogunbiyi@uclan.ac.uk

Abstract

The UK Government has recognised the importance of the construction industry in achieving the overall goals of sustainable development. Therefore the Government has put several policies and strategies in place to achieve a more sustainable construction. Sustainable construction is considered as the application of sustainable practices and sustainable development principles to the activities of the construction sector. Lean construction is a new production philosophy which has the potential of bringing innovative changes in the construction industry. The Lean principles focus on the minimisation of both material and process wastes which in turn contribute to sustainable construction in terms of energy consumption and improvement in health and safety etc. This study aims at exploring the concept of sustainable construction and examines how the lean approach can impact on the sustainability practices within the construction industry. The study uses literature review to achieve the stated aim. The findings revealed that the application of lean construction principle, tools and methods have direct contributions to the attainment of sustainable practices within the construction industry. However, the study postulates that the better understanding of lean concept, proper implementation and integration of lean and sustainability concepts are required for lean construction to contribute to sustainable construction.

Keywords: *Lean construction, Sustainable Construction, Sustainability*

1.0 Introduction

The UK construction industry is noted for its economic contribution with an output worth over £100billion a year. It provides employment for over three million workers and accounts for eight per cent of gross added value [1]. Nonetheless, the construction industry is also noted for its poor safety record evident from high rate of accidents on construction sites leading to workers injury or loss of lives [2]. This suggests the reason why more attention is paid to the sector. However, there are other benefits to be gained from a more sustainable construction industry. The adoption of a sustainable approach was suggested to lead to important business benefits and address the shortcomings of the construction industry identified in the Rethinking Construction report. This reflects that becoming more sustainable could lead to efficiency, profit-orientated practice and achieving value for money, as it is about helping society and protecting the environment. There is a growing awareness as to the competitive advantages that can be convened by businesses taking a sustainable approach [3].

Lean construction is a new production philosophy which has the potential of bringing innovative changes in the construction industry. The concepts and principles of lean is to generally make the construction process leaner by removal of waste which is regarded as non-value generating activities [4]. The removal of waste (process and material) and value generation in terms of adding value to the customer are the major contributions of lean construction to sustainable development [5]. This is achieved by the use of lean principles: pull system, flow, value stream mapping, continuous improvement and involvement of employees.

There are several key factors to be taken into action by the construction industry. These factors have been suggested by the UK Government in its strategy for more sustainable construction [6]. These factors include:

1. Design for minimum waste
2. Aim for lean construction
3. Minimum energy in construction and use
4. Pollution reduction
5. Preservation and enhancement of biodiversity
6. Conservation of water resources
7. Respect for people and local environment
8. Setting targets
9. Monitoring and reporting in order to benchmark the performance

Among several factors, the lean construction principles will be focused on, as the main area of this study is to critically review the concept of Sustainable Construction (SC), and examine how the application of lean principles can impact on the sustainability practices within the construction industry. Accordingly, this study pulls from two main bodies of literature: i.e. the literature on sustainable development and lean construction in the broader context of the construction industry (see Figure 1). As earlier mentioned, the construction industry is considered as a key sector for achieving sustainable development goals because it plays a vital role in the drive to promote sustainable growth and development.

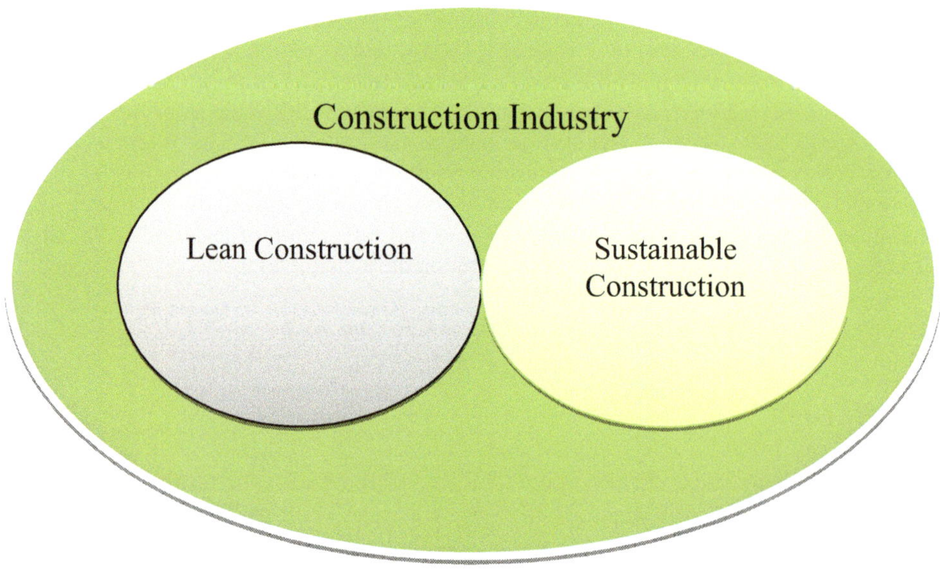

Figure 1: Literature review focus

The potential of lean to contribute to sustainable construction has been raised for discussion [5]. Therefore, it is of utmost importance to examine the possibilities of lean contributing to sustainable construction. Several studies have been carried out on lean and its application within construction at project level with great benefits achieved and there are many studies that have investigated lean construction and sustainability separately [7, 8]. However, studies that highlight the contributions of lean construction towards sustainability are few. The insufficiency of literature addressing this issue and the absence of research-based papers are assumed as a lack of awareness of the potential of lean construction as a means of achieving sustainability and an unrecognised relationship between sustainability and lean construction objectives. For instance, Forbes et al.[9] proposed a framework for providing technical support for lean methods application in some environments in developing countries. Sacks et al. [2] developed a research framework for analysis of the interaction between lean and BIM. However, there has been little or no study done to look at the impact of lean on sustainable construction in terms of developing a framework at the organisational level. Against this background, this study aims to examine the contributions of the implementation of the lean approach in sustainable construction.

2.0 Sustainable Construction

It is difficult to describe sustainable construction without defining or describing sustainable development. There are several definitions of sustainable development given in the literature [10, 11]. Sustainable development is a broad concept which has been adopted and interpreted in numerous contexts. For example many authors have seen the concept as vague and fuzzy [12, 13]. According to Sage [14], sustainable development refers to the fulfilment of human needs through simultaneous socio-economic and technological progress and conservation of the earth's natural systems. However, the most popular definition of sustainable development is the one given in the Brundant report "development that meets the needs of the present without compromising that ability of future generations to meet their own needs" [15]. Nevertheless, there are some areas of agreement in the various definitions. This reflects that the goal of sustainable development is to enable humanity all over the world to satisfy their basic needs and enjoy a better quality of life without compromising the quality of life of future generations. The concept of sustainable development has been described with three dimensions: economic, social and environmental aspect. Sustainable development and social responsibility have become increasingly important strategic issues for companies in virtually every industry [16].

The term sustainable construction means different things to different people as there are multiple definitions, and variance in terms of scope and context as well as practices [11, 12, 17]. Bourdeau et al. [17] stated that sustainable construction practices are widely different depending on how the concept of sustainable construction is developed in various countries. Therefore, simply put, sustainable construction is the response of the building sector to the challenge of sustainable development [5].

The implementation of sustainable construction is still under explored. The decision making process and the actors as well as the inter-relationship has to be understood when implementing sustainable construction [18]. The issues of sustainable construction are divided into 3 aspects: the environmental, economic and the social issues. CIEF [19] suggests sustainable construction as a solution for significant cost savings, to bring innovations and to enhance competitiveness for long time survival of any organisation. Sustainable construction practices not only provides increased market share and profitability but also brings many other intangible benefits such as visible brand name to the organisation in the industry, quality in construction, employee motivation and satisfaction, improved customer's satisfaction, and complements / awards from regulatory authorities and improved shareholder relations [19, 20].

3.0 Lean Thinking in Construction: Lean Construction

The application of lean thinking in construction was pioneered by Koskela who suggested that construction production should be seen as a combination of conversion and flow processes for waste removal. The concept of lean is attributed to the manufacturing industry and was introduced to construction [4].The use of lean concept has been advocated in the UK, several seminars and initiatives have been undertaken in an effort to encourage its uptake. The Construction Industry Research and Information Association (CIRIA), Construction Productivity Network (CPN), Construction Lean Improvement Programme (CLIP) and the Lean Construction Institute UK (LCI-UK) are some of the examples of institutions established. Seminars and conferences have been organised to tease out the main issues in the development and awareness of lean construction principles with real life case studies of some construction organizations presented [19]. In spite of these efforts, there seems to be some barriers to the successful implementation of lean construction. Generally the rate of lean implementation within the UK construction industry is relatively low and the application of lean in sustainable construction is still under explored [21]. Some studies have identified the barriers to the implementation of lean construction. These barriers need to be overcome in order for construction industry to reap the benefits of implementing lean construction. The application of lean principle to construction has been presented to result in benefits such as improved quality, improved safety, waste reduction, increased productivity, more client satisfaction, increased reliability, and improvements in design.

A study carried out by Sarhan and Fox [22] reveals that there seems to be positive trends in the development of a lean culture among UK construction organisations. Lack of understanding of how to successfully apply lean thinking principles to specific construction processes was also revealed. This study of lean culture within the UK construction organisations was carried out after the study of Common et al., and Johansen and Walter [22]. Lean thinking has become an important concept within the UK construction industry following the Egans report. There has been significant improvement in the agenda for change in the UK construction industry. Few studies have been carried out in order to establish the current levels of awareness and implementation of lean thinking within the UK construction industry. An example of such studies is the application of the Last Planner into a UK construction project. Last Planner is one of the lean tools and techniques and perhaps the most developed tool. The tool was applied to a UK construction project to ascertain its value and its possible barriers. However, the study raised a number of important structural and cultural problems for the success of Last Planner in the UK [23].

Shah and Ward [24] pointed out that it is essential to differentiate between those studies considering lean from a philosophical perspective related to guiding principles or overarching goals, and those analysing the concept from a practical perspective as a set of management practices, tools, or techniques that can be observed directly. The implementation of lean construction have been targeted towards some specific tools and principles without a full integration on different aspects such as supply chain, safety, planning and control, production design and management, culture and human aspects [25-27]. Framing an encompassing definition that covers all aspects of lean is seen as a difficult task [28]. Alves et al., [26] stated that there are many meaning of lean when applied to construction. Therefore, this study deems it fit to scrutinize various definition of lean as applied to construction. Table 1 presents various definition of lean.

Lean offers significant benefits in terms of waste reduction and increased organisational and supply chain communication and integration. The elimination of waste leads to cost benefits advantage, however these are pre-requisite for creating a lean process. The lean implementation effort stage one focus on waste elimination from a technical and operational perspective [29]. Process Mapping, Value Stream Mapping, and 5S (Workplace Organisation) are some of the tools for achieving such processes. There are 7 types of waste identified under lean: overproduction, overstocking, excessive motion, waiting time, delay and transportation, extra-processing, defect and rework. In the same manner, there are various methodologies for attaining lean production: just in time (JIT), total quality management, concurrent engineering, process redesign, value based management, total productive maintenance and employee involvement.

Table 1: Definitions of Lean

Sources	Definition
Manrodt[30]	Lean is a systematic approach to enhancing value to the customer by identifying and eliminating waste (of time, effort and materials) through continuous improvement, by flowing the product at the pull of the customer, in pursuit of perfection
Ballard et al. [31]	Lean is "a fundamental business philosophy – one that is most effective when shared throughout the value stream"
Lean Construction Institute [32]	Lean construction is a production management-based project delivery system emphasising the reliable and speedy delivery of

	value
Radnoret al. [33]	Lean is a philosophy that uses tools and techniques to create a change of organisational culture in order to implement the 'good practice of process/operations improvement that allows the reduction of waste, improvement of flow, more focus on the needs of customers and which takes a process view'
Construction Industry Institute [34]	"The continuous process of eliminating waste, meeting or exceeding all customer requirements, focusing on the entire value stream and pursuing perfection in the execution of a constructed project."
Shah and Ward [24]	''an integrated socio-technical system whose main objective is to eliminate waste by concurrently reducing or minimising supplier, customer, and internal variability.''

3.1 Lean Approach in Sustainable Construction

Lean construction is one of the strategies for improving the sustainability of construction, in other words one method of achieving sustainable construction. Lean approach in sustainable construction focuses on the removal of all forms of wastes from construction processes to allow more efficiency. Existing studies have suggested theories to support that lean is a method for optimising resources, improving safety, productivity, working condition and overall, the social, environmental and the economic bottom line [35]. There are several forms of waste under the lean terminology: processes, material and poor safety are considered as a potential wastes that hinder flow of value to the client. Construction should be seen as flow processes (consisting of both waste and conversion activities), not just conversion processes [4]. The promotion of health and safety practice can contribute to sustainable construction by enhancing workers' social life and minimising direct and indirect cost of accidents. Material waste elimination has been identified as the most efficient and cost effective approach to promote sustainable practice on construction sites. Similarly, the principles of lean construction focus on creating a sustainable change by stressing on efficient, waste-free and safe flow, storage and handling of materials to minimise cost, energy and resource consumption, and provide value for clients and end users [7].

Some of the key issues of sustainability identified in the literature include: global warming and climate change which is seen as one of the main threats to the environment as a whole [36]. Peng and Pheng [37], investigated the contribution of the lean concept to achieve low carbon installation in the construction sites using precast concrete products and found that the lean concept can be adopted to reduce carbon emission in terms of re-designing the site layout, improving the supply chain and installation work flow. Many studies have highlighted the contributions of lean construction towards the environmental aspect of sustainability. For example Huovila and Koskela [5] presented minimisation of resource depletion, pollution and matching business and environmental improvement as the contribution of lean construction to sustainable development. However, the contribution of lean construction to sustainable development is not limited to the environmental aspect but also to the social and economic aspect. The different lean applications might have different results on the three pillars of sustainable development.

The lean impact has been described to cover the economic, social and environmental aspect of sustainable construction. This include more value to client with less waste of time and resources, process improvement and overall project delivery, productivity improvement, cost reduction, improved quality and safety as well as promotion of continuous improvement. A good example of this is the case study of the modular home building by Nahmens [29] which was carried out to evaluate the use of lean construction to improve sustainability. Lean construction strategies serves as a platform for improvement in the delivery of the sustainable modular houses. Figure 2 presents the main effect of the application of the lean concept for the purpose of sustainability in the aforementioned example.

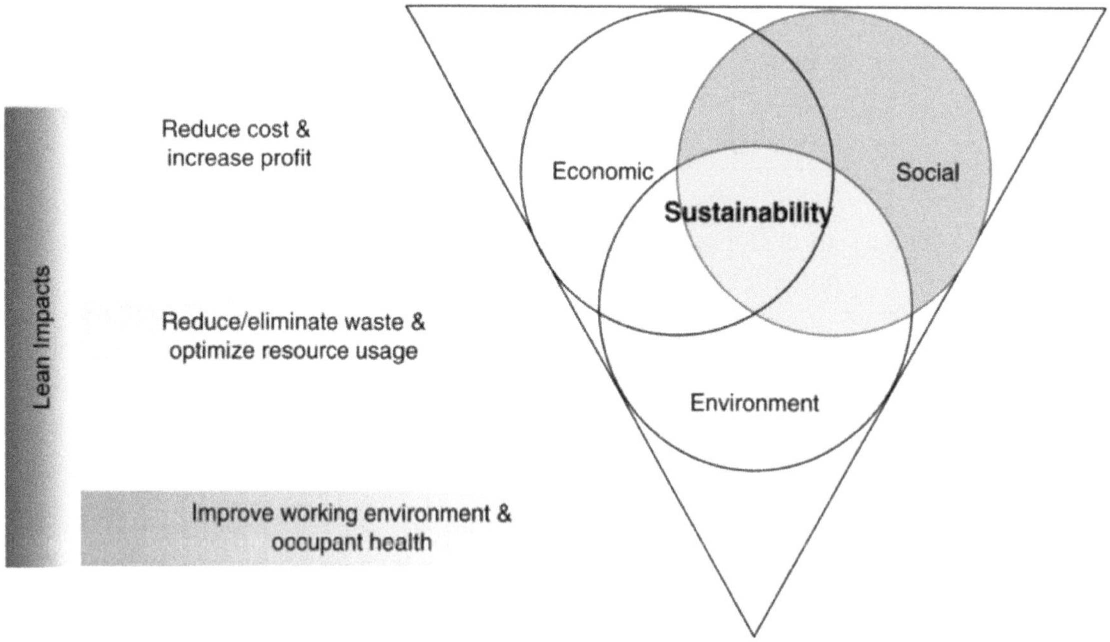

Figure 1: Conceptual model: effect of lean on sustainability
Source: (Nahmens and Ikuma, 2009)

As much as adopting lean concept has been attributed to positive influence on sustainable construction in terms of improved safety, many research works have shown both negative and positive effects of lean on safety. However, in terms of sustainability, lean and safety influence economic sustainability by reducing costs and increasing productivity, environmental sustainability by reducing or improving materials and social sustainability by affecting the well-being of workers.

3.2 Sustainable Practice and lean concept

According to Tan et al., [38], Sustainable construction practices include five major areas: compliance with sustainability legislation, design and procurement; technology and innovation; organisational structure and process; education and training; and measurement and reporting. The successful implementation of lean and sustainable concepts by an organisation depends on the level of commitment and knowledge. The implementation of sustainability throughout the organisation including the organisation's project will yield more result than when implemented only on the project [39]. Different company characteristics can influence the choices in sustainable construction practices. The selected sustainable construction practices should be consistent with the overarching strategy. The benefits of implementing sustainable practices include improved regulatory compliance requirements; reduction of liability and risk; enhanced reliability among customers and peers; reduction of harmful impacts to the environment; prevention of pollution and waste (which can result in cost reduction); improvements in site and project safety (by minimising injuries related to environmental spills, releases and emissions); improved relationships with stakeholders such as government agencies, community groups, and clients [40].

The benefits of implementing sustainable practices in construction can be grouped under environmental, economic and social aspects. Hall and Purchase [41] stated that numerous sustainability and lean practices, such as productivity, safety, efficiency, and waste minimisation, are interconnected. The conceptual relationship between lean and sustainability has been presented in the literature. Lean practices can be adopted in a construction project at design phase to reduce costs and enhance sustainability [42, 43]. Few studies have been carried out to investigate the application of sustainability and lean concept. Despite the pressure on the construction industry to adopt the concept of sustainability to improve the current unsustainable pattern of project delivery, its uptake is relatively slow i.e. the adoption of sustainable practice in construction project. Koranda et al., [8] developed a framework for implementing lean techniques and sustainability in a construction project as shown in figure 3. This framework captured the major sustainability issues at project level.

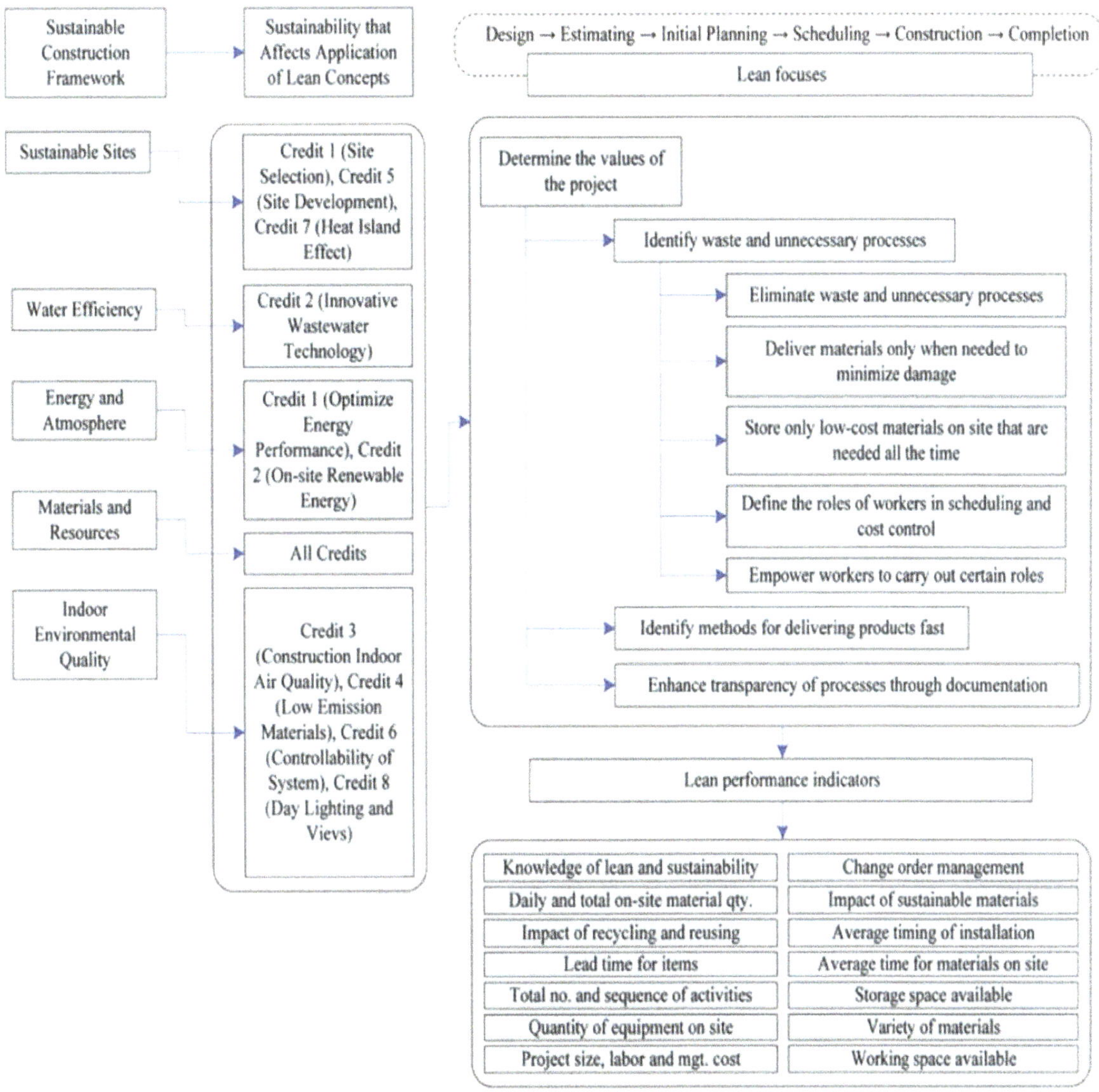

Figure 3: Framework for implementing lean techniques and sustainability in a construction project (Source: Koranda et al. 2012)

There is need for leadership participation in the quest for attaining a more sustainable construction as the leadership role in construction organisation is one of the paramount factors that can provide an overall vision, direction and vision towards the attainment of a sustainable construction. Therefore, it is highly essential that leaders have full knowledge of the concept of sustainability to be able to guide their organisations effectively [44]. Likewise, top level leadership commitment has been identified as one of the success factors for the implementation of lean. This suggests that thorough understanding of lean and sustainability concepts as well as principles are necessary for proper application on a construction project.

3.3 Lean Tools and Methodologies for Sustainable Construction

Various lean tools and techniques for enabling sustainability have been discussed by several authors. Some studies have explored various issues of sustainability by means of lean initiatives and established the benefits that can be derived by applying the lean principles/tools [42, 45]. Lean design methods such as Integrated Design, Design for Maintainability (DFM), Set-based Design, Target Costing and 3D Modelling can be used during the construction of sustainable project. Many studies have suggested integrated design method to be one of the most critical methods for sustainable construction [46-48]. Just-in-time (JIT) is a major component of the lean construction concept, the principle of just in time is to ensure that the correct quantities of materials are delivered as at when needed in the right quantity to the exact location in good condition [49-51]. Bae and Kim [43] carried out the quantitative assessment of lean methods and sustainability impacts of construction project. This was based on the lean project delivery phases which include: lean project definition, lean design, lean supply, lean assembly and whole delivery process. It was revealed that most lean construction methods provide positive economic impacts for sustainable projects while there are few negative impact as well as the combination of both impact (positive and negative) on the social and the environmental aspects.

There are many lean tools and techniques/principles among which 5S, value stream mapping, just in time, visualisation tool, last planner, value analysis, pull approach and continuous improvement appears to be the commonly adopted lean tools and techniques/principles [45]. Value stream mapping (VSM) is the mapping of wastes throughout the organisation. 5S and value stream mapping are commonly noted for environmental improvement. 5S helps companies to look at their workplace in a new dimension. Companies use 5S to clean and streamline areas within their works, removing unwanted parts, tools and general debris and setting a new standard for cleanliness and tidiness. It also helps in organising construction site, thereby resulting to environmental improvement and health and safety improvement.

4.0 Conclusions

The study has drawn from literature on both lean and sustainability reflecting the principles of lean and how it impacts on sustainable construction. Better understanding of lean concepts by the construction industry can contribute to improvement in all aspect of sustainable construction. The concept of lean and sustainable construction both seeks to minimise waste, but this is achieved through different approaches. There is need for construction stakeholders to set their priorities before the start of a project for better integration of the two concepts. More emphasis should be laid on lean approach in sustainable construction framework. There should be more level of commitment and knowledge by an organisation in order to successfully implement and derive maximum benefits from the concept of lean and sustainability. However, the application of lean in sustainable construction is not only possible on the operational level; it could also be applied at the strategic level. Therefore, this study will go on to further present the application of lean and sustainability at the strategic level and also explore the benefits that can be achieved.

References

[1] Construction Industry Research and Information Association CIRIA, "Guide to sustainable procurement in construction. London" Construction Industry Research and Information Association C695, 2011.

[2] R. Sacks, B. Dave, L. J. Koskela, and R. L. Owen, "Analysis framework for the interaction between lean construction and building information modeling," Proceedings of the 17th annual conference of International Group for lean Construction, 2009.

[3] Construction Industry Research and Information Association CIRIA "Sustainable construction: company indicators", London: CIRIA, C563, 2001.
[4] L. Koskela, "An exploration towards a production theory and its application to construction," D. Tech. thesis, Helsinki University of Technology, 2000.
[5] P. Huovila, and L. Koskela, "Contribution of the Principles of Lean Construction to Meet the Challenges of Sustainable Development," In Formoso, C.T., ed. Proceedings of IGLC-6. Guaruja, Brazil, 1998.
[6] Department of the Environment, Transport and the Regions DETR, "Sustainable development: what it is and what you can do", DETR Green Ministers Report. 2000
[7] I. Nahmens, and L. H. Ikuma, "An Empirical Examination of the Relationship between Lean Construction and Safety in the Industrialised Housing Industry," Lean Construction Journal, vol. 5(1), pp.1-12, 2009.
[8] C. Koranda, W. Chong, C. Kim, J. Chou, and C. Kim, "An Investigation of the Applicability of Sustainability and Lean Concepts to Small Construction Projects," KSCE Journal of Civil Engineering, vol. 16(5), pp. 699-707, 2012.
[9] L. H. Forbes, S. M. Ahmed, and M. Barcala, "Adapting lean construction theory for practical application in developing countries," Proceedings of the first CIB W 107 International Conference: Creating a Sustainable Construction Industry in Developing Countries (Eds. Division of Building Technology, CSIR), Stellenbosch, South Africa, 11-13 November, 2000.
[10] P. Glavic, and R. Lukman, "Review of sustainability terms and their definitions," Journal of Cleaner Production, vol. 15 (18), pp. 1875-1885, 2007.
[11] D. P. Wyatt, "Recycling and Serviceability: The Twin Approach to Securing Sustainable Construction," In Proceedings of First International Conference of CIB TG 16 on Sustainable Construction, Tampa, Florida, 6-9 November, pp. 69- 78, 1994.
[12] R. C. Hill, and P.A Bowen, "Sustainable Construction Principles and a Framework for attainment," Construction Management and Economics, vol. 15(3), pp. 223-39, 1997.
[13] P. S. Brandon, "Sustainability in Management and Organisation: The key issues?," In Procedings, Conference on Cities and Sustainability:Sustainaing our Cultural Heritage, Kamdalama, Sri Lanka, Februaury 11-16, 2000.
[14] A. P. Sage, "Risk management for sustainable development," IEEE International Conference on Systems, Man and Cybernetics, vol. 5, pp. 4815-4819, 1998.
[15] G. H. Brundtland, "Our common future: Report of the World Commission on Environment and Development," Oxford University Press, Oxford, 1987.
[16] J. Fiskel, "Sustainabilty and resilience: towards a system approach," Sustainability: science, practice, and policy, vol. 2, pp. 1-8, 2006.
[17] L. Bourdeau, P. Huovila, R. Lanting, and A. Gilham, "Sustainable Development and the Future of Construction: A Comparison of Visions from various Countries," CIB Report 225, Rotterdam. 1998.
[18] Y. Rydin, U. Amjad, and M. Whitaker, "Environmentally sustainable construction: Knowledge and learning in London planning departments," Planning Theory and Practice, vol. 8 (3), 363-380, 2007.
[19] Construction Industry Environmental Forum CIEF, "Lean construction for sustainable business," Joint CIEF and CPN seminar held at the Centre for Construction Innovation, CUBE, 113-115 Portland Street, Manchester M1 6DW, 2009.
[20] WRI Report, "Hot climate, cool commerce: A service sector guide to green house gas management," World Resources Institute, Washington D.C, http://pdf.wri.org/hotclimatecoolcommerce.pdfaccesed 3/2/13, 2006.
[21] A. Mossman, "Why isn't the UK construction industry going lean with gusto?," Lean Construction Journal, vol. 5 (1), pp. 24-36, 2009.
[22] S. Sarhan, and A. Fox, "Trends and challenges to the development of a lean culture among UK construction organizations," http://www.iglc20.sdsu.edu/papers/wp-content/uploads/2012/07/116%20P%20015.pdf accessed on 02/3/13, 2012.
[23] E. Johansen, and G. Potter, "An experience of introducing last planner into a UK construction project,"http://www.leanconstruction.dk/media/17682/An%20Experience%20of%20Introducing%20Last%20Planner%20into%20a%20UK%20Construction%20Project.pdf accessed 5/4/13, 2003.
[24] R. Shah, and P. T. Ward, "Defining and developing measures of lean production," Journal of Operations Management, vol. 25 (4), pp. 785-805, 2007
[25] F. Picchi and A. Granja, "Construction sites: using lean principles to seek broader implementations," Proceedings of International Group of Lean Construction, 12th annual conference, Copenhagen, Denmark, August 3th – 5th, 2004

[26] T. Alves, and C. Tsao, "Lean Construction – 2000 to 2007," Lean Construction Journal, vol. 3 (1), pp. 46-70, 2007
[27] I. Pavez, and L. F. Alarcón, "Lean Construction Professional's Profile (LCPP): implementation in Chilean contractor organizations," Proceedings of International Group for Lean Construction 16th Annual Conference, Manchester, UK, July 14th – 20th, 2008,
[28] J. Petterson, "Defining lean production: some conceptual and practical issues," The TQM Journal, vol. 21 (2), pp.127-142, 2009.
[29] S. Green, and S. May, "Lean construction: arenas of enactment, models of diffusion, and the meaning 'leanness'," Building Research & Information, vol. 33 (6), pp.498- 511, 2005.
[30] K. B. Manrodt, K. Vitasek, and R. H. Thompson, "Lean practices in the supply chain, Jones Laselle 2008 Report," http://www.joneslanglasalle.com/Documents/JLL-LeanPracticesInSupplyChain.pdf accessed 4/4/13. 2008.
[31] H. G. Ballard, Y. W. Kim, J. W. Jang, and M. Liu, "Roadmap for Lean Implementation at the Project Level," CII Research Report No. 234-11 Construction Industry Institute, The University of Texas, Austin, TX, 409pp, 2007.
[32] Lean Construction Institute, http://www.leanconstruction.org.uk/what-is accessed 11/12/12, 2012.
[33] Z. Radnor, P. Walley, A. Stephens, and G. Bucci, "Evaluation of the lean Approach to Business Management and its use to Public sector". http://www.scotland.gov.uk/Resource/Doc/129627/0030899.pdf accessed 9/10/12, 2006
[34] Construction Industry Institute, "The Application of Lean manufacturing Principles to Construction". https://www.construction-institute.org/scriptcontent/more/rr191_11_more.cfm accessed 08/01/13, 2013.
[35] I. Nahmens, and L. H. Ikuma, "Effect of lean construction on sustainability of modular homebuilding," Journal of Architectural Engineering, vol. 18 (2), 2012
[36] IPCC, "Climate Change 2007: Synthesis Report", Intergovernmental Panel on Climate Change, available from http://www.ipcc.ch/ accessed10/05/2011, 2007.
[37] W. Peng, and L. Pheng, "Lean and green: emerging issues in the construction industry – a case study. EPPM, Singapore, 20-21 September, 2011.
[38] Y. Tan, L. Shen, and H. Yao, "Sustainable Construction Practice and Contractor's Competitiveness: A preliminary Study," Habitat International, vol. 35, pp. 225-230, 2011.
[39] S. M. A. Beheiry, W. K. Chong, and C. T. Haas, "Examining business impact of owner commitment to sustainability," Journal of Construction Engineering and Management, vol. 132 (4), pp. 384-392, 2006.
[40] G. Christini, M. Fetsko, C. Hendrickson, "Environmental Management systems and ISO 14001 certification for Construction Firms," ASCE Journal of Construction Engineering and Management, pp. 330-336, 2004.
[41] M. Hall and D. Purchase, "Building or bodging? Attitudes to sustainability in UK public sector housing construction development," Sustainable Development, vol. 14 (3), pp. 205-218, 2006.
[42] O. Ogunbiyi, A. Oladapo, and J. Goulding, "Lean procurement: the use of lean construction techniques in project value enhancement," Joint CIB W070, W092 & TG72 International Conference on Facilities Management, Procurement Systems, 23-25 January, 2012, University of Cape Town, Cape Town , South Africa, 2012.
[43] J. Bae, and Y. Kim, "Sustainable Value on Construction Projects and Lean Construction," Journal of Green Building, vol. 3 (1), pp. 156-167, 2008.
[44] A. Opoku, and C. Fortune, "The implementation of sustainable practices through leadership in construction organizations," In: Egbu, C. and Lou, E.C.W. (Eds.) Procs 27th Annual ARCOM Conference, 5-7 September 2011, Bristol, UK, Association of Researchers in Construction Management, 1145-1154, 2011.
[45] S. Vinodh, K. R. Arvind, and M. Somanathan, "Tools and techniques for enabling sustainability through lean initiatives," Journal of Technology and Environmental policy, vol. 13, pp. 469-479, 2010.
[46] A. Lapinski, M. Horman, and D. Riley, "Lean processes for sustainable project delivery," Journal of Engineering and Management, vol. 132 (10), pp. 1083-1091, 2006.
[47] D. Riley, V. Sanvido, M. J Horman, M. McLaughlin, and D. Kerr, "Lean and Green: The Role of Design-Build Mechanical Competencies in the Design and Construction of Green Buildings," In Tommelein, T. (ed.), Proceedings of the 2005 Construction Research Congress, April 5-7, San Diego, CA, 2005
[48] M. J Horman, D. Riley, M. H. Pulaski, and C. Leyenberger, "Lean and Green: Integrating Sustainability and Lean Construction," CIB World Building Congress, Toronto, Canada, 2004.

[49] K. Birdi, C. Clegg, M. Patterson, A. Robinson, C. B. Stride, T. B. Wall, and S. J Wood, "The impact of human resource and operational management practices on company productivity: A longitudinal study," Personal Psychology, vol. 61 (3), pp. 467-501, 2008.

[50] X. Mao, and X. Zhang, "Construction process reengineering by integrating lean principles and computer simulation techniques," Journal of Construction Engineering and Management, vol. 134 (5), pp. 371-381, 2008.

[51] P. E. Eriksson, "Improving construction supply chain collaboration and performance: a lean construction pilot project," Supply Chain Management: An International Journal, vol. 15 (5), pp. 394-403, 2010

THE EFFECT OF AMORPHOUS SILICA RESIDUE IN THE PRODUCTION OF CONCRETE

Elvis M. Mbadike[1*], N. N. Osadebe[2]

[1] Department of Civil Engineering, Michael Okpara University of Agriculture, Umudike, Umuahia, Nigeria
[2] Department of Civil Engineering, University of Nigeria, Nsukka

*Corresponding E-mail: elvis_mbadike@yhaoo.co.uk

Abstract

In this research work, the effect of amorphous silica residue (ASR) in the production of concrete was investigated. A mix proportion of 1:1.9:3.9 with water/cement ratio of 0.48 was used. The percentage replacement of Ordinary Portland Cement (OPC) with amorphous silica residue was 0%, 5%, 10%, 20% and 30%. Concrete cubes of 150mm x 150mm x 150mm and concrete beams of 150mm x 150mm x 600mm of OPC/ASR were cast and cured at 3, 7, 28, 60 and 90 days. At the end of each hydration period, the three concrete cube and beams for each hydration period were crushed and their average compressive and flexural strength recorded. A total of seventy five (75) concrete cubes and seventy five (75) concrete beams were cast. The result of the compressive strength test for 5-30% replacement of cement with amorphous silica residue ranges from 12.78-38.16N/mm^2 while the control test (0% replacement) ranges from 10.86-26.04N/mm^2. The result of the flexural strength test for the same replacement level of cement with amorphous silica residue ranges from 2.29-11.69N/mm^2 while the control test ranges from 2.14 – 7.80N/mm^2. The initial setting time of OPC/ASR for 5-30% replacement level of cement with amorphous silica residue ranges from 37-53mins while the final setting time ranges from 408-573mins. The initial and final setting time of the control test is 58mins and 580mins respectively. Relevant literature has been cited to justify this research work. The main objective of this work is to determine the effect of amorphous silica residue on the setting time, compressive strength and flexural strength of concrete produced with it.

Keywords: *Compressive strength, flexural strength, amorphous silica residue, setting time*

1.0 Introduction

In the last few decades, many researches have been directed towards the utilization of waste materials in cement and concrete [1]. Not surprisingly, there has been increasing interest in the use of amorphous silica residues as supplementary cementing materials, even though little literature is available on this topic [2].

The amorphous silica residue is a waste material from the technological process used in the manufacture of amorphous silica product. The chemical reactions occurring in the process are complex, but one of the reactions includes the formation of amorphous silica slurry, some of which passes through a filler press in the form of very tiny particles. This amorphous silica residue which is thus a waste material of the process, is then sent to a setting tank to flocculate and finally to landfill disposal.

Initial work conducted by the civil and environmental department of the Louisiana State University, USA was primarily concerned with the characteristics of amorphous silica residue in comparison with a typical silica fume [3]. It is envisaged that the pozzolanic activity of amorphous silica residue would be much higher due to its higher specific surface areas and amorphous characteristics.

This study contributes to the development of a methodology for assessing concrete manufactured from waste. The methodology is based on the study of concrete containing experimental waste (Amorphous Silica Residue) [4].

The durability and the environmental impact of concrete are closely connected to its transport properties which control the kinetics of the penetration of water and aggressive agents into concrete [5]. The movement of chemical species within the material and the leaching of certain chemicals are also closely linked to concrete diffusivity [6]. Finally, the strength characteristics of concrete containing increasing levels of amorphous silica residue were studied to identify the effect of amorphous silica residue on concrete produced with it.

2.0 Literature Review

The chemical composition is an important factor that affects the activity of pozzolanic material. Potential pozzolanic materials contain substantial amount of SIO_2, AL_2O_3 and Fe_2O_3. [7] reported that the total SIO_2, AL_2O_3 and Fe_2O_3 were found to be a good indication of the pozzolanic activity. [8] has set a minimum value of 70% for the sum of SIO_2 plus AL_2O_3 plus Fe_2O_3 of the total compounds that make up the material. The Indian standard stated that for good pozzolans, the CaO content should be not be greater than 10%, and the in addition, the sum of SIO_2 plus AL_2O_3 should be greater than 50% [9]. Generally, amorphous silica reacts with calcium hydroxide more rapidly than does silica in the crystalline form [10].

Bulk chemical compositions and several comparative fundamental properties of both materials are listed in table 2 and 1 (Refer to section 4.0). The chemical properties of both AS residue and SF shows that the materials contain mainly SIO_2 and others in small percentage as can be seen in table 2. Table 1 shows some of the physical properties of both materials. The percentage amorphousness of AS residue is 71 while SF is 23. This shows the extent of formation of calcium silicate hydrate when ASR is used in concrete production. The amorphous characteristic (reactive property) makes ASR to have a very high pozzolanic property and therefore can be used to replace cement at a certain replacement level during concrete production. Amorphous Silica Residue (ASR) can be used where high compressive and flexural strength in concrete is required. The advantage of using the material in concrete is that it reduces the setting time of cement paste. Another advantage is that it can be used in a cold weather concreting. The disadvantage is that the concrete produced with it may have premature setting.

3.0 Methodology

Concrete mixtures with five levels of amorphous silica residue ranging from 5-30% and concrete mixtures with no amorphous silica residue were investigated to determine their effect on compressive and flexural strength. The mixtures were labeled M0, M5, M10, M20 and M30 with the different amorphous silica (SR) residue replacement percentages represented by the final digits in the label. The mixtures were proportioned for a target cube strength of $39N/mm^2$ and had a cementitious material content of $320kg/m^3$, a fine aggregate content of $615kg/m^3$, a coarse aggregate content of $1260kg/m^3$, and a water cement ratio of 0.48. The coarse aggregate used is granite and a clean river sand is used as fine aggregate. Both aggregates conforms to [11] and [12] respectively for coarse and fine aggregates, while the cement conforms to [13].

Tests to determine setting time, compressive and flexural strength were carried out in this study. For the tests, amorphous silica residue was used to replace 0 to 30% of cement by weight. For the compressive strength test, 150mm x150mmx150mm concrete cube specimen were used while 150mmx150mmx600mm beam moulds were used for flexural strength. A total of 75 specimens for concrete cubes and 75 specimens for concrete beams were cast and cured in water at room temperature in the laboratory for 3,7,28,60 and 90 days. At the end of each hydration period, three specimen for each were tested for compressive and flexural strength and the average recorded.

The setting time was determined in the laboratory using Vicat apparatus. For the control test (0% replacement), 200g of cement was used with 96g of water during the experiment to form a cement paste. The paste was then placed inside the Vicat mould and finally placed on the Vicat apparatus. Before the placement of the paste on the apparatus, the initial setting pin was fixed on the apparatus for the initial setting time. The apparatus is calibrated in millimeters. For the initial setting time, the initial setting pin was dropped on the paste to 5±1mm calibration mark on the apparatus. The initial setting time was then recorded starting from the time water was added to cement to the time the dropping of the pin was 5±1mm mark on the apparatus [14, 15].

Similarly, the final setting time was recorded using the final setting pin. The final setting time was taken when only the inner pin makes a mark on the paste when allowed to drop freely. The final setting time was then recorded starting from the time water was added to the cement to the time the inner pin of the final setting pin makes a mark on the paste. The experiment was repeated with 5%, 10%, 20% and 30% replacement of cement with amorphous silica residue.

4.0 Results and Discussions

Table 1 shows the result of the comparison of the amorphous silica residue and silica fume (Crystalline). The comparison was carried out in the Civil Engineering laboratory of the University of Nigeria, Nsukka.

Table 1: Comparison of Amorphous silica Residue and Silica Fume (Crystalline)

Properties	AS Residue	SF (crystalline)
Colour	White	Dark grey
Partial size (mm)	26	192
Specific surface Area (m^2/kg)	94700	23000
Pozzolanic Reactivity index (%)	88	30
Amorphousness (%)	71	23

Note: The comparison was carried out in the Civil Engineering Laboratory, University of Nigeria, Nsukka.

Table 2 shows the chemical analysis (characterization) of AS residue. The analysis was carried out in the Project Development Institute, Enugu State, Nigeria. The result shows that AS residues contain mainly SIO_2 (97.5%) and other oxides. P_2O_5 was not detected during the analysis. the analysis of silica fumes shoes also it contains mainly SIO_2 (95%) and other oxides. P_2O_5 was detected to be 1.1%.

Table 2: Chemical Analysis of AS and SF

Oxide (%)	AS Residue	SF
SIO_2	97.5	95
AL_2O_3	0.18	1.31
CaO	0.07	0.91
Fe_2O_3	0.05	0.18
MgO	0.12	0.39
Na_2O	0.74	0.19
K_2O	0.04	0.87
SO_4	1.3	-
P_2O_5	-	1.1

Note: The analysis was carried out in Project Development Institute, Enugu State, Nigeria.

The effect of AS residue on setting time can be seen in table 3, setting of the paste was accelerated by AS residue which results in the shortening of the setting time. The higher the setting time, the lower the strength of concrete and vice versa. For example, the time for initial and final setting of concrete with 30% AS residue was shortened to about 1½ that of the control test (0% replacement). the setting time of 5 – 30% replacement of cement with ASR ranges from

37-53mins and 408 – 573 for initial and final setting time respectively while the control test is 58mins and 580mins for initial and final setting time.

Table 3: Initial and final setting time of OPC/ASR paste

% Cement replacement	Initial setting time (mins)	Final setting time (mins)
0	58	580
5	53	573
10	49	522
20	46	496
30	37	408

The initial and final setting time of the control test (0% replacement) is 58mins and 580mins respectively; while 5-30% replacement of cement with amorphous silica residue ranges from 37-53mins for initial setting time and 408-573mins for the final setting time. The setting time decreases with the increase in the percentage replacement of cement with amorphous silica residue.

The compressive strength of concrete produced at each age (3,7,28,60 and 90 days) is shown in the tables 4-8 (Appendix-A) and that of the flexural strength is shown in table 9-13 (Appendix-B) compared to the control test, those made with AS residue showed greater strength both for the compressive and flexural strength respectively. The increasing strength was observed at curing periods as early as three days. Accelerated strength development started from seven days to ninety days for all concrete tested for both compressive and flexural strength.

In table 4, the compressive strength obtained when 100% cement (0% replacement) is used and cured at 3, 7, 28, 60 and 90days ranges from 10.56 – 26.04 N/mm^2. Table 5 shows that the compressive strength obtained by replacing cement with 5% of ASR ranges from 12.78 – 28.03N/mm^2 at the same hydration period. Table 6 shows the 10% replacement of cement with ASR. The result of the compressive strength ranges from 15.05 – 32.83 N/mm^2 while table 7 ranges from 18.64 – 35.16 N/mm^2 for 20% replacement of cement with ASR at 3, 7, 28, 60, and 90 days hydration period. The compressive strength of concrete obtained at 30% replacement of cement with ASR ranges from 23.69 – 38.16N/mm^2 at the same hydration period. Table 9 shows that the flexural strength of concrete obtained at 0% replacement of cement with ASR ranges from 2.14 – 7.80N/mm^2 while that of 5, 10, 20, and 30% replacement of cement with ASR in table 10, 11, 12 and 13 ranges from 2.29 – 8.12N/mm^2 , 2.92 – 10.20N/mm^2, 3.21 – 10.69 N/mm^2 and 4.16 – 11.69 N/mm^2 respectively for 3,7, 28,60 and 90 days hydration period.

The result of the compressive strength for 5-30%, replacement of cement with amorphous silica residue ranges from 12.78 – 38.16N/mm^2 while that of control test ranges from 10.86-26.64N/mm^2. The result shows that there is increase in the strength of concrete produced as the percentage replacement level of cement with amorphous silica residue increases.

Table 9-13 shows the result of the flexural strength of concrete with the same percentage replacement level of cement with amorphous silica residue. The result shows that flexural strength increases with increase in hydration period. The result of the flexural strength for 5-30%, replacement of cement with amorphous silica residue ranges from 2.29-11.69N/mm^2 while that of the control test ranges from 2.14-7.80N/mm^2.

The result shows that flexural strength also increases with increase in the replacement level of cement with amorphous silica residue (ASR).

5.0 Conclusions

The conclusion of this study can be summarized as follows:

a. Due to their high specific surface area and amorphous characteristics, amorphous silica residues have remarkably high pozzolanic activity and can be used as supplementary cementing materials.

b. With the addition of amorphous silica residues the setting time of the paste is shortened. This behaviour may be due to the early formation of a large amount of calcium silicate hydrate gel from hydration of tricalcium silicate and from the reaction between amorphous silica residues and calcium hydroxide.

c. The addition of amorphous silica residues can increase both the compressive and flexural strength of concrete produced.

d. The increased compressive and flexural strength may be due to the fact that calcium hydroxide diminished or wholly disappear and calcium hydrate increased with the addition of the amorphous silica residues.

e. Strength development in concrete increases with increase in hydration period.

References

[1] R. Helmth, “Fly ash in cement and concrete, Portland Cement Associate,” Journal of Cement and Concrete Research, vol. 30, pp. 201-204, 1987.

[2] Malhorta, “Superplasticizers and other chemical admixtures in concrete,” ACI Journal, vol. 32, pp. 100-101, 1987.

[3] L. Wang, R. K. Seals, and A. Roy, “Investigation of utilization of amorphous silica residues as supplementary cementing materials,” Journal of Concrete Research, vol. 13, pp. 85-88, 2001.

[4] B. Kessler, M. Rollet, and F. Sorrentino, “Microstructure of cement paste as incinerator ash host,” Proceedings of 1st international symposium on cement industry solution to waste management, Calgary, pp. 235-251, 1992.

[5] P. Pimienta, S. Remond, N. Rodrigues, and J. P. Bournazel, “Assessing the properties of mortars containing municipal solid waste incineration fly ash,” International congress, creating with concrete, University of Dundee, pp. 319-326, 1999.

[6] S. Remond, P. Pimienta, and D. P. Bentz, “Effects of the Incorporation of municipal solid waste fly ash in cement paste and mortars,” Journal of Cement and Concrete Research, vol. 10, pp. 12-14, 2002.

[7] A. A. Al-Rawas, A. W. Hago, T. C. Corcoran, and K. M. Al-Ghafri, “Properties of Omani Artificial Pozzolans,” Applied clay sciences, vol. 13, pp. 275-292, 2008.

[8] American Society for Testing and Materials, ASTM C618, “Specifications for fly ash and Raw or calcined natural pozzolan for use as a mineral admixture in Portland cement concrete,” 2003.

[9] R. S. Vashney, “Concrete Technology,” Oxford IBH Company, 2006.

[10] O. R. Werner, “Report on the Use of Natural pozzolans in concrete,” ACI Materials Journal, Report of ACI, Committee 232, vol. 91 (4), pp. 410-426, 2009.

[11] British Standards institution BS 877, “Foamed or expanded blast furnace slag lightweight aggregate for concrete,” London, 1967.

[12] British Standards Institution BS 3797, “Lightweight aggregate for concrete,” London, 1964.

[13] British Standards institution BS12, “Specification for Ordinary Portland Cement,” London, England, 1978.

[14] A. M. Neville, “Properties of Concrete,” 4th edition, Pitman Publishing Company Ltd., 1995.

[15] A. M. Neville, and J. Brooks, “Concrete technology,” 3rd edition, Pearson Publishers, India, 1995.

APPENDIX-A

Table 4: Result of compressive strength obtained with 0% replacement of cement with amorphous silica residue

Cube size (mm)	Age of Cube(days)	Test load (kN)	Compressive strength (N/mm^2)	Average compressive strength (N/mm^2)
150x150x150	3	248	11.02	
150x150x150	3	215	9.56	10.86
150x150x150	3	270	12.00	
150x150x150	7	341	15.16	
150x150x150	7	330	14.67	14.64
150x150x150	7	317	14.09	
150x150x150	28	378	16.80	
150x150x150	28	426	18.93	17.58
150x150x150	28	383	17.02	
150x150x150	60	401	17.82	
150x150x150	60	490	21.78	19.87
150x150x150	60	450	20.00	
150x150x150	90	500	22.22	
150x150x150	90	570	25.33	26.04
150x150x150	90	688	30.58	

Table 5: Result of compressive strength obtained with 5% replacement of cement with amorphous silica residue

Cube size (mm)	Age of Cube(days)	Test load (kN)	Compressive strength (N/mm^2)	Average compressive strength (N/mm^2)
150x150x150	3	273	12.13	
150x150x150	3	315	14.00	12.78
150x150x150	3	275	12.22	
150x150x150	7	340	15.11	
150x150x150	7	300	13.33	14.67
150x150x150	7	350	15.56	
150x150x150	28	426	18.93	
150x150x150	28	406	18.04	18.83
150x150x150	28	439	19.51	
150x150x150	60	409	18.18	
150x150x150	60	500	22.22	22.18
150x150x150	60	588	26.13	
150x150x150	90	496	22.04	
150x150x150	90	684	30.40	28.03
150x150x150	90	712	31.64	

Table 6: Result of compressive strength obtained with 10% replacement of cement with amorphous silica residue

Cube size (mm)	Age of Cube(days)	Test load (kN)	Compressive strength (N/mm^2)	Average compressive strength (N/mm^2)
150x150x150	3	350	15.56	
150x150x150	3	299	13.29	15.05
150x150x150	3	367	16.31	
150x150x150	7	413	18.36	
150x150x150	7	501	22.27	19.84
150x150x150	7	425	18.89	
150x150x150	28	530	23.56	
150x150x150	28	591	26.27	24.31
150x150x150	28	520	23.11	
150x150x150	60	596	26.49	
150x150x150	60	549	24.40	26.07
150x150x150	60	615	27.33	
150x150x150	90	694	30.84	
150x150x150	90	789	35.07	32.83
150x150x150	90	733	32.58	

Table 7: Result of compressive strength obtained with 20% replacement of cement with amorphous silica residue

Cube size (mm)	Age of Cube(days)	Test load (kN)	Compressive strength (N/mm^2)	Average compressive strength (N/mm^2)
150x150x150	3	442	19.64	
150x150x150	3	396	17.60	18.64
150x150x150	3	420	18.67	
150x150x150	7	480	21.33	
150x150x150	7	500	22.22	21.55
150x150x150	7	475	21.11	
150x150x150	28	580	25.78	
150x150x150	28	630	28.00	26.30
150x150x150	28	565	25.11	
150x150x150	60	682	30.31	
150x150x150	60	720	32.00	32.77
150x150x150	60	810	36.00	
150x150x150	90	800	35.56	
150x150x150	90	897	39.87	35.16
150x150x150	90	676	30.04	

Table 8: Result of compressive strength obtained with 30% replacement of cement with amorphous silica residue

Cube size (mm)	Age of Cube(days)	Test load (kN)	Compressive strength (N/mm^2)	Average compressive strength (N/mm^2)
150x150x150	3	540	24.00	
150x150x150	3	549	24.40	23.69
150x150x150	3	510	22.67	
150x150x150	7	580	25.78	
150x150x150	7	591	26.28	27.43
150x150x150	7	680	30.22	
150x150x150	28	690	30.67	
150x150x150	28	578	26.69	29.49
150x150x150	28	700	31.11	
150x150x150	60	820	36.44	
150x150x150	60	760	33.78	36.74
150x150x150	60	900	40.00	
150x150x150	90	905	40.22	
150x150x150	90	784	34.80	38.16
150x150x150	90	888	39.47	

APPENDIX-B

Table 9: Result of flexural strength obtained with 0% replacement of cement with amorphous silica residue

Bean size (mm)	Age of beam (days)	Test load (kN)	Modulus of rupture (N/mm^2)	Average modulus of rupture (N/mm^2)
150x150x600	3	8.52	1.54	
150x150x600	3	12.84	2.28	2.14
150x150x600	3	14.68	2.61	
150x150x600	7	18.90	3.36	
150x150x600	7	18.60	3.31	3.20
150x150x600	7	16.56	2.94	
150x150x600	28	35.00	6.22	
150x150x600	28	26.88	4.78	5.43
150x150x600	28	29.80	5.30	
150x150x600	60	35.10	6.24	
150x150x600	60	41.00	7.29	6.52
150x150x600	60	34.00	6.04	
150x150x600	90	44.40	7.89	
150x150x600	90	45.10	8.02	7.80
150x150x600	90	42.10	7.48	

Table 10: Result of flexural strength obtained with 5% replacement of cement with amorphous silica residue

Bean size (mm)	Age of beam (days)	Test load (kN)	Modulus of rupture (N/mm^2)	Average modulus of rupture (N/mm^2)
150x150x600	3	12.88	2.29	
150x150x600	3	11.00	1.96	2.29
150x150x600	3	14.80	2.63	
150x150x600	7	16.90	3.00	
150x150x600	7	14.50	2.58	2.86
150x150x600	7	16.91	3.01	
150x150x600	28	35.66	6.34	
150x150x600	28	31.00	5.51	5.91
150x150x600	28	33.10	5.88	
150x150x600	60	39.80	7.16	
150x150x600	60	42.80	7.61	7.37
150x150x600	60	41.30	7.34	
150x150x600	90	50.00	8.88	
150x150x600	90	47.00	8.36	8.12
150x150x600	90	40.00	7.11	

Table 11: Result of flexural strength obtained with 10% replacement of cement with amorphous silica residue

Bean size (mm)	Age of beam (days)	Test load (kN)	Modulus of rupture (N/mm^2)	Average modulus of rupture (N/mm^2)
150x150x600	3	15.10	2.68	
150x150x600	3	15.60	2.77	2.92
150x150x600	3	18.70	3.32	
150x150x600	7	18.96	3.37	
150x150x600	7	23.00	4.09	4.04
150x150x600	7	26.20	4.66	
150x150x600	28	41.00	7.29	
150x150x600	28	39.30	6.99	7.01
150x150x600	28	38.00	6.76	
150x150x600	60	49.30	8.76	
150x150x600	60	43.00	7.64	8.19
150x150x600	60	46.00	8.18	
150x150x600	90	56.10	9.97	
150x150x600	90	55.00	9.78	10.20
150x150x600	90	61.00	10.84	

Table 12: Result of flexural strength obtained with 20% replacement of cement with amorphous silica residue

Bean size (mm)	Age of beam (days)	Test load (kN)	Modulus of rupture (N/mm^2)	Average modulus of rupture (N/mm^2)
150x150x600	3	15.80	2.81	
150x150x600	3	19.60	3.48	3.21
150x150x600	3	18.80	3.34	
150x150x600	7	22.40	3.92	
150x150x600	7	28.00	4.98	4.14
150x150x600	7	19.80	3.52	
150x150x600	28	40.76	7.25	
150x150x600	28	44.10	6.95	7.08
150x150x600	28	39.60	7.04	
150x150x600	60	49.70	8.84	
150x150x600	60	54.00	9.60	9.39
150x150x600	60	54.80	9.74	
150x150x600	90	59.60	10.60	
150x150x600	90	57.00	10.13	10.69
150x150x600	90	63.80	11.34	

Table 13: Result of flexural strength obtained with 30% replacement of cement with amorphous silica residue

Bean size (mm)	Age of beam (days)	Test load (kN)	Modulus of rupture (N/mm^2)	Average modulus of rupture (N/mm^2)
150x150x600	3	23.10	4.11	
150x150x600	3	25.30	4.50	4.16
150x150x600	3	21.80	3.88	
150x150x600	7	36.20	6.44	
150x150x600	7	39.70	7.06	6.76
150x150x600	7	38.10	6.77	
150x150x600	28	47.20	8.39	
150x150x600	28	41.00	7.29	8.21
150x150x600	28	50.30	8.94	
150x150x600	60	52.96	9.42	
150x150x600	60	53.00	9.42	9.63
150x150x600	60	56.60	10.06	
150x150x600	90	61.30	10.90	
150x150x600	90	67.00	11.91	11.69
150x150x600	90	68.90	12.25	

17

Shear Behaviour of RC T-Beams with Externally Bonded Discrete CFF Strips – A Experimental and Finite Element Study

J. Jayaprakash[1,*], Abdul Aziz Abdul Samad [2], Mustafasanie M. Y[3], Anwar M. P[4], Mohamed A. A[5]

[1, 4 5] Department of Civil Engineering, The University of Nottingham, Malaysia
[2] Faculty of Civil and Environmental Engineering, Universiti Tun Hussein Onn, Malaysia
[3] School of Civil Engineering, Universiti Sains Malaysia, Malaysia

*Corresponding E-mail: J.Jayaprakash@nottingham.edu.my (or) j_jaypee2000@yahoo.com

Abstract

The application of fibre reinforced polymer (FRP) composites for retrofitting and strengthening of existing reinforced concrete (RC) structures has fascinated the attention of researchers and engineers in the recent decades. This paper presents the results of experimental and finite element (FE) investigation of shear behaviour of reinforced concrete T-beams repaired with externally bonded bi-directional discrete carbon fibre fabric (CFF) strips. The reinforced concrete T-beams were tested under four point bending system to investigate the performance of CFF shear strengthening scheme in terms of ultimate load carrying capacity. These beams were modelled using LUSAS software. To evaluate the behaviour of the simulated models, the predicted results were compared with the experimental results. The experimental results show that the gain in shear capacity of the CFF repaired beams ranged between 20% and 40% over the control beam. Thus, it can be concluded that the externally bonded CFF strips significantly increased the shear capacity of CFF repaired beams. It was generally observed that the developed FE model shows better agreement with the experimental results. The results of load-deflection profile, cracking pattern, modes of failure, and strain distribution in discrete CFF strips are presented.

Keywords: *CFF Strips, Concrete, Beam, Shear, Repair*

1.0 Introduction

Fibre Reinforced Polymer (FRP) has been effectively used as an external reinforcement to enhance the structural performance of deficient or damaged concrete structures. The demand to use the FRP fabrics or sheets is due to its better characteristics such as high strength in the direction of fibres, non-corrosive, high stiffness, and resistance to acids, and chemicals [1-2]. This plate bonding technique can be used to increase the load carrying capacity of the reinforced concrete members such as columns, beams, slabs, and masonry walls.

Shear failure of reinforced concrete beam is catastrophic and could occur abruptly without any advance warning. However, the addition of Fibre Reinforced Polymer composites is very complex because of the brittle nature of the material. A number of experimental investigations [3-8] have been studied the shear strengthening behaviour of reinforced concrete beams with the application of continuous carbon fibre sheets. It was found that the externally bonded carbon fibre reinforcement significantly increased the shear capacity and stiffness of the beam. Some researchers [2, 9-11] have investigated the effectiveness of the strengthened beams using discrete carbon fibre strips. The result showed that the discrete carbon shear strips increased the shear capacity of the beams.

Limited researchers have attempted to simulate the behaviour of FRP strengthened beams using finite element numerical analysis [12]. Arduini et al [13] simulated the behaviour and failure mechanism of strengthened RC beams using finite element analysis. The FRP plates were modelled using two dimensional plate elements, however the crack patterns were not predicted. Kachlakev et al [14] performed non-linear finite element analysis to model the uncracked reinforced concrete FRP strengthened beams using ANYSIS software. The smeared crack model was used to predict the non-linear behaviour of the concrete material. The stress-strain relationship of concrete was assumed to be linearly elastic up to the ultimate tensile strength. The steel rebars were assumed to be perfectly elastic-plastic material and identical in tension and compression. The fibre sheet was taken as orthotropic and transversely isotropic material. He also assumed that the bond between concrete and materials was perfect without any bond slip. From the analysis, it was found that the predicted results showed good agreement in comparison to the experimental results. Similarly, Santhakumar and Chandrasekaran [15] have developed a finite element model to study the shear behaviour of rectangular CFF strengthened beams using ANSYS software. The predicted load deflection profile was validated with the experimental results of Norris et al [16]. The assumptions were similar to the Kachlakev et al [14] finite element model. In the analysis, Newton-Raphson Equilibrium Iteration was employed at each increment of load. The numerical results agreed well with the experimental results.

Literature review reported that most of numerical analysis have dealt with shear behaviour of the strengthened rectangular beams with continuous wrapping system [14-15]. However, the reinforced concrete T-beam strengthened with discrete CFF strips has been limited. Moreover, some worthwhile parameters such as strain in fibres and cracking patterns have not been widely addressed.

This paper presents the results of experimental and finite element (FE) study of shear behaviour of reinforced concrete T-beams repaired with externally bonded bi-directional discrete Carbon Fibre Fabric (CFF) strips. The analysis used LUSAS Finite Element software. To evaluate the behaviour of the simulated models, the modelled results were compared with the experimental results. The numerical results of load-deflection profile, cracking pattern, modes of failure, and strain distribution in discrete CFF strips are presented.

2.0 Experimental Beams and Material Properties

This study used the experimental results of control (TS1a) and repaired CFF beams (TS1-1 and TS1-2) of the author [9]. These specimens were of 2980mm span with flange and web cross sections of 400mm x 100mm and 240mm x 120mm, respectively. Fig. 1 shows the cross section and internal longitudinal and transverse reinforcement details of the reinforced concrete T-beams. Beams TS1-1 and TS1-2 were precracked and repaired externally with discrete CFF strips spaced at 150 mm and 200 mm centre to centre, respectively. The width of CFF strip was 80mm with a orientation of 90 degree to the longitudinal axis of the beam. The external CFF reinforcement details are shown in Fig. 2. Tables 1 and 2 show the material properties of concrete & internal steel rebars and epoxy and discrete CFF strip reinforcement, respectively.

These beams were tested under four point bending system. The control beam was subjected to loading to develop precracks followed by unloading to zero and then reloaded for failure. However, the repaired beams were loaded for two cycles of loading to develop and widening of precracks prior to the application of external discrete CFF strip reinforcement (i.e. precracking phase). Subsequently, the repaired specimens were loaded to failure without any intervention.

Table 1. Details of material properties of steel and concrete

Specimen	Span (mm)	Compressive strength of concrete (N/mm^2)	Tensile strength of steel rebars (N/mm^2)		
			Stirrup 8mm	Tensile 16mm	Compression 12mm
TS1a, TS1-1 and TS1-2	2980	16.70	620.31	311.22	565.42

Table 2. Material properties of CFF composites [17]

External Discrete CFF Strips	Thickness (mm)	Tensile strength (N/mm^2)	Modulus of Elasticity (N/mm^2)
U-strips (0/90) Bi-directional	0.09	3,800	230,000
Epoxy	-	30	3800

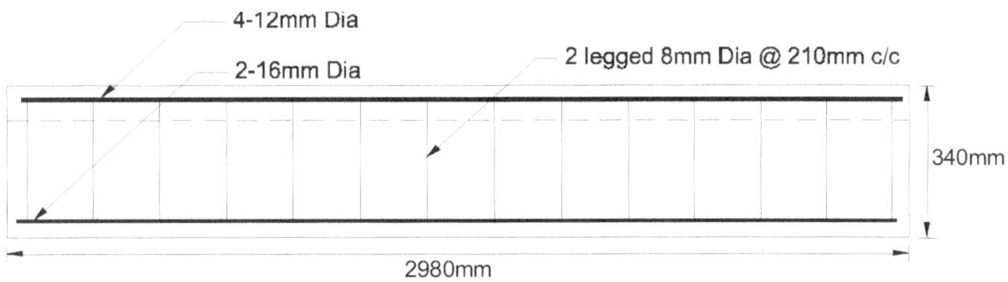

(a) Reinforcement details of T-beams

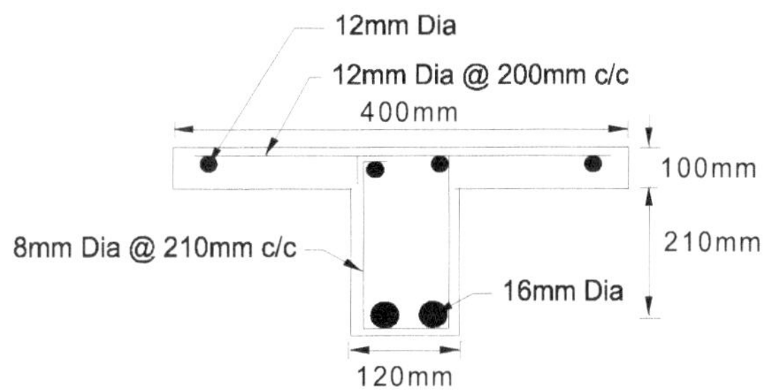

(b) Cross-section

Figure 1. Reinforcement and cross-section details of T- beam

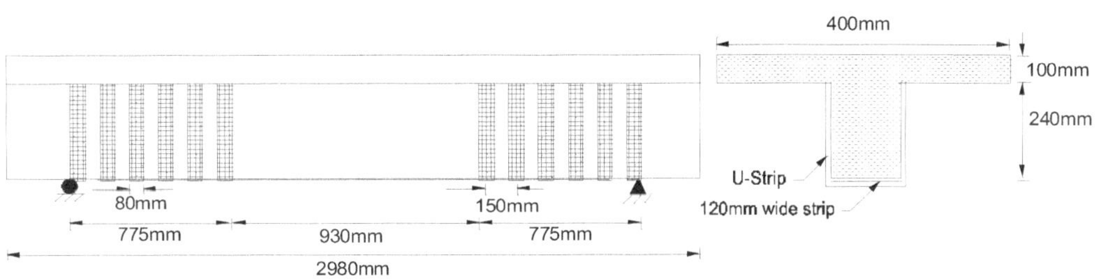

(a) CFF discrete U-strip with spacing of 150 mm for repaired beam TS1-1

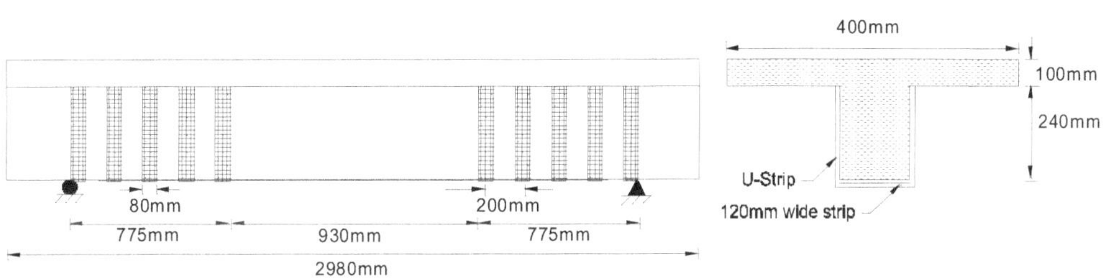

(b) CFF discrete U-strip with spacing of 200 mm for repaired beam TS1-2

Figure 2. External CFF discrete strip reinforcement details of repaired beams

3.0 Finite Element Analysis

Concrete: Solid 82 elements were used to model the concrete. The elements are hexahedral in shape with quadratic interpolation. The stress strain curve of concrete in compression was linearly elastic up to 30 percent of maximum compressive strength. After the maximum compressive strength, the curve descends into a softening region and eventually crushing failure occurs at an ultimate strain [18]. This study used a perfectly plastic relationship instead of compressive strain-softening curve [14, 19]. Under uniaxial tension, the material was assumed to be linearly elastic up to the tensile strength. The stress-strain relationship of concrete is shown Fig. 3(a). The poison's ratio and ultimate strain values for concrete were taken as 0.2 and 0.035, respectively. The elastic modulus and tensile strength of the concrete were calculated using Eqns. 1 and 2 as recommended in AS3600: 2001 [20] (cited in Pham and Al-Mahaidi [21]).

$$E_c = 5050\sqrt{f_c'} \tag{1}$$

$$f_{ct}' = 0.4\sqrt{f_c'} \tag{2}$$

The variables E_c, f'_c, and f'_{ct} represent modulus of elasticity, compressive strength, and tensile strength of concrete, respectively.

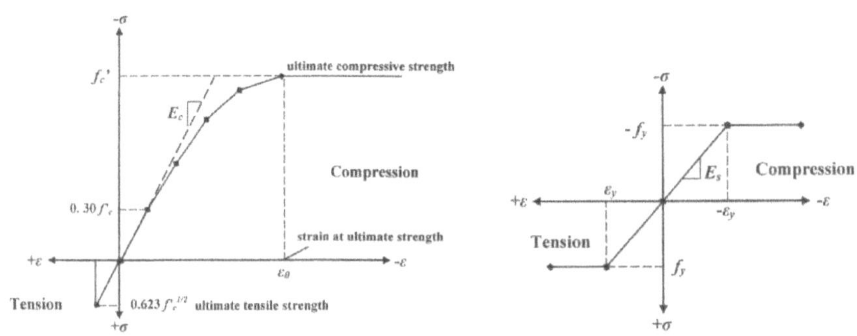

(a) (b)

Figure 3. Stress-strain curve for (a) concrete [14, 19] and (b) steel rebar [19]

Steel Reinforcement: The steel rebar was modelled as line element. Each division of line element was same as mesh element. This rebar was assumed to be elastic-perfectly plastic and identical in tension and compression [19, 22]. The stress-strain relationship of steel rebar is shown in Fig. 3(b). Furthermore, the bond between the rebar and concrete was assumed to be perfect (i.e.) no loss of bond slip [22]. The values of elastic modulus and poison's ratio of steel rebars were taken as 200,000MPa and 0.3, respectively.

Discrete CFF Strip Reinforcement: The CFF composite reinforcement used surface element. A linear elastic isotropic material was adopted for external discrete CFF strip reinforcement [23]. The bond between the discrete CFF strips and concrete was assumed to be perfect, whereas the bond interface was not considered in this model [14 15].

A half of the full beam was modelled by taking the advantage of symmetry of section and loading pattern with proper boundary condition [14]. It was modelled as uncracked beam but the precracking phase was ignored. A lagrangian isoparametric equation was employed to represent the geometric non-linearity of the model. Fig. 4 portrays the typical finite element model of the repaired beams with discrete CFF strips.

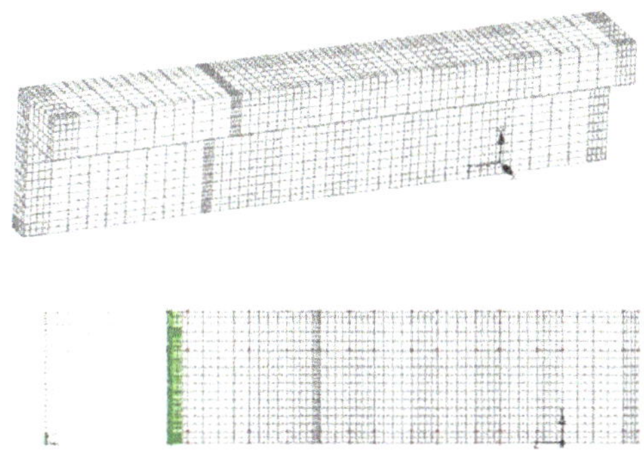

Figure 4. Elevation view of geometry model of repaired beam TS1-1

4.0 Results and Discussion

4.1 Experimental Observation

The control beam TS1a observed a diagonal shear crack at a load of 68kN. As the applied increased, these observed shear cracks propagated along the bottom of the flange towards the loading point. Moreover, some flexural cracks from the constant moment zone grew into the flange. The failure mode of the control specimen was controlled by shear at a peak load of 134.74 kN. Fig. 5(a) portrays the shear failure pattern of the control beam TS1a.

Beams TS1-1 and TS1-2 were precracked similar to the control specimen TS1a. The observed flexural and shear cracks were almost identical as in the beam TS1a. Subsequently, these beams TS1-1 and TS1-2 were repaired with CFF strips at a spacing of 150mm and 200mmc/c, respectively. The shear crack in the beam TS1-1 exhibited between the CFRP strip gaps (i.e. unwrapped zone). As the applied load increased, the CFF strips were ruptured approximately at a load of 148kN and followed by shear cracks emerging at the supports prior to the failure. However, in beam TS1-2, there was no sign of rupture in any CFF strip, however the flexural crack under the loading point grew up to bottom of flange. Both beams TS1-1 and TS1-2 were failed in flexure at a peak load of 187.95kN and 161.34kN, respectively. The flexural failure of these beams occurred under the loading points. The observed propagation and distribution of shear cracks in TS2-2 were relatively more as compared to beam TS1-1 due to the increase in spacing of CFF strips. These beams TS1-1 and TS1-2 attained a shear enhancement of approximately 40% and 20% over the control beam TS1a. Fig. 5(b) portrays the flexural failure pattern of the CFF repaired specimen TS1-1.

4.2 Cracking and Failure Patterns – Numerical Observation

The cracking patterns of control and CFF repaired beams have been generated using LUSAS finite element software for different loading cases. Fig. 5(c) portrays the cracking patterns of the control beam TS1a. At the early stage of loading, the local crushing of concrete was observed near the support of the beam as the support was not modelled in this study. The flexural cracks were then initiated in the flexural zone and propagated towards the flange zone. With further increase of load, the diagonal shear cracks began at the shear span and extended along the bottom of the flange. Some compressive cracks were also developed in the flange at the point of application load. The tested control beam was also accomplished similar observations except the local crushing at the support (see Fig. 5(a)).

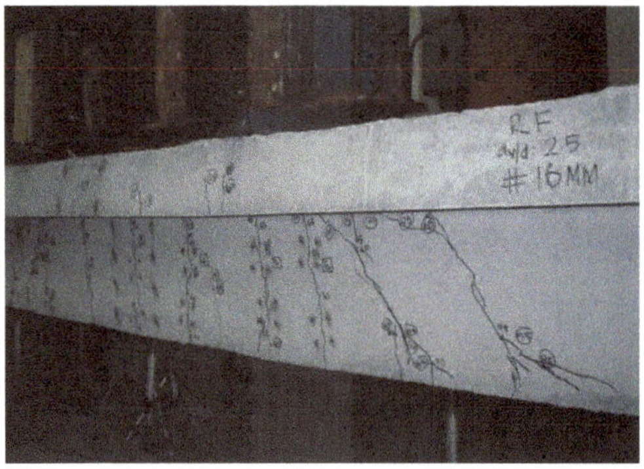

(a) Control beam TS1a

(b) CFF repaired beam -TS1-1

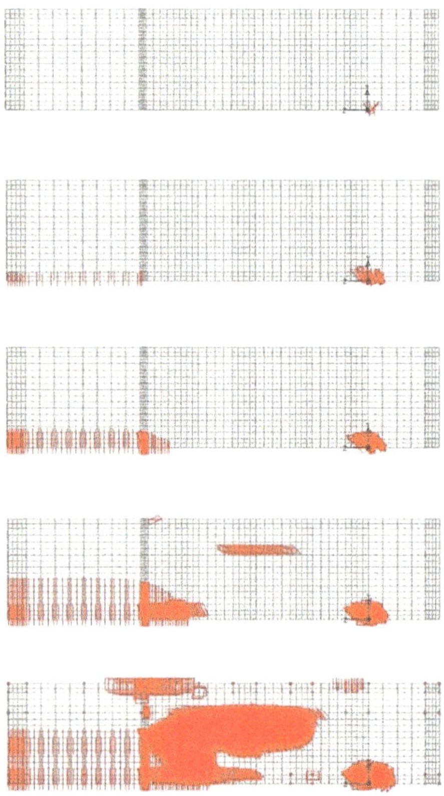

(c) Numerical study – Beam TS1a

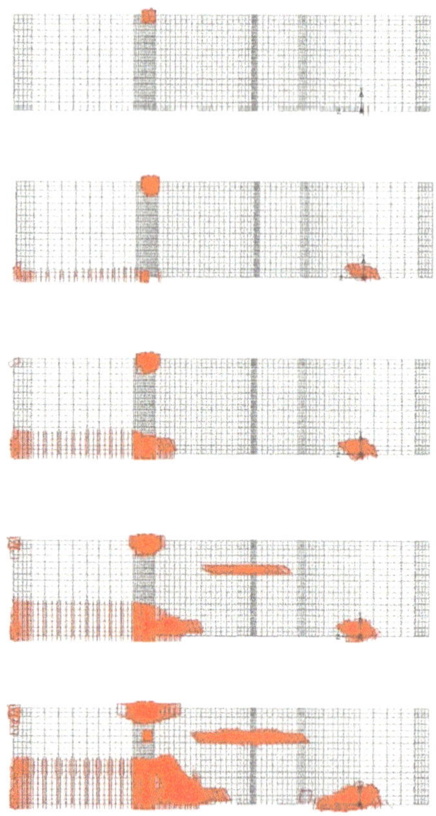

(d) Numerical study – Beam -TS1-1

Figure 5. Numerical and experimental cracking patterns for (a) control (TS1a) and repaired (TS1-1) beams

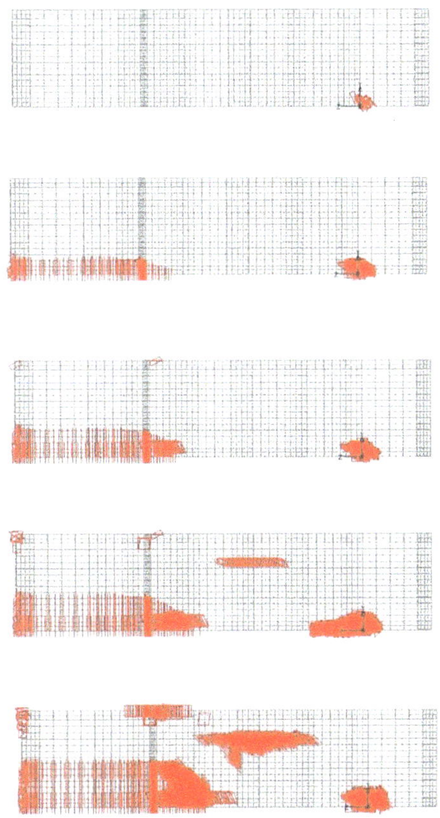

Figure 6. Cracking patterns for repaired beam TS1-2

The cracking patterns of repaired beams TS1-1 and TS1-2 at various loading stages are depicted in Figs 5(d) and 6, respectively. These CFF repaired beams had local crushing of concrete and flexural cracks similar to the control beam. The shear cracks were also emerged in the shear zone and propagated along the bottom of flange. Moreover, the flexural cracks were observed under the loading points. It can be seen from figures, the distribution of cracks in the CFF repaired beams were not same as control beam due to the application of external CFF shear reinforcement. All these observations (i.e. shear and flexural cracks) were also achieved in the experimentally repaired beams (e.g. see Figure 5(b)) prior to the failure load. From the overall discussion, it can be concluded that the cracking pattern of the modelled control and repaired beams was almost similar to the experimental beams except the local crushing failure at the support. It was also observed that the modelled control TS1a failed in shear, however the CFF repaired beams TS1-1 and TS1-2 were failed in flexural mode. The predicted failure modes of the numerical models were same as observed in the tested control and CFF repaired beams.

4.3 Load Deflection Profile

The typical deformation of the modelled beam is shown in Fig. 7. Figs. 8(a) - (c) show the comparison of experimental and predicted load-deflection profile for control and CFF repaired beams. In control beam TS1a, the stiffness of the predicted deflection curve was similar to the experimental results up to a load approximately of 110kN. However, in the later stage, there was an unexpected drop in deflection value of the predicted deflection curve. Comparing the experimental and predicted results of repaired beam TS1-1, the numerical results show better agreement with the experimental results, whereas the beam TS1-2 was relatively less agreement in comparison to the beam TS1-1. The deflection profile of beam TS1-2 shows similar trend until the load 100kN but the stiffness increased as the applied load increased. Table 3 shows the comparison of experimental and predicted FE results of control and CFF repaired T-beams. It was observed that the predicted results were varied between 25.92% and 31.76% over the experimental results. The discrepancy in predicted results is probably due to the several reasons as reported by Kachlakev et al [14]. At first, this effect is due to the presence of micro cracks in concrete which could be produced by drying shrinkage in the concrete and handling of the beams. The microcracks probably reduce the stiffness of the tested beams, whereas the finite element models do not include the microcracks. Secondly, this finite element analysis assumed that the bond between the concrete and steel is perfect but it would not be true in the actual beam. As bond slip occurs, there is no composite action between the concrete and steel reinforcement [14]. Furthermore, the experimental beams were precracked before applying the CFF shear reinforcement, however this FE analysis ignored the effect of precracking phase of the modelled beams. In due course, the assumption prefect bond between concrete surface and CFF in the model may cause the failure to simulate the premature debonding failure of the CFF. This could also lead to the overestimating the member stiffness and the ultimate load bearing capacity [24]. As a consequence, the overall stiffness and load carrying capacity of the numerical results was expected to be greater than the experimental results.

(a)

(b)

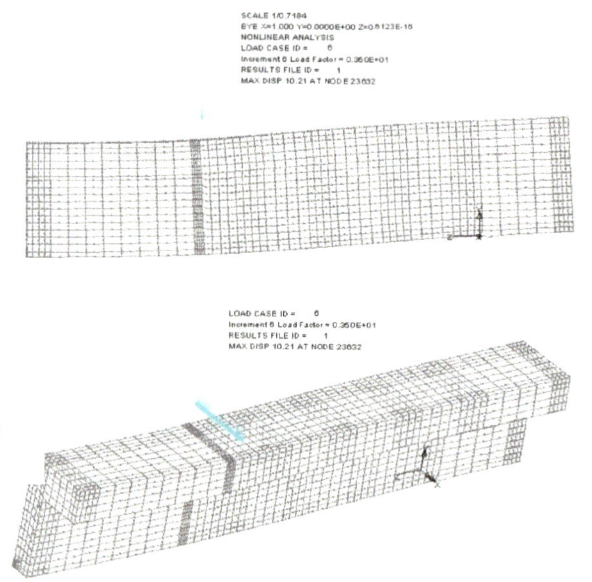

Figure 7. Typical deformation of modelled beam (a) Elevation view and (b) Isometric view

Table 3. Comparison of experimental and predicted FEM results for control and CFF repaired T-beams

Beam	External CFF Reinforcement		Failure load		
	Width (mm)	Spacing (mm)	EXP (kN)	FEM (kN)	(%) Difference [(FEM-EXP)/EXP] x 100
TS1a (Control)	---		134.74	170.0	25.92
TS1-1 (Repaired)	80	150	187.98	247.7	31.76
TS1-2 (Repaired)	80	200	161.34	224.0	27.97

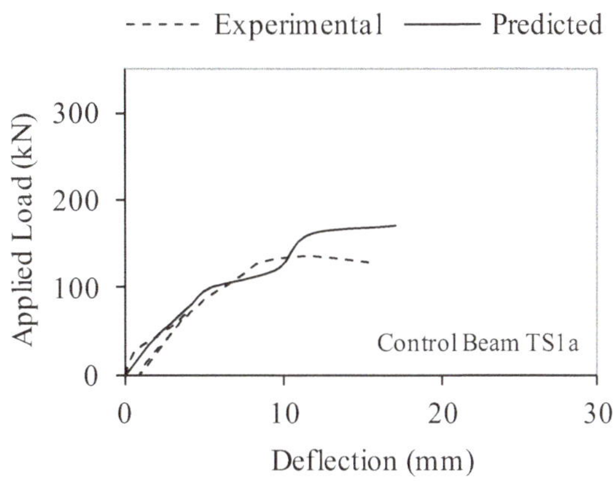

(a) Control beam TS1a

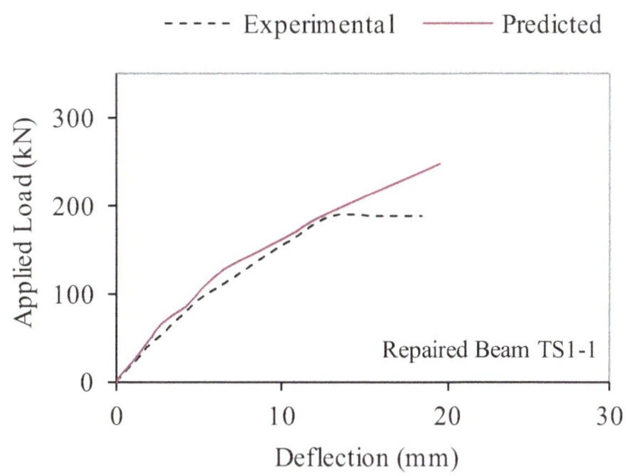

(b) CFF repaired beam TS1-1

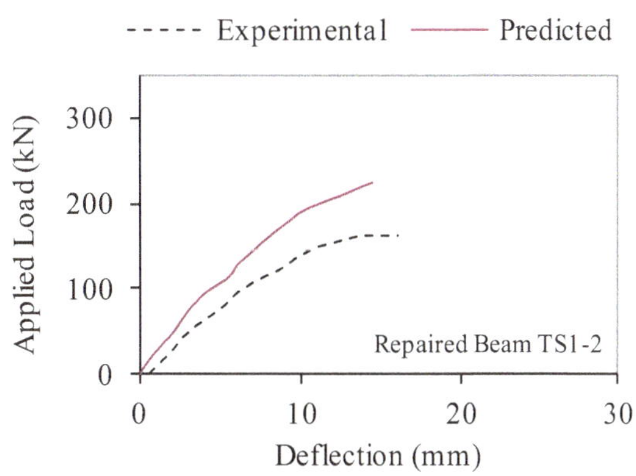

(c) CFF repaired beam TS1-2

Figure 8. Load – deflection profile for control and CFF repaired T-beams

4.4 Strain Distribution in Discrete CFF Strips

The distribution of fibre strains in the CFF repaired T-beams predicted the location of failure zones such as debonding or fracture of discrete CFF strips where the stress is highly concentrated. Figs. 9(a) and (b) depict the development of strain distribution in discrete CFF strips of the modelled beams TS1-1 and TS1-2, respectively. From the figures, it can be seen that the discrete CFF strips placed proximity to the mid-shear span of the beams TS1-1 and TS1-2 were stressed. This was probably due to the propagation of shear crack in the concrete band (i.e. between CFF strips). It also confirms that the discrete CFF strips only began to carry the load after the formation of crack in the concrete. Moreover, the negative strain values were probably caused by the fracture of discrete CFF strips.

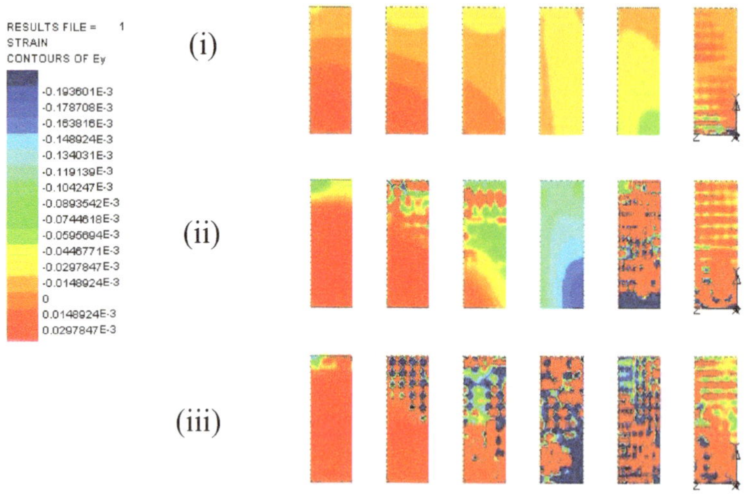

Figure 9(a). Development of strain distribution in discrete CFF strips for repaired beam TS1-1 at different stages of loading (i) 27.7 kN; (ii) 47.7 kN; (iii) 107.7 kN

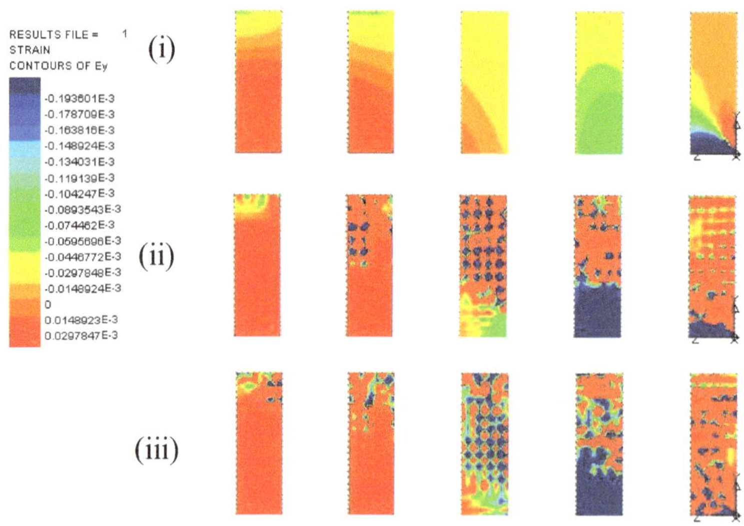

Figure 9(b). Development of strain distribution in discrete CFF strips for repaired beam TS1-2 at different stages of loading (i) 20.0 kN; (ii) 48.0 kN; (iii) 112.0 kN

5.0 Conclusions

This paper presents the experimental and finite element results of shear behaviour of reinforced concrete T-beams repaired with discrete CFF strip composites. The following conclusions can be drawn from this experimental and numerical investigation:

- The experimental results show that the gain in shear capacity of the CFF repaired beams ranged between 20% and 40% over the control beam. Thus, it can be concluded that the externally bonded CFF strips significantly increased the shear capacity of CFF repaired beams.
- The numerical results have shown that the application of external shear enhancement can enhance the shear behaviour of the repaired beams that is similar to the experimentally tested beams. The load deflection behaviour of the predicted finite element model shows better agreement with the experimental results. However, the difference in stiffness was slightly higher as assumptions taken in the numerical study may not be fully applied in the real specimens.
- This numerically predicted model can be used to depict the initiation and propagation of cracks in the CFF repaired beams but it is partially able to observe the cracking pattern in the tested beams with discrete CFF strips. Besides that, the numerical model assists to predict and identify the high risk of failure zone for discrete CFF strips.

References

[1] A. Khalifa, J. G. William, A. Nanni, and M. I. Abdel Aziz "Contribution of externally bonded FRP to shear capacity of flexural members," ASCE, Composites for Construction J, Vol. 2, No. 4, pp. 195-203, 1998

[2] J. Jayaprakash, A. A. Abdul Aziz, A. A. Ashrabov, and A. A. Abang Abdullah, "Repair of precracked RC rectangular shear beams using CFRP strip technique," An Int. J Structural Engineering and Mechanics, Vol. 26, No. 4, pp. 427-439, 2007.

[3] A. Khalifa and A. Nanni, "Improving shear capacity of existing RC T-section beams using CFRP composites," Cement and Concrete Composites J, Vol. 22, No. 2, pp. 165-174, 2000.

[4] H. A. Raghu, J. J. Myers, and A. Nanni, "Shear performance of RC beams strengthened in situ with composites," Report, University of Missouri, Rolla, 2001.

[5] A. Khalifa and A. Nanni, "Rehabilitation of rectangular simply supported RC beams with shear deficiencies using CFRP composites," Construction and Building Materials J, Vol. 16 pp. 135-146, 2002.

[6] P. Alagusundramoorthy, I. E. Harik, and C. C. Choo, "Shear strength of R/C beams wrapped with CFRP fabric," Research report, KTC-02-14/SPR 200-99-2F, 2002.

[7] B. Taljsten, "Strengthening concrete beams for shear with CFRP sheets," Construction and Building Materials, Vol. 17, pp. 15-26, 2003.

[8] A. B. Adhikary and Mutsuyoshi, "Behaviour of concrete beams strengthened in shear with carbon fibre sheets," Composites for Construction J, Vol. 8, No. 3, pp. 258-264, 2004.

[9] J. Jayaprakash, A. A. Abdul Aziz, and A. A. Ashrabov, "Investigation on effects of variables on shear capacity of pre-cracked RC T-beams with externally bonded bi-directional CFRP discrete strips," Composite Materials J, Vol. 44, No. 2, pp. 241-261, 2010.

[10] Z. Zhang and C. T. T. Hsu, "Shear strengthening of reinforced concrete beams using carbon fibre reinforced polymer laminates," Composites for Construction J, Vol. 9, No. 2, pp. 158-169, 2005.

[11] J. G. Teng, G. M. Chen, J. F. Chen, O. A. Rosenboom, and L. Lam, "Behaviour of RC beams shear strengthened with Bonded or unbonded FRP wraps," Composites for Construction J, Vol. 13, No. 5, pp. 394-404, 2009.

[12] K. W. Neale, U. A. Ebead, H. M. Abdel Bakky, W. E. Elsayed, and A. Godat, "Analysis of the load deformation behaviour and debonding for FRP strengthened concrete structures," Advances in Structural Engineering, Vol. 9, No. 6, pp. 751-763, 2006.

[13] M. Arduini, A. Di Tommaso, and A. Nanni, "Brittle failure in FRP plate and sheet bonded beams," ACI Structural, Vol. 94, No. 4, pp. 363-370, 1997.

[14] D. Kachlakev, M. Thomas, and Y. Solomon, "Finite element method analysis of a concrete bridge repaired with fibre reinforced plastic laminates," Computers and Structures, Vol. 72, pp. 379-407, 1999.

[15] R. Santhakumar and E. Chandrasekaran, "Analysis of retrofitted reinforced concrete shear beams using carbon fibre composites," Electronic J. Structural Engineering, Vol. 4, pp. 66-74, 2004.

[16] T. Norris, H. Saadatmanesh, and R. Ehsani Mohmammad, "Shear and flexural strengthening of R/C beams with carbon fibre sheets," ASCE, Structural Engineering J, Vol. 123, No. 7, pp. 903-911, 1997.

[17] Sika Manufacturer's Manual sheet, Switzerland (Supplier: Sika Kima Sdn. Bhd.)

[18] M. Y. H. Bangash, "Concrete and Concrete Structures: Numerical Modeling and Applications," Elsevier Science Publishers Ltd., London, England, 1989.

[19] D. Kachlakev, M. Thomas, and Y. Solomon, "Finite element modeling of reinforced concrete structures strengthened with FRP laminates," Report for Oregon Department of Transportation, Final report SPR 316, 2001.

[20] AS3600 2001, Concrete structures.

[21] H. B. Pham and R. Al-Mahaidi, "Assessment of available prediction models for the strength of FRP retrofitted RC beams," Composite Structures, Vol. 66, pp. 601-610, 2004

[22] A. M. Ibrahim and M. Mahmood Sh, "Finite element modelling of reinforced concrete beams strengthened with FRP laminates" European Journal of Scientific Research, Vol. 30, No. 4, pp. 526-541, 2009.

[23] H. B. Pham and R. Al-Mahaidi, "Finite Element Modelling of RC Beams Retrofitted with CFRP Fabrics," Proceedings of 7th International Symposium on Fiber Reinforced Polymer Reinforcement for Reinforced Concrete Structures (FRPRCS 7), pp. 499-514, 2005.

[24] R. S. Y. Wong and F. J. Vecchio, "Towards modeling of reinforced concrete members with externally bonded fiber-reinforced polymer composites," ACI Structural J, Vol. 100, No. 1, pp. 47-55, 2003.

18

Construction Managers' Perception of the Factors Affecting Sustainability in Construction Projects

Reza Taheriattar[1], Morvarid Farzanehrafat[2]

[1] School of Civil and Environmental Engineering, University of New South Wales, Sydney, Australia

[2] School of Civil Engineering, Iran University of Science and Technology, Tehran, Iran

*Corresponding E-mail : rezataheria@gmail.com

Abstract

Sustainable construction is a comprehensive concept which requires long-term planning. Moreover, construction managers play a key role in leading, planning and scheduling of a construction project. As a result, sustainability of construction projects can be affected by construction managers' decisions. In addition, for greater development of sustainable construction, affecting factors should firstly be notified. Therefore, it seems necessary to investigate construction managers' perception of the factors affecting sustainability in construction projects. To quantify managers' perspective of construction sustainability, a survey involving 15 construction managers from Iran construction industry was employed. A statistical comparative analysis used to identify the most important factors affecting sustainability performance at project-level. The results show that most of the factors affecting sustainable construction can be addressed by management teams. The findings will be useful for managers to improve construction sustainability performance at project-level.

Keywords: *Sustainability Performance, Affecting Factors, Construction Managers, Project-Level*

1.0 Introduction

The Increased exploitation of natural resources caused by technology advancement and population growth besides limited resources, resulted in the increase of importance and concern for realizing sustainable development during recent decades [1]. In this regards, holding the United Nations Summit on Environment and Development in 1972, Agenda21, Rio Conference in 1992, followed by many other national and international meetings and conferences demonstrate the growing concerned for protecting the environment for the future generations by introducing sustainable development concept [2, 3].

On the other hand, according to CIB (1998), 54 percent of energy consumed in the U.S. is directly or indirectly related to the buildings and construction activities [4]. Also, the MOC report (1999) depicts that almost 25l of the energy consumed in China is directly caused by producing building materials and implementing construction activities. Meanwhile, the solid waste from destruction of buildings is 10 to 20 times by weight as much as the waste produced from the construction of new buildings [4]. Construction industry, which creates critical infrastructures to support other industries, provided over 6.4 million job opportunities in the United States in 2004 alone. Moreover, this industry as the largest employer in Europe (7.5l of total employment), has significantly contributed to the economy (9.7l of GDP and 47.6l of the Gross fixed capital formation in 1999) [5]. Considerable effects of construction activities on environmental, economic and social aspects, illustrate the potential to help implementing sustainable development. Thus, paying attention to the concept of sustainable development in construction projects seems necessary and justifiable.

As the first approach, basics and criteria must be notified. In terms of sustainable construction criteria and principles, some research has been conducted. In this regard, Kibert (1994) proposed six principles of sustainable construction based on which Miyatake (1996) provided a comprehensive perspective of sustainable construction and identified the important roles of technology development in realizing sustainable construction [6]. Also, Hill and Bowen

(1997) introduced a framework of key principles of sustainable construction for environmental management in the construction process [1]. Ofori (1998) provided a comment on the contents of Hill and Brown, and endorsed their stress on the importance of environmental considerations in construction [7]. Huang et al. (2001) created a framework for the appraisal indicators of sustainable construction at state-level in order to help a nation to pinpoint areas needing improvement [8]. In addition, Pulaski (2004) tried to express most important approaches in order to realize sustainable development in site operations. Shen et al. (2005) briefly identified variables affecting sustainable development ability as the basis to examine the sustainability of a construction project [4]. Also, a set of key assessment indicators were introduced by Shen et al. (2011) to evaluate the sustainability of an infrastructure project [9]. However, many other important indicators were not considered in their research.

Despite the existing studies on sustainable construction principles and indicators in different levels, no effort has been done to widely identify and prioritize the most significant factors affecting sustainable construction at project-level. However, sustainability performance improvement considerably requires identifying the main areas of impact. In addition, sustainable construction, which is a comprehensive and long-term concept, requires long-term thinking and planning. Thus, construction managers, who are usually responsible for making long-term decisions, play a key role in sustainability implementation. Therefore, this research aims to identify and prioritize the most important factors affecting sustainability performance of construction projects from construction managers' viewpoint.

2.0 Definitions

2.1 Sustainable Development

The concept of sustainability was inherently perceived by the first human civilizations. They understood the importance of using resources on a sustainable basis and the fact that life-support systems are required for the survival of human life on the earth. World Commission on Development and Environment expressed that, even though, new civilizations have recently commenced exploring sustainable forms of development, old civilizations have tried sustainable consumption of resources for thousands of years. In the present developed world, public awareness of environment increased in 1960s and this international concern was reflected in the United Nations conference on environment in Stockholm in 1972. Also, the idea of Eco-Development; which was the precursor of the concept of sustainable development, was emerged from this conference as 'an approach to development aimed at harmonising social and economic objectives with ecologically sound management' [1]. In the same year, the Club of Rome published "The limits to Growth", a document which emphasized that concerns about pollution, environmental degradation and natural resource depletion were crucial to the long-term future of humanity. The limits-to-growth perspective challenged the pro-growth perspective and since it threatened previous ideas and theories, was subjected to severe reactions. Finally, a synthesis of these conflicting perspectives was emerged in the perspective of sustainable development. The International Union for the Conservation of Nature and Natural Resources (IUCN)" published "World Conservation Strategy" in 1980. The strategy marked a significant shift in conservation, from focusing solely on the practice of fencing off nature reserves to viewing conservation and development as integrated concepts. This strategy defined development as 'modification to the biosphere to satisfy human needs' and conservation as 'the management of human use of the biosphere to yield the greatest sustainable benefit for present and future generations [1].

According to the definition presented by Bruntland (Table 1), development is not limited to a specific area, but it can be used in all areas and for everyone or everything in the present and future. Also, this definition is based on two concepts:

- The concept of "Needs" includes conditions to provide an acceptable standard life for all humans.

- The concept of "Limits" of environment capacity to meet the present and future needs.

Sustainable development is mainly managing the relationship between meeting human needs and protecting the integrity of the biosphere. Indeed, to satisfy the society demands in such a way that critical environmental limits are not exceeded [10]. For the common future, the best attitude is meeting needs along with preventing restrictions to increase or preferably causing to decrease.

Table 1: Several Definitions of Sustainable Development

Reference	Definition
-World Commission on Environment and Development (WCED), Bruntland, 1987	Development that meets the needs of the present without compromising the ability of future generations to meet their own needs
-Caring for the Earth, IUCN UNEP 1991	Improving the quality of human life while living within the carrying capacity of supporting ecosystems
-International Council for Local Environmental Initiatives, ICLEI 1996	Development that delivers basic environmental, social and economic services to all residences of a community without threatening the viability of natural, built and social systems upon which the delivery of those systems depends
-Amsterdam Treaty, 1997	Determined to promote economic and social progress for their peoples, taking into account the principle of sustainable development and within the context of the accomplishment of the internal market and of reinforced cohesion and environmental protection, and to implement policies ensuring that advances in economic integration are accompanied by parallel progress in other fields
-(Sage, 1998)	Refers to the fulfillment of human needs through simultaneous socio-economic and technological progress and conservation of the Earth's natural systems
-Forum for the Future, UK's Sustainable Development Association (Parkin, 2000b)	A dynamic process which enables human all people to realize their potentials and improve the quality of their life in ways that simultaneously protect and enhance the Earth's life-support systems

2.2 Triple Bottom Line

Sustainable development is defined based on three main themes of economic, social and environmental (Fig. 1) which is known as Triple Bottom Line:

2.2.1 Social Sustainability

The social sustainability is providing an appropriate level of prosperity; including physical and psychological, for community members. This field of sustainable development includes subgroups and issues such as peace and social security, trust, training, developing awareness, knowledge and skills, employees' health and safety, recruitment, communications, work environment efficiency, creating opportunities for leisure-cultural activities, reduction of crime commitment and providing services for disabled people.

2.2.2 Environmental Sustainability

The philosophy of environmental sustainability is to deliver the earth to future generations in a way better than what we had. According to this definition, human activities can result in a sustainable environment, if they live without destroying environment and over consuming natural resources. So, materials must be totally made from recycled materials or renewable resources, waste recycling must be done completely, renewable and clean energy resources such as wind, solar, nitrogen and geothermal energy must be used in order to conserve energy, natural diversity must be protected and the emission of greenhouse gases must be reduced.

2.2.3 Economic Sustainability

Economic sustainability returns to economic boom and reduction of financial losses, lowering expenses and accelerating investment return. Some issues such as investing on human resources and facilities to create competitive economy, job opportunities, dynamic local economy, enhancing added value to the earth and crating markets and new economic opportunities are related to the concept of economic sustainability.

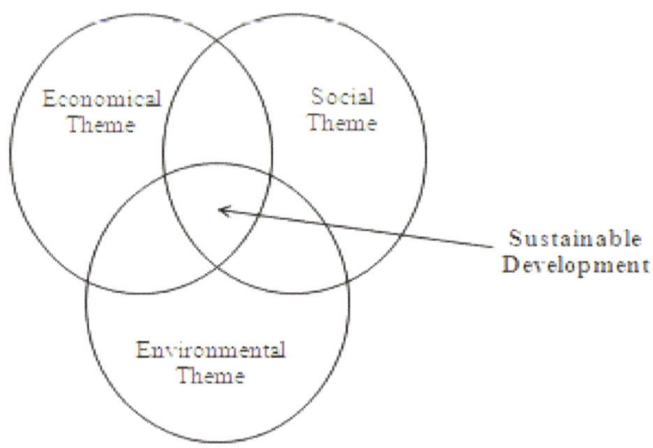

Figure 1: Sustainable Development Themes

2.3 Sustainable Construction

The term of sustainable construction is generally used to express the responsibility of construction industry in achieving sustainable development [1]. In other words, sustainable construction is a way of helping the achievement of sustainable development in construction industry and is based on three main bases of environmental protection, social welfare and economic prosperity [11]. Environmental sustainability is concerned with the extraction of natural resources and built environment which includes construction activities that can cause severe adverse impacts on the environment. Social well-being is related to human feelings: security, satisfaction, safety and comfort and human contributions: skills, health, knowledge and motivation. Finally, economic sustainability is relevant to project financial gains for the benefits of the clients, constructors, public and the government [11]. Table 2 illustrates several important definitions of sustainable construction.

Table 2: Several Definitions of Sustainable Construction

Reference	Definition
-1st International Conference on Sustainable Construction, Kibert,1994	The creation and responsible management of a healthy built environment based on resource efficient and ecological principles
-Huovila, 1999	In its own processes and products during their service life, aims at minimizing the use of energy and emissions that are harmful for environment and health and produces relevant information to consumers for their decision-making
-CIB, 1999	Official definition: a way of building which aims at reducing (negative) health and environmental impacts caused by the construction process or by buildings or by the built-up environment More precise definition: the reduction in the use of natural resources and the conservation of the life support function of the environment by construction processes, buildings and the built-up environment under the premise that the quality of life is maintained
-Chang et al. 2000	Low environmental impact, high contact with the environment, amenities and health
-Plessis et al. 2002 (Agenda 21 for Sustainable Construction in Developing Countries)	A holistic process aiming to restore and maintain harmony between the natural and the built environments, and create settlements that affirm human dignity and encourage economic equity
-Huang and Kou, 2002	Environmentally friendly construction for achieving sustainable coexistence with the natural environment throughout the stages of planning, design, construction and service life, stressing environmental ethics including consumption of minimal energy and resources, harmony with the environment and sharing with later generations
-Architecture and Building Research Institute , 2003	Architectural design geared towards human health and comfort, pursuing coexistence with the global environment, and fostering the sustainability of the people's living environment. Buildings should consume relatively few natural resources and manufacture relatively little waste

3.0 Methodology

3.1 Affecting Factors Identification

Previous investigations have not properly identified factors affecting the achievement of sustainable construction. Therefore, recent studies were precisely reviewed to identify factors affecting sustainable construction in order to become an appropriate reference for all the future studies. Some of the existing research studied and classified the sustainable construction criteria. These criteria may exactly be the affecting factors or may contribute to recognize and extract the factors by a little change in existing attitude. For example, the factor of "Utilizing Clean and

Renewable Energies" was understood from "Renewable Energies" as a sustainable construction criterion. Moreover, some of the other studies which provided some solutions for sustainable development implementation were used as a reference to extract the affecting factors. For instance, the factor of "Neighborhood Disturbance" has been derived from a research suggested pre-fabrication method. It has been suggested as an appropriate solution to achieve sustainable construction because the neighbors' comfort can be maintained as a benefit of this method. Also, some studies have presented the assessment indicators which can be literally considered as the factors affecting sustainable construction. Finally, through studying previous research and integrating the factors affecting sustainable construction, 64 factors identified and classified into 5 groups as illustrated in Table 4. Moreover, some of the factors identified by the experts' opinions denoted as Experts.

3.2 Relative Importance Determination

At this stage, in order to determine the most important factors, a questionnaire survey was set and given to a group of experts including construction managers in the construction industry of Iran.

3.2.1 Survey Setting

Initially, a comprehensive framework containing the affecting factors was set to ask experts' opinions. Construction Managers were invited to indicate the significance of each factor in improving sustainability performance of construction projects, using a five point Likert scale. The scores of 5, 4, 3, 2, and 1 considered to represent very high, high, medium, low and very low significance in achieving sustainable construction, respectively. At the end of the survey, a list of effective factors definition was provided to clarify researchers' purpose and uniform participants' views. Thus, possible problems and uncertainties in understanding effective factors could be eliminated by referring to the list.

3.2.2 Expert Group

The success of a survey clearly rests on the combined expertise of the participants who make up the expert panel. There are two key aspects to panel composition: panel size and panel member qualifications. A panel of 10–50 participants was previously recommended, whereas reasonable results can be obtained with a homogenous group of 10–15 experts. It is generally regarded that as the number of members increases, the reliability of a composite judgment increases. Thus, an expert panel of 15 construction managers with more than two decades experience from different construction projects were employed to indicate the significance of each indicator (Table 3). This number was felt to be sufficient with a composition of highly qualified panelists [12].

Table 3: Construction Managers' Area of Expertise

Area of Expertise	Number
Construction Management	5
Environmental Engineering	3
Geotechnical Engineering	2
Transportation Engineering	2
Water Resources Management	3
Total	15

3.2.3 Survey Distributing and Data Gathering

The surveys distributed among the experts through a two-round method. In the first round, experts' comments and consultations were received, some factors were eliminated or merged and some others were added to the list of factors. Then, the effective factors were again distributed among the experts in a format similar to the first round. At the end of second round, after collecting surveys and analyzing data, the average values were calculated as the relative importance of factors affecting sustainability performance of construction projects (Table 4).

Table 4: Relative Importance and References of the Identified Affecting Factors

Affecting Factor	Importance (1 to 5)	Ref No.*	Affecting Factor	Importance (1 to 5)	Ref No.
Workforce & Materials			**Project Control & Management**		
Motivation	4.4	13	Management Quality	4.5	4,24
Workers Commitment	4.3	10	Utilizing Clean and Renewable Energies	4.4	15,25
Sustainability Performance Monitoring and Controlling	4.1	4,12	Utilizing Reduction Techniques for Energy Consumption	4.4	15,25
Materials Durability and Service Life	3.9	14,18	Greenhouse Gases Emission and Amount of Dust Fallen	4.3	9,16,18,22
Materials Availability	3.4	14,19	Energy Efficiency	4.3	18
Using Multipurpose Materials	3.3	18	Cleaning up Contaminated Water and Land	4.1	8,19
Crew Size and Composition	3.1	14,15,16	Energy Consumption	4.1	22
Design Specifications & Engineering Documents			Using Clean Technologies, Tools and Materials	4.1	4,14,18,23
Considering Sustainability Principles in Design and Contract	4.3	12	Waste Management	4.0	14,23
Design Alteration during Execution	3.8	14	Job Satisfaction	4.0	11
Architectural Quality	3.8	18,20	Resources Consumption	4.0	14
Public Design	3.1	8,21	Working According to Schedule	3.9	Experts
Design Complexity	2.8	22	Use of Non-Drinkable Water	3.8	8,21
Project & Work Characteristics			Competitiveness	3.6	1,15
			Water Recycling	3.6	24
Environment Degradation	4.4	13	Maintenance Management	3.6	22
Productivity	4.3	22	Corporation's Capital	3.5	9
Product Quality	3.8	11,17	Job Stability	3.5	22
Project Profit	3.8	4	Efficient Use of Materials	3.5	22
Execution Complexity	3.6	22	Environmental Accounting	3.4	Experts
Harmony between Natural and Built Environments	3.6	8	Accessibility and Communication	3.3	4
Product Lifetime	3.5	17	Employing Sub-Contractors	3.1	1
Required Skills on Workforce	3.4	14	Corporation's Credit	3.0	Experts
Direct Costs	3.3	15,18	Employment Creation	3.0	18,22
Land Use	3.3	22	Site Discipline	3.0	18,20
Maintenance Costs	3.3	18	Developing Green and Open Space	2.4	23,24
Neighborhood Disturbance	3.3	14	**External Environment**		
Project Time	3.3	14	Bio-Diversity	4.3	4
Public Traffic Disruptions	3.3	14	Local Economy	4.1	4,8
Scale of Serviceability	2.9	9	Considering Sustainability Principles in Housing Policy	4	15
Contractors' Claims	2.8	Experts**	Investor & Operator Confidence	3.8	26
Flexibility and Lack of Monotony	2.6	23	Environmental Pollution Lawsuits	3.4	16
Air Quality	2.4	18	Global Warming	2.6	24
Work Space	2.1	18	Ozone Depletion	2.6	20

* According to the numbers assigned in the References section

** The affecting factors identified using Experts' opinions

4.0 Results and Discussions

Construction managers believed that the factors which exert most influence on sustainability performance of construction projects include management quality, utilizing clean and renewable energies, utilizing reduction techniques for energy consumption, motivation, environmental degradation followed by greenhouse gas emission, energy efficiency, bio-diversity, workers' commitment, considering sustainability in design and contracts, and productivity (Fig. 2). The highest importance for managers' quality which originates from managers' knowledge and experience, confirmed the considerable effect of managers on sustainability implementation. It is because managers can change the direction of a construction project by making specific decisions. Indicating factors such as utilizing clean and renewable energies, environmental degradation, greenhouse gases emission and bio-diversity demonstrate the necessity of concentration on environmental aspect of sustainability for better implementation of sustainability in construction projects. In addition, the factors of utilizing reduction techniques for energy consumption and energy consumption depicted the significance of energy management from the construction managers' viewpoint. The great relative importance for motivation and workers' commitment proves the considerable impact of workforce and social aspect of sustainability for improving sustainability performance of construction projects. Considering sustainability principles in design and contracts, as an important affecting factor, provides a proper framework for better implementation of sustainability principles during the construction process. Moreover, construction managers believed that productivity notably influence on sustainability performance of construction projects. It is because productivity, which directly influence on time and costs of a construction project, may improve the different aspects of sustainability especially economic sustainability by reducing time or costs of a construction project. On the whole, it can be gathered from Fig. 2 that the factors related to the project control and management registered a high importance in improving sustainability performance of construction projects from the construction managers' perception.

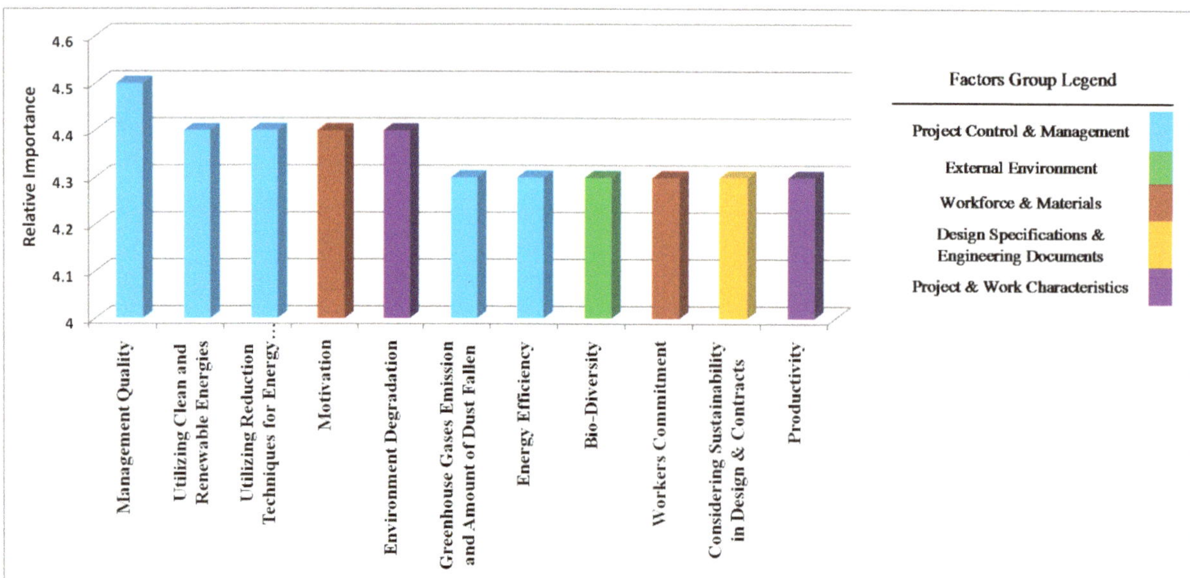

Figure 2: Most Important Factors

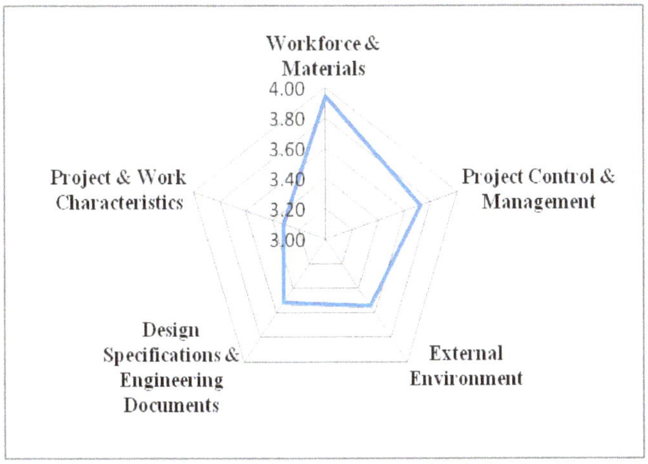

Factors Group	Relative Importance (1 to 5)
Workforce & Materials	3.94
Project Control & Management	3.72
External Environment	3.54
Design Specifications & Engineering Documents	3.52
Project & Work Characteristics	3.32

Figure 3: Relative Importance of the Factors Groups

5.0 Conclusion

The great change in people's demand for natural resources accompanied by technology advancement during recent decades resulted in the increase of importance and concern for realizing sustainable development. In addition, considerable effects of construction activities on environmental, economic and social aspects of sustainable development illustrate its potential to help implementing sustainability. Thus, many studies conducted to interpret the concept and principles of sustainable construction but the lack of developing a comprehensive framework for identification of most important areas affecting construction sustainability is seen. The researchers believe that managers need to recognize the factors affecting sustainability of construction projects based on which can appropriately select the best strategy. Therefore, this study has widely identified factors affecting sustainable construction from the construction managers' viewpoint.

Construction managers perceived that the factors related to "Workforce and Materials" and "Project Control and Management" make a great contribution to improving sustainability performance of construction projects (Fig. 3). Thus, it is rational to emphasize on these special areas in their decision-making for sustainability improvement. Totally, the represented framework can provide a richer understanding of construction sustainability and contribute to mitigate the management challenges in identifying strategies for sustainability improvement.

Acknowledgement

The authors would like to thank Mr. Shahin Dabirian, the PhD candidate at Iran University of Science and Technology, and Sustainability Department in Isfahan Municipality for their wholehearted assistance in developing the experts group and collecting data.

References

[1] Hill R. C. and Brown P. A. (1997), Sustainable Construction: Principles and a Framework for Attainment, Construction Management and Economics, Vol. 15, No. 3, pp. 223 – 239.

[2] Chaharbaghi K. and Willis R. (1999), Study and practice of sustainable development, Engineering Management Journal, Vol. 9, No. 1, Feb. 1999, pp. 41 – 48.

[3] Parkin S. (2000), Context and drivers for operationalizing sustainable development, Proceedings of ICE, Vol. 138, Nov. 2000, pp. 9 – 15.

[4] Shen L. Y., Wu Y. Z., Chan E. H. W. and Hao J. L. (2005), Application of System Dynamics for Assessment of Sustainable Performance of Construction Projects, Journal of Zhejiang University SCIENCE, 6A(4), pp. 339 – 349.
[5] Chong W. K., Kumar S., Haas C. T., Beheiry S. M. A., Coplen L. and Oey M. (2009), Understanding and Interpreting Baseline Perceptions of Sustainability in Construction among Civil Engineers in the United States , Journal of Management in Engineering, Vol. 25, No. 3, pp. 143 – 154.
[6] Miyatake Y. (1996), Technology development and sustainable construction, Journal of Management in Engineering, Vol. 12, No. 4, 1996, pp. 23 – 27.
[7] Ofori G. (1998), Sustainable Construction: Principles and a Framework for Attainment - Comment, Construction Management and Economics, Vol. 16, No. 2, pp. 141 – 145.
[8] Huang R. Y. and Hsu W. T. (2011), Framework development for state-level appraisal indicators of sustainable construction, Civil Engineering and Environmental Systems, Vol. 28, No.2, pp.143–164.
[9] Shen L., Wu Y. and Zhang X. (2011), Key Assessment Indicators for the Sustainability of Infrastructure Projects, Journal of Construction Engineering and Management, Vol. 137, No. 6, pp. 441 – 451.
[10] Plessis C. D. (2007), A Strategic Framework for Sustainable Construction in Developing Countries, Construction Management and Economics, Vol. 25, No. 1, pp. 67 – 76.
[11] Zainul Abidin N. (2010), Investigating the Awareness and Application of Sustainable Construction Concept by Malaysian Developers, Habitat International, Vol. 34, pp. 421 – 426.
[12] Rajendran S. and Gambatese J. A. (2009), Development and Initial Validation of Sustainable Construction Safety and Health Rating System, Journal of Construction Engineering and Management, Vol. 135, No. 10, pp. 1067 – 1075.
[13] Schianetz K. and Kavanagh L. (2008), Sustainability Indicators for Tourism Destinations: A Complex Adaptive Systems Approach Using Systemic Indicator Systems, Journal of Sustainable Tourism, Vol.16, No.6, pp.601-628.
[14] Chen Z., Li H., Ross A., Khalfan M. M. A. and Kong S. C. W. (2008), Knowledge-Driven ANP Approach to Vendors Evaluation for Sustainable Construction, Journal of Construction Engineering and Management, Vol. 134, No. 12, pp. 928 – 941.
[15] Yunus R., Yang J. and Too E. (2011), Sustainability Factors in Industrialized Building System, World Sustainable Building Conference, 18-21 October, Helsinki.
[16] Ho Y. and Wang S. (2005), System Dynamics Model for the Sustainable Development of Science City, The 23rd International Conference on the System Dynamics Society, 17-21 July, Boston.
[17] Jaillon L. and Poon C. S. (2008), Sustainable Construction Aspects of Using Prefabrication in Dense Urban Environment: A Hong Kong Case Study, Construction Management and Economics, Vol. 26, No. 9, pp. 953 – 966.
[18] Chen Y., Okudan G. E. and Riley D. R. (2010), Sustainable Performance Criteria for Construction Method Selection in Concrete Buildings, Automation in Construction, Vol.19, pp.235-244.
[19] Bourdeau L. (1999), Sustainable development and the future of construction: a comparison of visions from various countries, Building Research & Information, Vol. 23, No.6, pp. 354 – 366.
[20] Hiete M., Kuhlen A. and Schultmann F. (2011), Analysing the Interdependencies between the Criteria of Sustainable Building Rating Systems, Construction Management and Economics, Vol.29, No.4, pp.323-328.
[21] Hakkinen T. (2007), Assessment of Indicators for Sustainable Urban Construction, Civil Engineering and Environmental Systems, Vol.24, No.4, pp.247-259.
[22] Alansar N. W., Flanagan R. and Alansar W. E. (2008), Model for Calculating the SustainableBuilding Index (SBI) in the Kingdom of Bahrain, Energy and Buildings, Vol. 40, pp. 2037 – 2043.
[23] Ross N., Brown P. A. and Lincoln D. (2010), Sustainable Housing for Low-income Communities: Lessons for South Africa in Local and Other Developing World Cases, Construction Management and Economics, Vol. 28, No. 5, pp. 433 – 449.
[24] Tae S. and Shin S. (2009), Current Work and Future Trends for Sustainable Buildings in South Korea, Renewable and Sustainable Energy Reviews, Vol. 13, pp. 1910 – 1921.
[25] Manoliadis O., Tsolas I. and Nakou A. (2006), Sustainable Construction and Drivers of Change in Greece: A Delphi Study, Construction Management and Economics, Vol.24, No.2, pp.113-120.
[26] Beheity S. M. A., Chong W. K. and Haas C. T. (2006), Examining the Business Impact of Owner Commitment to Sustainability, Journal of Construction Engineering and Management, Vol.132, No.4, pp.384-392.
[27] Sage A. P. (1998), Risk management for sustainable development, Proceedings of the IEEE International Conference on Systems, Man and Cybernetics, Vol. 5, 1998, pp. 4815 – 4819.
[28] Parkin S. (2000), Sustainable development: the concept and the practical challenge, Proceedings of ICE, Vol. 138, Nov. 2000, pp. 3 – 8.

[29] C. A. and Ding, G. K. C. (Eds.), Sustainable practices in the built environment, Langston, Butterworth-Heinemann, Oxford, 2001.
[30] Matar M. M., Georgy M. E. and Ibrahim M. E. (2008), Sustainable Construction Management: Introduction of the Operational Context Space (OCS), Construction Management and Economics, Vol. 26, No. 3, pp. 261 – 275.
[31] Sterman J. D. (2000), Business dynamics: systems thinking and modeling for a complex world, McGraw-Hill.
[32] Park M., Nepal M. P. and Dulaimi M. F. (2004), Dynamic Modeling for Construction Innovation, Journal of Management in Engineering, Vol.20, No.4, pp.170-177.

19

LOCAL BAMBOO AND EARTH CONSTRUCTION POTENTIALS FOR PROVISION OF AFFORDABLE STRUCTURES IN NIGERIA

Nwoke ,O.A[1] and Ugwuishiwu ,B .O[2]

Department of Agricultural and Bioresources Engineering
Faculty of Engineering, University of Nigeria Nsukka.

Email:nwoke.oji@unn.edu.ng[1], boniface.ugwuishiwu@unn.edu.ng[2]

Abstract

In Nigeria and other developing countries where reinforced concrete in construction is widely used, the high and steadily increasing cost of steel has made construction very expensive. This, coupled with the political will, usually christened “Nigerian Factor” has made any conceived affordable mass housing program by successive governments a mirage. This development has triggered off the search for alternative and suitable replacement for steel reinforcement in concrete works. This search for a cheaper alternative has led to the exploration of abundant, naturally occurring materials such as bamboo, coconut fibres, sisal and oil palm fibres which can be obtained locally at low cost and low levels of energy using local manpower and technology. The use of these locally available materials as substitute for the conventional materials in reinforced concrete elements can cut construction costs by as much as between 30% and 80%. Interest in these local materials is heightened by the facts that not only are they considered cheap; they are also “eco-friendly”. Also, the rising level of pollution in the construction industry has called for the adoption of “Eco-structures”, which are constructions that are in harmony with the surroundings and do not violate the environment neither through the chosen building materials nor through the construction methods. Several studies have shown that contemporary earth construction has the potentials to address the urban housing crisis in the developing countries. On the other hand there is a wrong perception among the users and the professionals that, ‘earth houses are only used by the poor people’. This paper investigates the information available on bamboo and earth material and their possible use as a low cost sustainable building material in Nigeria in the light of problem of affordable structure. The findings of this paper show that more research has to be done to come up with reasonable conclusion. It has however been found that this materials has been used to carry out some of the building activities. Finally the author thinks that if more information is found more can be done to put this material to use.

Keywords: *bamboo, earth, structure, construction, affordability, housing, materials*

1.0 INTRODUCTION

In most third world, tropical and Asian countries, investigation are on at finding suitable substitutes for steel reinforcement in concrete that would be cheap, strong, durable, possess good tensile characteristic and environmental friendly. Experts have evaluated natural fibre such as bamboo, oil palm fibre, palm kernel fibre and coconut palm fibre that are locally available, relatively less expensive, low energy demand during production, environmental friendly and 100% bio-gradable as possible substitutes for steel reinforcement and steel mesh. If bamboo is stripped into thin pieces and woven into a mesh it could be impregnated in the middle of a 10cm thick cement mortar to construct walls\ for low cost houses. The system being suggested will be factory made walls where door and window openings are pre-formed in the bamboo mesh and embedded in a 10cm (4inch-) thick cement mortar. The idea of a low cost house to my mind should not connote an inferior construction where inferior materials are used. Our ancestors have used strips of bamboo embedded in mud to build houses. If we can go a bit further to precast wall sections with bamboo fibres reinforced cement we shall have vindicated our modern education. Bamboo has been extensively studied as an alternative reinforcing material for concrete [1-2]. Ref [3] reported that natural fibres of the agave family have significant mechanical properties that make them suitable as potential reinforcement of cementitious matrix. Ref [4] conducted various tests to show the natural fibre reinforced concrete is stiffer, stronger and more ductile compared to plain concrete or even conventional steel reinforced concrete when subjected to abnormal cyclic loads.

Ref [5] noted that earth has been the most essential of the building materials, since the dawn of Man. Ever since mankind first congregated in villages almost 10,000 years ago. Unbaked earth has been one of they principal building materials, used in every continent. It is probably the very ubiquity of earthen buildings as expressed by [6] that has resulted in the low esteem with which it is held A cliché of television media reporting proclaims that they live in Mud huts? The phrase intended to convey poverty, squalor and misery of those in need of aid including new housing. In much of the world architects engineers and the government or authorities they served have assimilated the message and advocated building in permanent materials, especially unyielding concrete. Like other national materials such as palm bamboo or timber, earth building is associated with the disadvantage and the unprogressive, while concrete, steel frame sheet metal and plate glass are recognized as modern and prestigious even if they are often climatically inappropriate and expensive to produce and use. The rising cost and limited availability of modern construction materials such as block, cement and concrete make building an expensive proposition. Modern materials in building systems were developed because they are strong and longer lasting. The question remains for whom?

The whole world is in dilemma because of depletion of natural building material. Nigeria is therefore not an exception to the universal problem. This paper investigates the information available on bamboo, palm fibre and earth material and their potential use as a low cost sustainable building material in Nigeria in the light of problem of affordable structure.

1.1 HISTORICAL BACKGROUND OF EARTH AS A CONSTRUCTION MATERIAL IN AFRICA

In Africa, the Egyptian civilisation provides abundant evidence of the use of earth in building as found in the early human settlements at the Merimd and Fayum sites in the Nile delta, which dates from the fifth millennia before Christ. The dominance of the Egyptian dynasty promoted buildings of prestigious structures made of brick from the Nile clay, desert sand and straw from the grain fields. These bricks were made by hand and dried in the sun before the development of the mould. The excavation at Saggarah and Bbydos shows the use of bricks which were covered by stone. The art of brick vaulting was also developed in the lower Nubia, between Luxor and Aswan [7].

It is essential to look at historical evidence of the success of earth construction. It is currently estimated that over one third [8] to over one half [9] of the world's population lives in some type of earthen dwelling. The history of earth building lacks documentation, because it has not been highly regarded compared to stone and wood [10]. There are cities built of raw earth, such as: Catal Hunyuk in Turkey; Harappa and Mohenjo-Daro in Pakistan; Akhlet-Aton in Egypt; Babylon in Iraq; [11].

"*30% of the world's population, or nearly 1,500,000,000 people, live in a home built in unbaked earth. Roughly 50% of the population of developing countries, the majority of rural populations, and at least 20% of urban and suburban populations live in earth homes*" [10].

According to [12], *"earth architecture should not of course be considered a miraculous solution to neither all our housing problems, nor one which can be applied successfully anywhere, everywhere."* Before any building is constructed with earth, it is essential to identify the soil to be used. The identification process involves various tests, which need the use of a laboratory. Apart from the laboratory identification process, local knowledge of the soil and traditional skills are necessary. In Africa, suitable soil is found in most of the countries.

Weathered soils are composed mostly of iron and aluminium oxide and when dominated with iron oxide, they are called

2.0 BAMBOO TECHNOLOGY

According to our discovery bamboo has been the construction material used in about 70% of the houses found in the rural areas of the eastern part of Nigeria. However it was evident to us that those houses were not done properly and have not followed the accepted scientific standards of construction technology – walls were not straight not reinforced and the bamboo is exposed to the effects of external elements such as sun and rain with no foundation to stand on steadily. As a result the houses looked poor and had to be repaired every year and therefore the age old bamboo technology has been looked down upon as inferior and suitable only for temporary shelter. With simple drawings and constant follow up and finally with model houses we can fill the technological gap by adding the scientific knowledge and technological expertise to enhance the traditionally accepted, user friendly, environmentally sound technology to make the locally available material- bamboo more accepted and the construction to be ***stronger, durable*** *and more* ***practical*** Then to make bamboo durable, and therefore to be accepted by the community the treatment of bamboo has to be introduced. The end result will be able to build cost effective, affordable yet decent houses in Nigeria. Starting with a pilot program and expanding on the same, with families the new bamboo technology is user-friendly, community-driven, environmentally sustainable, durable and conducive in different climatic conditions. Additionally it will also provide alternative income generation activities through growing and harvesting bamboo.

2.1 IMPACT AND CHALLENGES OF USING THESE BUILDING TECHNOLOGIES

Impact

1. The immediate impact is on the number of families that will be served.
2. The program will become a 'community owned' program as the technology will be known and owned by the community being it a heritage of the communities especially the under privileged, neglected and the deprived communities.
3. Added pride and dignity to them promoting their own heritage and tradition.
4. Construction of houses will result in the increase of awareness of the need to protect the environment. The increased use of bamboo in construction will produce a 70% reduction of plantation timber used thus helping to reduce deforestation, encouraging new and existing bamboo cultivators to grow more bamboo, utilise wasteland, unused land and river banks, which will result in better soil conservation and mitigation of flood disasters.
5. This initiative will increase the options for income generation as bamboo mats can be made and marketed as more and houses are built.
6. As demand for bamboo increases more people will start cultivating bamboo.

Challenges:

1. The biggest challenge will be to change the mindset of the people to accept that bamboo construction is durable. Constant orientation and awareness program, distribution of educational materials and community group meetings will be helpful in addressing this issue.
2. Acceptance of "cement" and 'burnt bricks' as the sole durable construction material by the majority. Research findings and the model houses have been able to make a change on this though still it remains a challenge with all media and construction world being supportive of the same.

3. Less cooperation from the village skilled labourers in promoting the bamboo technology.
4. The prevailing old houses depict a poor picture of bamboo technology. Therefore, people tend to see all bamboo construction as poor technology though treated bamboo is durable for more than 30 years according to research done.

2.2 A WAY FORWARD IN MAKING HOUSING AFFORDABLE

There is a myth to accept burnt bricks, cement and reinforced concrete as the *only* material that helps constructing durable houses. In general concept, use of cement and sand in maximum quantity are the reason for durability of house. Nevertheless, they are costly and not so affordable to the poorer families. On the other hand when we look back into our history and traditions we see a difference. Clay, lime and stones have been the best used materials to build massive structures and palaces in the past and those have lasted centuries and withstood natural calamities. What do we learn from this scenario? We are challenged, especially those who are involved and shares expertise in construction, to come out with some innovative yet scientific methods, techniques and alternatives to overcome the housing problem that has direct impact on Nigerians. Some demonstration projects using bamboo technology are shown in figures 2.1 and 2.2. These are suggested ways:

1 Propagate more locally available bamboo
2 Investigate other species favourable to Nigeria climate
3 Find other ways of improving the success of bamboo to ensure ready availability
4 Find more application of bamboo as low cost building materials
5 Test the locally available bamboo strength for other application.
6 Find out how existing spaces can be used e.g leaves, seeds, and sprouts are food for both people and animals. Jointed stems in different sizes are used for pipes, furniture and household utensils.

Fig.2.1: *280 Sq.ft house, Stone foundation, Bamboo wall plastered with cement* (source:www.habitat.org/ap) [13]

Fig.2.2: 280 Sq.ft. house, stone foundation, Bamboo wall in timber frame (source:www.habitat.org/ap) [13]

Alternative construction materials

There is a myth to accept burnt bricks, cement and reinforced Concrete as the *only* material that helps constructing durable houses. In general concept, use of cement and sand in maximum quantity are the reason for durability of house. Nevertheless, they are costly and not so affordable to the poorer families. On the other hand when we look back into our history and traditions both in Nigeria and other African countries we see a difference. Clay, lime and stones have been the best used materials to build massive structures and palaces in the past and those have lasted centuries and withstood natural calamities. What do we learn from this scenario? We are challenged, especially those who are involved and shares expertise in construction, to come out with some innovative yet scientific methods, techniques and alternatives to overcome the housing problem that has direct impact on the majority of the population who live in inadequate housing in Nigeria.

Other guidelines for Selection of construction materials are (source:www.habitat.org/ap) [13]

1. Study of prevailing condition of house which is going to renovate or, new construction.
2. Keep record of materials which may be used after dismantling (if old house is going to dismantled)
3. Survey for different kind of materials which are locally available for construction.
4. Short-list of available materials and record their prevailing rates.

2.3 SOME OF THE ALTERNATIVE MATERIALS WE DO WORK WITH FOR COST EFFECTIVE HOUSING

1. Bamboo
2. Sunburn Bricks
3. Stone Boulders
4. Local wood

2.3.1 BAMBOO

Bamboo is a non timber evergreen plant that is mainly distributed in subtropical and tropical zones. Bamboo grows fast and has a high regenerating rate. Once planted, bamboo has new shoots every year. Usually bamboo can be harvested for culms purpose in the third and following years. Bamboo is both a decorative and a structural member. For shelter purposes bamboo culms between 60 to 100 mm diameters are generally adopted, which means that most of the bamboo spices available are within this range. Bamboo is widely used as a basic timber for rural housing. It is used as poles, purlins, trusses, rafters, mats for wall/ceilings/roof, frame of doors & windows and fence posts, especially in tropical countries. Though it is a natural product, the presence of large amount of hemi cellulose, starch and abundance of moisture makes it highly susceptible to biodegrading agent like white-rot, soft-rot, brown – rot, group of decay fungi and stain fungi and insects like borers and termites. In rural areas people are using bamboo without treatment. So it doesn't long last and this has created a myth in the minds of the people that bamboo is not a durable construction material. But with a simple treatment on bamboo the strength and duration of bamboo could be as long as 25 -30 years [14].

Harvesting considerations

1. Choose only those bamboo which are more than 3 years of old and it can simply determine by observing the colour of the culms has changed from clear and shiny green to greyish green and white bands at each node that have almost disappeared with replacement of gray bands.
2. As chances of attack by fungi will be more while moisture content is large so it is better to harvest bamboo in dry season.
3. Always cut the bamboo 20 to 30 cm above the ground or just above the first node.
4. Always use sharp tool or saw to avoid damaging of bamboo.
5. Keep the bamboo aright with support of other trees or on such arrangement for two to three weeks which helps to reduce the starch content of culms and as a whole it helps to reduce the chances of attack by insects or Bostrichidae and Lyetidae beetle.

Treatment of Bamboo

Bamboo can be treated mainly two ways)

1. Non chemical Treatment
2. Chemical Treatment

1. Non-Chemical Treatment

Traditionally the following treatment methods are adopted.

a) Water immersion treatment
b) Smoking
c) Using diesel *(Source: National Building Code-1994)*

a) Water immersion treatment

Normal practice followed in many bamboo producing areas is to immerse freshly cut bamboos in running water for about three to six weeks and keep it them for at least one week. By this treatment starch from bamboo is leached out making the culms immune to insect attack.

b) Smoking

This method is simple and effective .In this, bamboo are kept in a temporary smoking chamber to destroy starch by allowing smoking until they have a slightly dark color on its surface.

c) Using Diesel

Dry bamboos, both whole and split could be treated with light diesel oil by dipping in to.

2. Chemical treatment

Several permanent treatment method are available to protect bamboos and from insect attack and fungal attack. These methods are both preventive and remedial. For such kind of treatment many kind of chemicals are in use like creosote oil, Borax and Boric Acid.

a) Internodes Injection

Internodes injection for whole bamboo (poles and columns) 20-25ml. Creosote oil poured through 6mm hole in each inter node, the hole is closed with wax or putty. It is rolled once in two days for a week before it is put to use. The oil gets absorbed through the inner walls through the cross structural area, this treatment is used for dry bamboo.

b) Diffusion process

In diffusion process freshly felled culms or bamboos having moisture content of above 50% are kept submerged in preservative solutions for 1-2 days followed by stacking under shade for 10 – 15 days. To treat dry bamboo culms, they are be wetted before content will be about 50%. Split bamboos are also could be treated in this way and this way this treatment is more effective in split bamboos and slivers. By puncturing or drilling holes near the internodes before submerging them in preservative solution, treatments could be carried out for whole bamboos. The hole thus made can plugged with putty after taking the bamboos from the treating solution before they are stacked. This method will enhance penetration of treatment solution in whole bamboos. Preservatives which could be included – Boron, compounds, Copper Chrome – Arsenic (CCA), Copper Chrome – Boric (CCB), Acid copper chrome (ACC) [14].

Possible building structures and points to be remembered

Mainly bamboo are used as various construction material in housing as from foundation to the top. It may either be used as reinforcement on sub soil foundation, as horizontal tie member for plinth, lintel, and top of wall as well as rafter and purlins of roofing work. From ancient time different architectural buildings are made up of bamboo. Generally from technical point of view, bamboo is stronger in tensile member but it may also be used as column with some of engineering techniques. It is recommended that it is not wise to use bamboo as horizontal member on more than 10-12 feet without middle support and only to be used after treatment as mentioned above. By special care and treatment, it prolongs the age of structural bamboo up to 30 to 40 years.

Fabrication: - cutting, joints and finishing

The dimension and size of the bamboo are mainly depends on structural component and loading. While designing any shelter or house, the built up area as well as the spacing between two partitions is more important. 11 feet to 15 feet between two partitions generally are suitable without having a support in between. In bamboo houses generally, the slant of the roof of GI or corrugated bamboo roofing sheet are used so the height will have to be maintained not more than 8 feet on lower side and 11 feet on top side. For the different purposes on housing bamboo can be fabricated as follows figure 2.3.

Only basic carpentry, Masonry tools and skills are necessary for the construction of bamboo houses. Figure 2.4 below shows end profiles of bamboo poles for use in construction. For the joints and other clamping points these kind of cutting or fabricating is preferred. With clear instruction and proper workmanship we may get such a kind of joint arrangement with using steel clamps and nut arrangement.

Fig 2.3: Bamboo Column
(source:www.habitat.org/ap) [13]

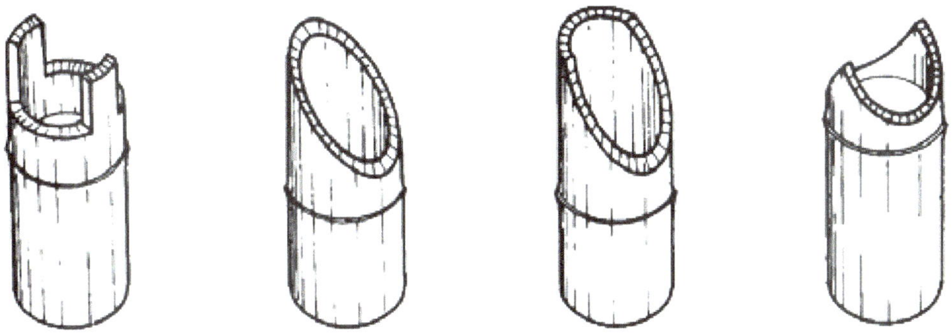

Fig.2.4: End profiles of bamboo poles for use in construction
(source:www.habitat.org/ap) [13]

In the proposed bamboo house the columns are constructed from whole bamboo poles. The poles are fixed at the level of foundations/plinth. For this purpose the end 500 mm of the pole is injected with cement based mortar and reinforced with a steel bar- see Figure 2.5 .The bamboo should be reinforced in this way before mounting them in the foundations.

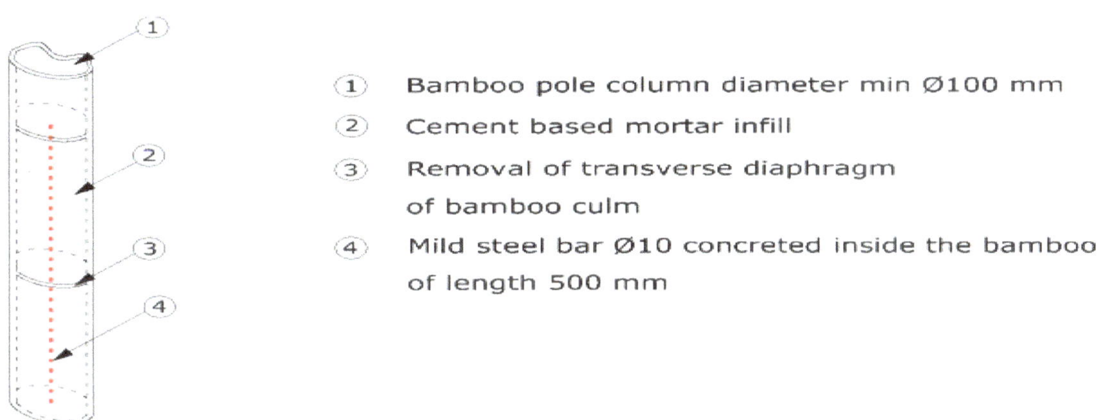

Fig.2.5: End 500 mm of the pole is injected with cement based mortar and reinforced with a steel bar (source:www.habitat.org/ap) [13]

All bamboo columns are connected at the top to a timber or bamboo with some steel plate. This beam has a rectangular cross section for timber and hooking arrangement of reinforcement from bamboo. The bamboo pole column is connected to this beam through an anchor bolt. The anchor bolt is embedded in the pole where cement based mortar is injected- see Figure 2.6.

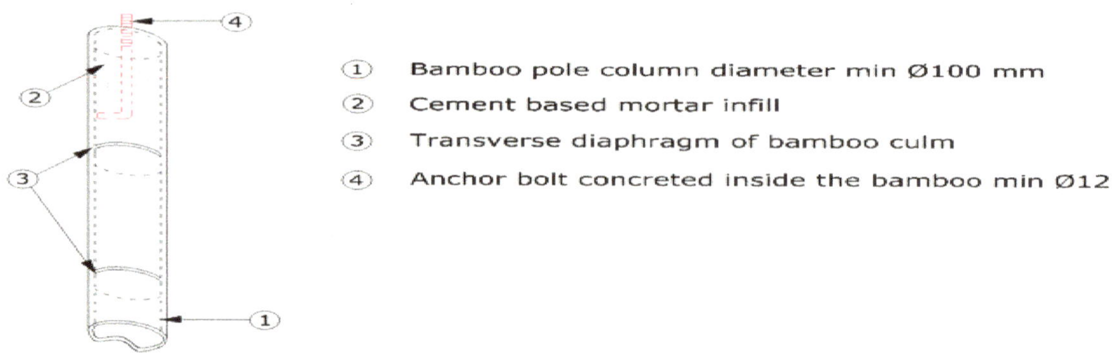

Fig.2.6: Bamboo pole column is connected to beam through an anchor bolt (source:www.habitat.org/ap) [13]

3.0 EARTHEN CONSTRUCTION

Earth as building materials is available everywhere, existing in many different compositions that can be processed in a myriads of ways.

3.1 TYPES

The following are among the more popular types of earthen construction [15]

1. Adobe
2. Rammed earth
3. Compressed earth blocks

3.1.1 ADOBE (LATERITE BLOCK)

Adobe is air/sum dried brick from and composed of inorganic soil and sand. The soil must have minimum clay content of 10%. Fibers such as straw may also be added to increasing the stability of the block as work well as binders. An adobe brick is typically 250 to 300mm and weighs between 13 -22kg [16]. Some of the steps that can be taken to protect adobe structures are [15]:

- Placing walls on concrete or stone foundations.
- Using water inhibiting additives
- Plastering adobe walls with stucco
- Providing substantial overhangs.

3.1.1 RAMMED EARTH

Rammed earth involves the compacting of moist soil between rigid forms to create monolithic earth walls with similar properties as that of adobe walls. The soil for rammed earth construction must have about 30% clay and 70% sand and small gravel. It is critical to ensure that the moisture content of the wall is just right as if the mixture is too dry. Then it will lead to a weak and crumbly wall [16].

3.1.2 COMPRESSED EARTH, BLOCKS

Compressed earth blocks (CEBs) are a relatively recent technology and combine the nest characteristics of traditional earthen technology and modern brick making process. Earth is poured into moulds and compressed either manually or mechanically.

3.2 BENEFITS

The most obvious advantage of earth construction is the abundance of the raw material earth. Other advantages of earthen construction include [17-24]:

- High thermal insulating properties
- High sound insulation
- Not susceptible to insects or rodents
- No waste generated during construction
- Inert- contains no toxic substances
- Construction is inexpensive and simple
- High workability and flexibility
- High fire resistant.
- Earth construction is economically beneficial.
- It requires simple tools and less skilled labour.
- It encourages self-help construction.
- Suitable for very strong and secured structure.
- It saves energy (low embodied energy).
- It balances and improves indoor air humidity and temperature.
- Earth is very good in fire resistance.
- Earth construction is regarded as a job creation opportunity.
- Earth construction is environmentally sustainable.
- Loam preserves timber and other organic materials.
- Earth walls (loam) absorb pollutants.
- Easy to design with and high aesthetical value.
- Earth buildings provide better noise control.
- Earth construction promotes local culture, heritage, and material.
- Earth is available in large quantities in most regions.

3.3 CONSTRAINTS OR DRAWBACKS

The in spite of the well established technology and also of the application areas [25], laterite blocks has not really found favour in housing programmes taken up in the country either in the public or in the private sector level. The constraints are mainly due to [19-24, 26]:

- Acceptability of the material in view of bias and stigma attached to the material. Otherwise considered to be the material which is and also for use of only the poor man
- The lack of institutional arrangement for market availability of the blocks as normally available for sandcrete blocks through block industries and suppliers.
- The lack of support through code provisions either in the national codes or through the Public Works Department (PWD) codes.
- The lack of references of materials in most of the schedule of rats in the country as an alternative to sandcrete blocks.
- The over-emphasizing use of materials like sandcrete blocks as of higher strength in comparison to stabilized laterite blocks.
- Tendency to adopt a safer pattern of using established materials.

- Lack of demonstration project that will infuse confidence among the people for using the laterite blocks technology for housing and building.
- Lack of sufficient machinery to be distributed all over the country of proven capabilities and training for entrepreneur and workers for manufacture of stabilized laterite blocks of required quality.
- Less durable as a construction material compared to conventional materials.
- Earth construction is labour intensive.
- Mud houses behave poorly in the event of earthquakes.
- Structural limitations.
- Need high maintenance.
- Professionals make less money from earth building projects.
- Special skills needed for plastering.
- Loam is not a standardised building material.
- Need higher wall thickness.
- Suitable only for in situ construction.

3.4 AREAS OF SUPPORT ASSISTANCE NEEDED

Based on experience [25], the following are major areas of support assistance needed to make laterite blocks making technology popular in housing project both at the public sector and private sector.

- They should be increased access to block making machines.
- Training of people in this field
- Codes support
- Schedule rate
- Public building using earth materials (demonstration projects)
- Exposure on the subject to students of architecture and engineering.
- Public awareness

Location of clay deposits in Nigeria is shown in Figure 3 1. Laterite usually contains a good quantity of clay particles. A systematic survey of clay deposits by the Nigeria

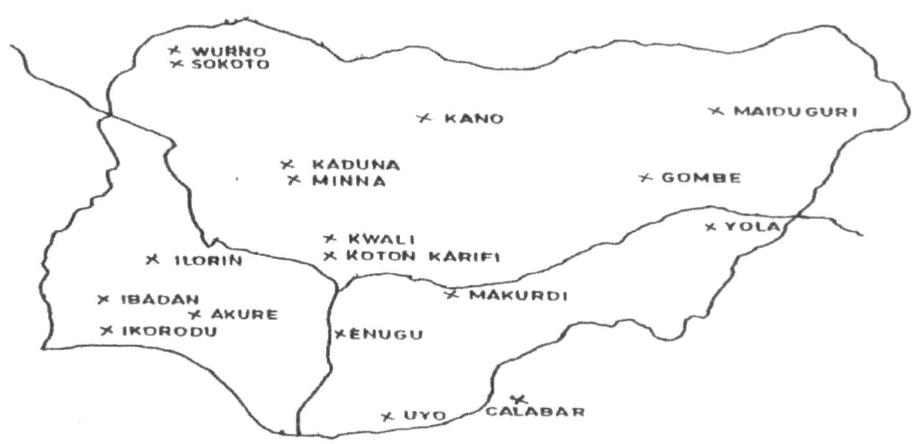

Fig 3.1: Location of Clay Deposits in Nigeria
(Source: Salau 2008) [27]

Geological Survey Department is given in Figure 3.1 and their preferred use, suggested in Table 3 1.

Table 3.1: Clay Deposit Suitable for Bricks in Nigeria (Source: Salau 2008) [27]

S/No	CLAY	LOCATION	STATE	ROUTE/AREA	SUGGESTED INDUSTRIAL USE
1	CLAY	Agbassah, Kingbo, Uyode	Delta	12m, E.W.E., Ughelli, 16km from Ughelli-Warri Road 4 and a half km N. W., Ughelli & 7km W.S.W, Ughelli	Bricks, Pipes (Floor) Foilet Bricks, pipes etc.
2	CLAY	Kiagbodo	Delta	Delta, Ughelli, 16km East of Ughelli, 30km N.W.W. Sapele on Ossiomo River	Bricks, pipes, etc
3	BALL CLAY	Ebutte Orega	Lagos	Ikorodu	Pottery, tiles and bricks
4	CLAY	Ifo	Ogun	Ijegun Junction at 36km on railway	Bricks, tiles, etc
5	CLAY	Oke-Seri	Ogun	2km East of Ibadan Rd. and 5km. North of Ijebu-Ode, Ijebu bextensive to Imope village.	Bricks Manufacturing Floor and roof tiles
6	CLAY	Ijaiye, Ake, Ijegun, Ibara	Ogun	Abeokuta, Ake, quarry Itoko, quarry near Own African Central School and Lafenwa quarry	Bricks, Pottery, etc.
7	BALL CLAY	Agbaignara Osu	Imo	Okigwe, Owerri	Refractory, Bricks etc
8	CLAY	Ondo	Ondo	Tributary of Oluwa River North Agbaje	Tiles, Bricks
9	CLAY	Rimi, Dakata Ganyawa	Kano	Kano	Bricks
10	BALL CLAY	Ekpene Obom	Akwa Ibom	Uyo near Catholic School	Pottery, Bricks, tiles, etc
11	BALL CLAY	Mbiafor, Ikot-Ekpene	Akwa Ibom	Uyo	Pottery, Bricks, tiles, etc
12	FIRE CLAY	Obuga	Enugu	Enugu, Escarpment to West of Enugu from Obuga to Nyaba River.	Refractory Bricks

3.5 CONSEQUENCES OF UNAFFORDABLE STRUCTURES

The after-effect of the unaffordable cost of construction is intentional and unprofessional economy in construction materials which normally lead to collapsed buildings and structures. Recently, there have been cases of collapsed buildings and structures but only few are reported. The Nigerian building professionals, especially the Structural Engineers have also been indicted. The incidences have led to questioning their competence and capability. Majority of the collapse of concrete structures are during the construction period due to the use of materials. Among the well publicized structural collapse of buildings are [27]:

i) 8-Storey Hotel building on Akinwunmi Street, Okupe Estate, Mende Maryland, Lagos State in 1989. (Collapsed on the Eve of Official Opening)

ii) Saque Comprehensive Primary and Secondary School, Port-Harcourt, River-State on the 15th June, 1990.
iii) Two-storey building at Maryland, Lagos State on the 1st Jan. 1999
iv) Uncompleted church building at Mafoluku, Lagos State
v) 3-storey building at Abuja, 1996
vi) 3-storey building at No.3 Cheme Avenue, New Haven Enugu, Enugu State, on the 12th June, 1997
vii) Main entrance canopy of the main approach to the Eleganza Ball Pen Industry Estate, Ajah, Lagos State, 2000
viii) Collapse of completed and earlier certified fit, three-storey building at Ebute-Metta West, Lagos Mainland in 2006
ix) Collapse of uncompleted Bank building at No 1, Murtala Muhammed International Airport Road, Mafoluku, Oshodi, opposite Glass House office of "Construction New Digest" in 2006
x) 4 collapses in Port Harcourt between Aug. and Sept 2005 (Abacha Road, Rombia Str, Old GRA and Tourist Beach) at different stages of construction in a reclaimed mangrove swamp.
xi) Partial Collapse of Bank of Industries Building at Custom Street, Lagos in 2006
xii) Two-storey Residential House at Ibadan Street, Ebute-Metta East in Lagos Mainland in 2007
xiii) Residential Building at Okepopo, Lagos Island in 2008

It should be noted that materials, either locally or imported, do not fail but the wrong selection of materials at the conceptual, design and construction stages are often the cause(s) [27]. From the experience of the investigation of the failed structures, the lesson to be drawn there from can be expressed with this statement.

Finally, the responsibility of an Engineer is given in the delight book of Henry Pertoki, *To Engineer is Human."Structural design is more akin to coaching. However, the design engineer must do better than any coach, for he is expected to win every game he plays. That is a tough assignment when one mistake can often mean a loss. And when defeat occurs, all one can hope is to analyze the game film and learn from the mistakes so that they are less likely to be repeated the next time out."*

4.0 CONCLUSIONS

The conclusion is that the dream of housing for all by Year 2020 and United Nations' Millennium Development Goals (Year 2015) can only be achieved if the abundant local structural engineering materials in Nigeria, on which a lot of research works have been done, can be used to get affordable houses and structures.

The failure of the government in the provision of adequate houses for the population has made laterite blocks inevitable for the masses. However the low esteem with which the material is regarded can be overcome if some of these recommendations given above can be implemented. In this way government will not burden herself with the provision of housing as this will now shift to private developers and the government can concentrate on essential building services

The findings of this paper show that more research has to be done to come up with reasonable conclusion. It has however been found that this materials has been used to carryout some of the building activities. Finally the author thinks that if more information is found more can be done to put this material to use

5.0 RECOMMENDATIONS

1. Standard Organization of Nigeria, agency responsible for Quality Control of Construction Material should ensure that all brands of imported cement satisfy the relevant Nigerian codes and suitable for the environment.
2. For sustainable economic and industrial development, governments at all levels and organized private sectors should invest on the development and application of abundant local materials in Structural Engineering.
3. There should be continued focus on the development and usage of earthen construction as this will significantly reduce the usage, overdependence on Ordinary Portland Cement and stabilize its cost. This will contribute to the achievement of the Millennium Development Goals.
4. The usage of bamboo and other natural fibres such as oil palm fibre, palm kernel fibre and coconut palm fibre should be encouraged as possible substitutes for steel reinforcement and steel mesh.
5. Governments and Organized Private Sectors should have confidence in results of research findings in Nigeria and adequately support their practical applications.
6. Various recommendations on the possible preventive measures on collapse of structures, presented to the Government by the Nigerian Society of Engineers should be considered while the appropriate Government Regulatory Bodies should ensure their enforcement in the construction industry in Nigeria.

6.0 REFERENCES

[1] Youssef, M.A.R., (1976), "Bamboo as a Substitute for Steel Reinforcement in Structural Concrete", New Horizons in Construction Materials. 1, 525-554.

[2] Kankam, I. A., Perry. S. H and Ben-George. M. (1980), "Bamboo Reinforced Concrete One-Way Slabs subjected to Live Loading" International Journal for Development Technology.

[3] Castro, J and Naaman, A.C (1981), "Cement Mortar Reinforced with Natural Fibres" ACI Journal, Vol 28, No.1, 69-78.

[4] Lakshmipathy, K and Santhakumar, P.T. (1980), "The Planting, the Flora of West Tropical Africa" Millibark, London, Vol II, P 251-253.

[5] Rao Gundu, R (1990), "Conservation of Earth Architecture of India", in Indian National Trust for Arts and Cultural Heritage(INTACH), Municipal Institute of Technology(MIT), Housing and Urban Development Corporation (HUDCO), National Seminar on Mud in Habitat Manipal 7-8 December,1990, pages: 1-8

[6] Norton, J. (1997), "Building with Earth", A handbook, 2nd Edition, Intermediate Technology publications, London

[7] Rastorfer, D. (1985), "The man and his work: Hassan Fathy", A Mimar Book, Concept Media, Singapore, Architectural Press, London.

[8] Dethier, J. (1981), "Down to earth: adobe structure – an old idea, a new future", New York facts on file, USA.

[9] Smith, E. W. and Austin, G. S. (1989), "Adobe, pressed earth, and rammed earth industries in New Mexico", New Mexico Bureau of Mines and Mineral Resources, Bulletin 127, USA.

[10] Houben, H. and Guillaud, H. (1989), "Earth construction", Intermediate Technology publications 1994, London.

[11] Easton, D. (1998), "The Rammed Earth House", Chelsea Publishing Company, White River Junction, Vermont, USA.

[12] Denyer, S. (1978), "African Traditional Architecture", Heinemann, Nairobi, Kenya.

[13] www. habitat.org/ap

[14] www.caneandbamboo.org

[15] Kadiri Kabir O. (2005), "mass housing through earth construction technology in Nigeria", Pakistan journal of social sciences, Vol 3 (5), pages: 755-760

[16] Allen,E. and R.Thallon (2001), "Fundamentals of Residential Contruction", New York : John Wiley and Sons in chandra varun, Alternative Construction System for Residential Construction.http://archnt2.tamu.edu/cgraham/student_project.htm pp:2-4

[17] Zami, M. S. and Lee, A. (2007), "Earth as an alternative building material for sustainable low cost housing in Zimbabwe", The 7th International Postgraduate Research Conference. March 28 – 29, 2007, The Lowry, Salford Quays, Salford, Greater Manchester, UK.

[18] Morton, T. (2007), "Towards the development of contemporary Earth Construction in the UK: drivers and benefits of Earth Masonry as a Sustainable Mainstream Construction Technique", International Symposium on Earthen Structures, Indian Institute of Science, Bangalore, 22-24 August. Interline Publishing, India.

[19] Hadjri, K., Osmani, M., Baiche, B. And Chifunda, C. (2007), "Attitude towards earth building for Zambian housing provision", Proceedings of the ICE institution of civil engineers, engineering sustainability 160, issue ES3

[20] Minke, G. (2006), "Building with earth, design and technology of a sustainable architecture", Birkhauser publishers for architecture. Basel, Berlin, Boston.

[21] Walker, P. Keable, R. Martin, J. and Maniatidis, V. (2005), "Rammed earth: Design and Construction Guidelines", BRE Bookshop, UK.

[22] Maini, S. (2005), "Earthen architecture for sustainable habitat and compressed stabilised earth block technology", Programme of the city on heritage lecture on clay architecture and building techniques by compressed earth, High Commission of Ryadh City Development, The Auroville Earth Institute, Auroville Building Centre – INDIA.

[23] Adam, E. A. and Agib, A. R. A. (2001), "Compressed Stabilised Earth Block Manufacture in Sdan", Printed by Graphoprint for the United Nations Educational, Scientific and Cultural Organization. France, Paris, UNESCO.

[24] Morris, J. and Booysen, Q. (2000), "Earth construction in Africa", Proceedings: strategies for a sustainable Built Environment, Pretoria, 23-25 August.

[25] Suresh,V.(1990), "Use of stabilized mud blocks in housing and building projects .In: Indian National Trust for Arts and Cultural Heritage(INTACH), Municipal Institute of Technology (MIT), Housing and Urban Development Corporation(HUDCO)(1990).National Seminar on Mud in Habitat Manipal 7-8 December,1990,pp:1-24

[26] Blondet, M. and Aguilar, R. (2007), "Seismic protection of earthen buildings", International Symposium on Earthen Structures", Indian Institute of Science, Bangalore, 22-24 August, Interline Publishing, India.

[27] Salau, M.A and Ikponmwosa, E.E (2008), "Structural Characteristic of Bamboo Reinforced Concrete Slabs" Journal of Engineering, India

20

A Review of Sustainability Reporting Tools (SRTs) for Communities

Renard Yung Jhien Siew*[1]

[1]Centre for Energy and Environmental Markets, University of New South Wales, Australia

*Corresponding E-mail: rensiew10@gmail.com

Abstract

Sustainability reporting tools (SRTs) for Communities is part of a new paradigm to aid local governments and councils in their decision making process. These integrative tools have evolved from prior versions of SRTs that focus purely on a single building performance to consider the sustainability performance of communities. This paper provides a review of mainstream SRTs for Communities, in particular: BREEAM for Communities; LEED for Neighborhood Development; CASBEE for Urban Development; Green Star for Communities; Sustainability Tool for Assessing and Rating (STAR); EcoCity; and HQE2R. A critique of these tools is provided to better understand some of the existing limitations including the lack of clarity in the size of development that these SRTs are capable of assessing; lack of published reasoning behind the allocated scores or weightings for the criteria selected; inadequate account of the different sources of uncertainty; the adoption of a static perspective; and the lack of acknowledgement of possible interaction or correlation between criteria. This review will be of interest to practitioners, academics and developers who are concerned about ways to improve the sustainability of the built environment.

Keywords: *Sustainability reporting tools (SRTs), urban community, rating, urban sustainability, neighborhood, assessment, triple bottom line*

1.0 Introduction

The definition of sustainable development provided by the Brundtland Report (1987) [1]: 'development that meets the needs of the present without compromising the ability of future generations to meet their own needs' is perhaps the most cited definition among leading scholars. This concept encompasses a cleaner environment with efficient use of resources and a more inclusive society with shared benefits of increased prosperity. In the built environment, the development of sustainable communities has been a key topic of discussion across global conferences. There is clearly an increased acknowledgement by practitioners, local governments and policy makers that urban sustainability is critical and deserves much attention as it has a huge impact on the livelihoods of surrounding communities and their future generations.

Berardi (2013) [2] claims that communities are nowadays considered an appropriate scale to measure the sustainability of the built environment citing that 'over 50% of the world's population live in urban areas and that this figure is expected to rise to 70% by 2050; in Europe, 75% of the population lives in urban areas, and by 2020, the number is expected to reach 80%'.

To guide stakeholders in making informed plans and decisions, sustainability reporting tools (SRTs) have evolved from a primary focus on environmental issues of single buildings such as Green Star, BREEAM and LEED [3] to the assessment of the sustainability of communities. This is, in part, due to the criticism that SRTs for buildings are incapable of addressing the volume of sustainability challenges that are beyond environmental issues [3, 4]. The scaling up of SRTs to a community level is perceived as an effective way of tackling a range of sustainability issues: pollution, biodiversity, social needs, transportation, climate change and energy among others synergistically [4]. Some of the more established SRTs to assess the sustainability of communities include BREEAM for Communities, CASBEE for Urban Development, Green Star for Communities, LEED for neighbourhood Development, Sustainable Tools for Assessing and

Rating (STAR), EcoCity and HQE2R. This paper aims to review and discuss the differences between these tools. A critique will be provided followed by suggestions for improvement.

2.0 Background

A search was done on Google, Scopus and Science Direct for articles, reports as well as guidelines with the following combination of key words: 'sustainability assessment'; 'communities'; and 'neighbourhood'. The search was restricted to articles published in all languages within the last five years (2009-2013). The search was restricted to the last five years to avoid redundant data and to select more recent updates on sustainability assessment for communities. The search results revealed that the seven most commonly discussed tools that are used to measure the sustainability or communities/neighbourhood are BREEAM for Communities, CASBEE for Urban Development, Green Star for Communities, Sustainability Tool for Assessing and Rating (STAR); EcoCity; and HQE2R. The nature and characteristics of these tools are discussed in this section.

2.1 BREEAM for Communities

One of the main features of the BREEAM for Communities tool is that it allows developers, local authorities and master planning professionals to integrate sustainable design features into the master planning stage. This tool is aligned with the current UK planning process addressing all core planning principles found within the national planning policy. There are three progressive steps involved in this tool, each comprising of a list of criteria to be addressed. All criteria listed under Step 1 are mandatory and must be addressed by users of this tool while criteria listed under steps 2 (except for consultation and engagement) and 3 are not. The list of criteria relevant to each step and relevant scores in parentheses are highlighted in Table 1 [5].

Table 1: Criteria for BREEAM Communities [5]

Steps	Criterion	Scores available
Step 1: Establishing the principle of development	Consultation plan (1); economic impact (2); demographic needs and priorities (1); flood risk assessment (2); noise pollution (3); energy strategy (11); existing buildings and infrastructure (2); water strategy (1); ecology strategy (1); land use (3); transport assessment (2)	29
Step 2: Determining the layout of the development	Consultation and engagement (2); design review (2); housing provision (2); delivery of services facilities and amenities (7); public realm (2); microclimate (3); utilities (3); adapting to climate change (3); 'green' infrastructure (4); local parking (1); flood risk management (3); water pollution (3); enhancement of ecological value (3); landscape (5); safe and appealing streets (4); cycling network (1); access to public transport (4)	52
Step 3: Designing the details	Community management of facilities (3); local vernacular (2); inclusive design (3); light pollution (3); training and skills (3); sustainable buildings (6); low impact materials (6); resource efficiency (4); transport carbon emissions (1); rainwater harvesting (3); cycling facilities (2); public transport facilities (2)	38
Innovation	Any innovation within the design, planning and construction industry that are currently not addressed by BREEAM Communities	7

The percentage of each criterion score achieved out of the total available scores is multiplied by a fixed criterion weighting. The sum of all weighted percentages is mapped to one of the six BREEAM awards as shown in Table 2 [3].

2.2 CASBEE for Urban Development

The Japan Sustainable Building Consortium (JSBC) introduced the Comprehensive Assessment System for Building Environmental Efficiency- for Urban Development (CASBEE-UD) in 2006 to assess sustainability in neighbourhood developments [4]. This tool utilises a unique methodology where the environmental quality (Q) within the site boundary is divided by the environmental load (L) beyond the site boundary to calculate the neighbourhood's environmental efficiency. It considers six categories namely natural environment, service functions for the designated area, contribution to the local community, environmental impact on microclimates, façade and landscape, social infrastructure, and management of the local environment. Each of these categories are composed of criteria which in turn are divided into more specific subcriteria [6]. According to Sharifi and Murayama (2014) [4], scores of subcriteria are added up to give the total score of the higher level criteria. This is repeated until final scores for both environmental quality (Q) and environmental load (L) are obtained, following which the Building Environmental Efficiency of urban Development (BEE-UD) is calculated based on Equation 1:

$$BEE-UD = \frac{25X(Q-1)}{25X(5-L)} \tag{1}$$

The final BEE-UD score is mapped to a CASBEE award as shown in Table 2 (please refer Page 42).

2.3 Green Star for Communities

The Green Star for Communities tool was developed in 2012 by the Green Building Council of Australia (GBCA) in consultation with various industry stakeholders but only released in March 2014. A total of 100 points is distributed across 36 subcriteria with the possibility of getting another 10 bonus points through the innovation criterion. The breakdown of the criteria is depicted in Table 3. In a similar fashion to previous tools such as Green Star (Design and As-Built) which specifically rates the sustainability potential of buildings [3], the cumulative scores from each criterion is mapped to a Green Star award as described in Table 2. The Green Star Communities tool only certifies community projects that have achieved at least 4 Green Stars. A re-certification every five years is required to ensure that commitments made during the design and planning stages are delivered.

Table 3: Green Star Communities criteria [7]

Criterion	Subcriterion
Governance	Accredited professional (1); corporate responsibility (4); sustainability awareness (3); engagement (6); operational governance (2); adaptation and resilience (4); environmental management (1)
Design	Site and context analysis (2); site planning and layout (4); urban design (4);
Liveability	Access to amenities (2); community development (4); healthy and active living (5); access to fresh food (2); safe places (2); culture, heritage and identity (3)
Economic prosperity	Employment and economic resilience (2); education and skills development (2); return on investment (2); community investment (4); affordability (4); incentive programs (2); digital economy (2); peak electricity demand (2);
Environment	Site sensitivity (2); ecological enhancement(2); heat island effect(1); light pollution(1); greenhouse gas emissions (6); 'green' buildings (4); potable water consumption(4); stormwater (3); materials(3); waste management (2); transport(3)
Innovation	10 points available demonstration of innovation

Table 2: Mapping scores to award

Levels of award							SRTs for Communities
	Unclassified	Pass	Good	Very good	Excellent	Outstanding	BREEAM for communities
	< 25%	≥ 25%	≥ 40%	≥ 55%	≥ 70%	≥ 85%	
		Poor	Fairly good	Good	Very good	Excellent	
		BEE < 0.5	BEE= 0.5-1.0	BEE= 1.0-1.5	BEE= 1.5-3.0	BEE ≥ 3	CASBEE for Urban Development
	1 Green Star	2 Green Stars	3 Green Stars	4 Green Stars	5 Green Stars	6 Green Stars	
	<20 points	≥20 points	≥30 points	≥45 points	≥60 points	≥ 75 points	Green Star for Communities
			Certified	Silver	Gold	Platinum	
			40-49 points	50-59 points	60-79 points	80-100 points	LEED for Neighborhood Development
			Reporting STAR	3-STAR	4-STAR	5-STAR	
			<200	200-399	400-599	600+	Sustainable Tools for Assessing and Rating (STAR)
		Bad	Medium	Good	Very good	Best score	
		E	D	C	B	A	Eco-City
-3	-2	-1	0	1	2	3	HQE^2R

2.4 LEED for Neighbourhood Development

The LEED for Neighbourhood Development tool integrates principles of 'green' buildings, New Urbanism and smart growth into neighbourhood design. This tool was initially developed by the U.S. Green Building Council (USGBC) in collaboration with the Congress of New Urbanism (CNU) and the Natural Resources Defense Council (NRDC). The six main criteria are further divided into sub-criteria as shown in Table 4. Some criteria are prerequisites while others carry points if requirements are met.

Table 4: LEED Neighbourhood Development criteria [8]

Criterion	Subcriterion
Smart location and linkage- encourages communities to consider location, transportation alternatives, and preservation of sensitive lands	Smart location (pre); preferred locations (10); brownfield development (2);housing and jobs proximity (3); imperilled species and ecological communities (pre), wetland and water body conservation (pre), site design for habitat or wetland and water body conservation (1); agricultural land conservation (pre), floodplain avoidance (pre), steep slope protection (1), restoration of habitat or wetlands and water bodies (1); long term conservation management of habitat or wetlands and water bodies (1); locations with reduced automobile dependence (7); bicycle network and storage (1)
Connected neighbourhoods - focuses on vibrant, equitable communities that are healthy	Compact development (6); walkable streets (12); connected and open community (pre); reduced parking footprint (1); tree-lined and shaded trees (2); transportation demand management (2); mixed used neighbourhood centres (4); access to civic and public spaces (1); access to recreation facilities (1); mixed income diverse communities (7); community outreach and involvement (2); local food production (1); neighbourhood schools (1); street network (2); transit facilities (1); visitability and universal design (1)
'Green' infrastructure and buildings – promotes the design and construction of buildings that reduces energy and water use while promoting more sustainable use of materials,, reuse of existing and historic structures, and other sustainable best practices.	Certified 'green' building (5); building energy efficiency (2); building water efficiency (1); water-efficient landscaping (1); existing building reuse(1); historic resource preservation and adaptive use(1); construction activity pollution prevention (pre); stormwater management (4); light pollution reduction (1); heat island reduction (1); solar –orientation (1); on-site renewable energy sources (3); district heating and cooling (2); infrastructure energy efficiency (1); solid waste management infrastructure (1); wastewater management (2); recycled content in infrastructure (1); minimised site disturbance in design and construction (1)
Innovation and design process- recognises innovative practices that go beyond the existing criteria in the tool	Innovation and exemplary performance (5); LEED accredited professional (1)
Regional priority- encourages projects to focus on meeting criteria that have a significance to the local environment	Regional priority credit (4)

pre = prerequisite

Unlike the other residential LEED tool, LEED for Neighbourhood Development comprises the following three stages of certification:

Table 5: Three stages in the LEED Neighbourhood Development tool [8]

Stages	Description
Stage 1- Conditionally approved plan	This stage provides the conditional approval for a LEED Neighbourhood Development plan before the completion of entitlements of the public review process. Completing this stage is envisioned to help the project garner support from the community and local government.
Stage 2-Pre-certified plan	Pre-certifies a LEED Neighbourhood Development plan and is applicable to either fully entitled projects or projects under construction. Completing this stage might help projects secure financing or attract tenants.
Stage 3- Certified Neighbourhood Development	Once the project is completed, it can apply for the certification process which recognises that the project has achieved all prerequisites and met certain level of sustainability performance.

The LEED Neighbourhood Development award is determined based on the aggregated scores for each criterion shown in Table 2.

2.5 Sustainable Tools for Assessing and Rating (STAR) Communities

STAR Communities was developed between 2008 and 2012 by a joint steering committee and technical advisory group [9]. The purpose of STAR Communities is to provide a common tool for local governments to inform the progress on sustainability initiatives using a set of performance-based criteria. This tool includes seven criteria, where each criterion is further divided into 5-7 subcriterion (total up to 44 subcriteria- see Table 6). Scores are given to neighbourhood communities for successfully meeting the requirements of each subcriterion. The aggregated subcriteria scores are then mapped to a STAR Communities award as shown in Table 2.

Table 6: STAR Communities criteria [9]

Criterion	Subcriterion
Built environment	Ambient noise and light (5); community water systems (15); compact and complete communities (20); housing affordability (15); infill and redevelopment (10); public spaces (15); transportation choices (20)
Climate and energy	Climate adaptation (15); greenhouse gas mitigation (20); greening the energy supply (15); industrial sector resource efficiency (10); resource efficient buildings (15); resource efficient infrastructure (10); waste minimisation (15)
Economy and jobs	Business retention and development (20); green market development (15); local economy (15); quality jobs and living wages (20); targeted industry development (15); workforce readiness (15)
Education, arts and community	Arts and culture (15); community cohesion (15); educational opportunity and attainment (20); historic preservation (10); social and cultural diversity (10)
Equity and empowerment	Civic engagement (15); civil and human rights (10); environmental justice (15); equitable services and access (20); human services (20); poverty prevention and alleviation (20)
Health and safety	Active living (15); community health and health system (20); emergency prevention and response (15); food access and nutrition (15); indoor air quality (5); natural and human hazards (15); safe communities (15)
Natural systems	'Green' infrastructure (20); invasive species (10); natural resource protection (20); outdoor air quality (15); water in the environment (20); working lands (15)

2.6 EcoCity

The European Commission in its 5th framework program initiated the EcoCity project which aims to develop a common concept and design model settlements across seven countries (Austria, Finland, Germany, Hungary, Slovak Republic, Italy and Spain). The developed tool focuses on main criteria such as: urban structure; transport; energy flows; material flows; socio-economic issues; and processes [10]. It uses relative benchmarks in its evaluation, that is the measured value of a criterion is compared against the given benchmark criterion value to derive its score between A (best) and E (worst). If the criterion meets the benchmark value, a score of D- normal practice is given [10]. The five award levels for EcoCity are detailed in Table 2.

2.7 HQE2R

The HQE2R project led by Charlot-Valdieu and funded by the European Commission aims to provide decision making tools for municipalities and their local partners focussing on the needs of residents and users of neighbourhoods [11]. The five main criteria identified under this project are: to preserve and enhance heritage and conserve resources; to improve the quality of the local environment; to ensure diversity; to improve integration; and to reinforce social life. These criteria are further divided into a total of 21 subcriteria detailed in Table 7. Each criterion is measured based on a -3 to +3 scale.

Table 7: HQE^{2}R criteria [11]

Criterion	Subcriterion
To preserve and enhance heritage and conserve resources	To reduce energy consumption and improve energy management; to improve water resource management and quality; to avoid land consumption and improve land management; to reduce the consumption of materials and improve their management; to preserve and enhance the built and natural heritage
To improve the quality of the local environment	To preserve and enhance the landscape and visual comfort; to improve housing and building quality; to improve cleanliness hygiene and health; to improve safety and risk management; to improve air quality; to reduce noise pollution; to minimise waste
To ensure diversity	To ensure the diversity of population; to ensure the diversity of functions; to ensure the diversity of housing supply
To improve integration	To improve the levels of education and job qualification; to improve access for all residents to employment, to services and to facilities; to improve the attractiveness of the neighbourhood by creating living and meeting place for all inhabitants of the city; to avoid unwanted mobility and to improve environmentally sound mobility
To reinforce social life	To reinforce local governance; to improve social networks and social capital

3.0 Comparison between SRTs for Communities

In contrast to SRTs for buildings and infrastructure [3], SRTs for communities have done well in terms of their coverage of sustainability issues encompassing economic, environmental and social dimensions as can be observed from Table A1 (Appendix A). It was particularly encouraging to note that all mainstream SRTs for communities have placed emphasis on the social needs of the communities calling for higher engagement levels with this group of stakeholder. There is also acknowledgement of the existence of 'green' buildings/ infrastructure. This is notable since the World Economic Forum [12] identifies the building sector as an area that needs to be addressed, accounting as it does for 40% of the world's energy use, 40% of carbon output and consuming one-fifth of available water [13]. Even though there appears to be more consistency in the themes and criteria selected in SRTs for Communities, yet, Table 8 shows that these different tools still have varying emphasis on the selected criteria. On the other hand, every criterion for EcoCity and HQE2R are measured on a predefined scale. No aggregation of scores is done for both of these tools.

Table 8: Different emphasis across selected criteria (available points for the criterion out of total points in percentage)

Criterion	**BREEAM for Communities (%)**	**Green Star for Communities (%)**	**LEED for Neighbourhood Development (%)**	**STAR (%)**
Climate and energy	11.1	6	4	14.3
Design and layout/Consultation plan	4.76	10	13	12.1
Transportation	7.14	3	3	2.86
Demographic needs/community/ social	15.87	18	42	14.3
Ecology and biodiversity	0.79	4	3	11.4
Noise pollution	2.38	0	0	0.71
Water strategy	5.55	4	8	0.71
Land use	6.35	0	2	1.43
Economic impact/ Business growth	15.9	20	0	14.3
'Green' buildings/ infrastructure	7.9	4	8	5
Innovation	5.55	10	5	0
Others	30.95	21	12	22.9

Of all the SRTs for Communities analysed, only Green Star has made an attempt to interlock with the Global Reporting Initiative (GRI), allocating points under its governance criterion for any demonstration of reporting in line with the GRI. The GRI is a globally accepted guideline for corporate sustainability reporting [3, 13].

A majority of the mainstream SRTs have started the process of engaging councils by publishing reasons behind the selection of the criteria although this can be done better by providing a more thorough justification behind the selection of criteria weightings and allocated scores [7].

4.0 Critique of SRTs for Communities

There are a few limitations with SRTs for Communities. First, a large majority of these mainstream tools do not clearly define the dimension or size of development that can be assessed with the exception of CASBEE-UD: where the minimum standard of assessment is defined as consisting of a group of buildings on two or three adjoining plots while the maximum standard of assessment is defined as consisting of a combination of tens, hundreds, or thousands of building plots and non-built land such as roads and parks [4]. This is problematic especially since some of these tools have started to introduce benchmarking in determining criteria scores such as EcoCity. It would be questionable as to whether the benchmarks selected are valid if the size of development assessed differs significantly.

Second, there is a lack of published reasoning behind the allocated scores or weightings for the criteria selected. Without such information, users may find it difficult to understand the rationale as to why certain criteria are weighted more heavily than others. This leads to heavy speculations or accusations of these tools being a misrepresentation of what should constitute appropriate sustainability assessment. Berardi (2013) [2] argues that SRTs for Communities misrepresents economic sustainability as low importance is given to the ability of a community to promote business and economic opportunities. As well, Berardi (2013) [2] adds that other important economic criteria such as 'industrial vitality, the amount of economic exchanges and financial viability in the community are ignored'.

Third, these tools do not sufficiently account for the different sources of uncertainty (see Table B1, Appendix B). From the analysis in Table B1, BREEAM Communities acknowledges uncertainty in criteria weighting depending on geographical locations. Other tools ignore uncertainty in criteria weighting and scores. Green Star for Communities has removed the process of applying weightings to criteria altogether, that is, the number of scores available for a criterion simply reflects on the weighting for that criterion (i.e. one point for the energy criterion is equivalent to one point for the transport criterion). This is potentially problematic as there is a tendency for a criterion which has the higher the number of subcriteria to be weighted more heavily. According to Siew (2014b) [14] there are four different types of uncertainty (conceptual, contextual, mathematical and input data) which should be acknowledged. Some of the criteria selected by these tools are non-deterministic and uncertainty in measurements may arise because of variations in the perception of users [14]. Ignoring the need to account for uncertainty is not scientific as prior studies have established that doing so alters the resultant ranking of alternatives [3, 15]. As well, it would be difficult to establish benchmarks without truly understanding the potential of measurement errors. There is a possibility of uncertainty propagating due to the nature of most SRTs for Communities where criteria or subcriteria scores are added and mapped to an award (see Section 2).

Fourth, it is argued that SRTs for Communities are not a true reflection of the state of sustainability because they adopt a static perspective; that is measurements are only realized at the beginning of the community development process [2] and not on a continuous basis. This static perspective prevents the analysis of trends which may potentially convey useful information about the sustainability performance of a community.

In addition, because of this static perspective, most SRTs are limited in that they only incorporate stakeholders' views at the start of the project and not on a long term basis.

Fifth, there is hardly any acknowledgement of interaction or correlation between criteria measured. The inclusion of interacting criteria might lead to 'double-counting' [16, 17] hence skewing the end result. This is certainly a problem when using a weighted sum or weighted average, because models of this form assume that all criteria are independent, which is rarely the case in practice. For example, two HQE2R subcriteria: preserving and enhancing the landscape and visual comfort; and improving the attractiveness of the neighbourhood may be highly dependent on each other.

5.0 Opportunities

Given the limitations highlighted, there are various opportunities to improve the current state of SRTs for communities as depicted in Table 9.

Table 9: Opportunities to address limitations in current SRTs for communities

No.	Limitations	Opportunities
1	Lack of clarity on the size of development assessed/benchmarks established	• Include both absolute and relative measurements as benchmarks. This will involve a number of iterations based on in -depth case studies of existing projects. • Include size of development that was used to establish benchmarks and propose recommended size of development that should be assessed with SRTs
2	Lack of published reasoning behind the allocation of scores and weightings	• SRT developers need to provide justification on why certain criteria are weighted more heavily than others; this is especially important so that users are able to better comprehend the rationale behind the allocation of scores
3	Do not sufficiently account for sources of uncertainty	• SRT developers need to encourage the acknowledgement of uncertainty in criteria measurements where appropriate
4	Static perspective	• Rather than having a static measurement at the beginning of the project, it might be worthwhile considering a half-yearly assessment of sustainability performance. This might require some tweaking of the criteria proposed to ensure that the criterion is not merely a checklist (input criteria) but rather a performance measure (output criteria)
5	The negligence of correlation between criteria	• Applying the 'Choquet integral' [18,19]which is an alternative aggregation method accounting for criteria interaction or • Adopt a criteria breakdown structure to minimise criteria overlapping each other.

6.0 Conclusion

This paper has presented and discussed existing SRTs for communities. Seven mainstream SRTs which include BREEAM Communities, CASBEE for Urban Development, Green Star for Communities, LEED for neighbourhood Development, Sustainable Tools for Assessing and Rating (STAR), EcoCity and HQE2R have been considered. This paper has provided a critique of these tools and highlighted a few limitations such as lack of clarity in the size of development that these tools are capable of assessing; lack of published reasoning behind the allocated scores or weightings for the criteria selected; do not sufficiently account for sources of uncertainty; the adoption of a static perspective; and the lack of acknowledgement of possible interaction or correlation between criteria. These issues need to be addressed to ensure the precision of sustainability measures.

There is still much room for research in this area. For example, in –depth case studies focussing on the application of these tools would help further illustrate the limitations of SRTs. This is especially true since Green Star for Communities was just recently released in March 2014 and has not been widely adopted compared to other tools. Sources of uncertainty and how they can be captured by SRTs should also be a main focus of researchers. Acknowledging uncertainty helps project planners to be more transparent in their decision-making process. It may also facilitate better discourse and more accurately capture the sustainability performance of communities [13, 20].

References

[1] Brundtland Report, "Our Common Future", United Nations, New York, viewed on 16/9/2014, http://conspect.nl/pdf/Our_Common_Future-Brundtland_Report_1987.pdf, 1987

[2] U. Berardi, "Sustainability assessment of urban communities through rating systems", Environment, Development and Sustainability, Vol. 15, pp. 1573-1591, 2013

[3] R. Y. J Siew, M.C.A. Balatbat and D. G. Carmichael, "A review of building/infrastructure sustainability reporting tools (SRTs)", Smart and Sustainable Built Environment, Vol. 2, No. 2, pp. 106-139, 2013

[4] A. Sharifi and A. Murayama, "Viability of using global standards for neighbourhood sustainability assessment: insights from a comparative case study", Journal of Environmental Planning and Management (in print), DOI: 10.1080/09640568.2013.866077, 2014

[5] BRE, "BREEAM communities manual 2012", BRE, UK, viewed on 5/9/2014, <http://www.breeam.org/communitiesmanual/#_frontmatter/breeam_communities.htm%3FTocPath%3D1> ,2012

[6] JSBC, "CASBEE urban development", CASBEE, Japan, viewed on 5/8/2014, <http://www.ibec.or.jp/CASBEE/english/method2E.htm>, 2007

[7] GBCA, "Green Star communities: guide for local government", Sydney, viewed on 5/9/2014, <http://www.gbca.org.au/uploads/189/2749/Green_Star_Communities_Guide_for_Local_Government_For_Web.pdf> ,2013

[8] USGBC, "A citizen's guide to LEED for neighbourhood development: how to tell if development is smart and green", USGBC, CNU & NRDC, U. S., viewed on 6/9/2014 <https://www.nrdc.org/cities/smartgrowth/files/citizens_guide_LEED-ND.pdf>, 2013

[9] STAR Communities, "STAR community rating system version 1.1", STAR Communities, U. S., viewed on 9/8/2014, https://www.starcommunities.org/uploads/rating-system.pdf ,2014

[10] P. Gaffron, G. Huismans and F. Skala, "Ecocity book II: how to make it happen", EcoCity, Utrecht, Vienna, viewed on 9/5/2014, http://www.gea21.com/_media/proyectos/ecocity/ecocity_book_2.pdf, 2008

[11] C. Charlot-Valdieu, P. Outrequin, and C. Robbins, "The HQE2R toolkit for sustainable neighbourhood regeneration and European application overview", The European Commission, France, viewed on 6/9/2014, <http://www.eukn.org/France/fr_en/E_library/Housing/Housing_Quality/Sustainable_Construction_Methods/HQ2ER_Sustainable_renovation_of_buildings_for_sustainable_neighbourhoods>, 2004

[12] World Economic Forum, "A profitable and resource efficient future: catalysing retrofit finance and investing in commercial real estate", World Economic Forum, Geneva, Switzerland, October 2011, viewed on 6 January 2013, <http://www3.weforum.org/docs/WEF_IU_CatalysingRetrofitFinanceInvestingCommercialRealEstate_Report_2011.pdf>, 2011

[13] R. Y. J. Siew, "Evaluating and enhancing the impact of sustainability reporting tools (SRTs)", PhD Thesis, School of Civil and Environmental Engineering, University of New South Wales, Sydney, 2014a

[14] R. Y. J. Siew, "Alternative framework for assessing sustainable building funds: green building fund", Building Research & information (in print), <http://dx.doi.org/10.1080/09613218.2014.936170>, (2014b)

[15] K.M. Hyde, H. R. Maier and C. B. Colby, "A distance-based uncertainty analysis approach to multi-criteria decision analysis for water resource decision making", Journal of Environmental Management, Vol. 77, No. 4, pp. 278-290, 2005

[16] M. Lenzen, "Double-counting in life cycle calculations", Journal of Industrial Ecology, Vol.12, No. 4, pp. 583-599, 2008

[17] E. Lai, S. Lundie, and N. J. Ashbolt, "Review of multi-criteria decision aid for integrated sustainability assessment of urban water systems", Urban Water Journal, Vol. 5, No. 4, pp. 315-327, 2008

[18] C. Labreuche and M. Grabisch, "The Choquet integral for the aggregation of interval scales in multicriteria decision making", Fuzzy Sets and Systems, Vol. 137, No. 1, pp. 11-26, 2003

[19] P. Meyer, and M. Roubens, "On the use of the Choquet integral with fuzzy numbers in multiple criteria decision support", Fuzzy Sets and Systems, Vol. 157, No. 7, pp. 927-938, 2006

[20] R. Y. J. Siew, M. C. A. Balatbat and D. G. Carmichael, "The relationship between sustainability practices and financial performance of construction companies", Smart and Sustainable Built Environment, Vol. 2, No. 1, pp. 6-27, 2013

Appendix A

Table A1: Comparison of criteria across different SRTs

Criterion	BREEAM Communities	CASBEE for Urban Development	Green Star for Communities	LEED for Neighbourhood Development	Sustainable Tools for Assessing and Rating (STAR) communities	EcoCity	HQE^{2}R
'Green' infrastructure/ buildings	X	X	X	X	X	X	X
Innovation	X		X	X		X	
Climate and energy	X	X	X	X	X	X	X
Design and layout/ Consultation plan	X	X	X	X	X	X	X
Transportation	X	X	X	X	X	X	X
Demographic needs/ Community/ Social	X	X	X	X	X	X	X
Ecology and biodiversity	X	X	X	X	X	X	X
Noise pollution	X	X			X	X	X
Water strategy	X	X	X	X	X	X	X
Land use	X	X	X	X	X	X	X
Economic impact/ Business	X		X		X	X	X

Appendix B

Table B1: Acknowledgement of uncertainty-criteria weighting and scores

SRT for Communities	**Acknowledgement of uncertainty**	
	Criteria weighting	**Criteria scores**
BREEAM Communities	Yes- variation depending on geographical locations	No
CASBEE for Urban Development	No	No
Green Star for Communities	No	No
LEED for Neighbourhood Development	No	No
STAR	No	No
EcoCity	No	No
HQE^2R	No	No

21

The Challenges of Green Construction in Oman

Ayisha Powmya[1], Nazirah Zainul Abidin[2]
[1] Built Environment Engineering Department, Muscat College, Oman
[2] School of Housing, Building and Planning, Science University of Malaysia (USM), Penang, Malaysia

*Corresponding E-mail : ampowmya@yahoo.co.in

Abstract

Oman has seen rapid growth in construction industry with the blooming of many new infrastructure projects. The conventional way of approaching construction projects in the past makes introducing sustainable construction a major challenge. The country has a sound environmental law, however, the effectiveness of its implementation is unclear. Green or sustainable concept is still new to Omani people and as such many initiatives to promote the concept are necessary. In line with the green development of its neighbouring countries, Oman should act fast to be at par with them. This paper explores the progress of green construction in the industry to understand how far this concept has penetrated the industry. A survey has been conducted among construction practitioners in Oman focusing on the current progress and barriers of implementation. A total of 12 barriers of implementation have been identified of which the lack of demand for green construction and lack of pressure by government are the top two. The findings suggest that many more efforts are necessary to push green construction to the forefront and the government should play a major role in this development.

Keywords: *Oman, Green Construction and Barriers*

1.0 Introduction

Sustainability is placed high on many countries' agenda as an effort to preserve the environment for the future generation while using the resources to meet the needs of human today. Hence, different green building certification schemes are deployed worldwide to encourage the promotion of a sustainable built environment and the adoption of green building best practices. Sustainable technologies need to be implemented in the planning, construction and also in operation and maintenance stages.

Most of Gulf Cooperation Council (GCC) countries have experienced swift booming in construction sector in the past decade resulting in escalation of domestic energy consumption per capita [1]. However, poor regard to environmental issues has put these countries as among the highest contributor to CO_2 emission per capita [2]. Green construction approach has gained importance recently in GCC countries. Many argued that one of the key challenges currently facing the construction industry in the GCC is the need to incorporate green elements in the project [3-5].

Oman is growing at a steady pace as a result of aggressive government spending and strong economic performance. The infrastructures and construction markets of the country will grow substantially as the government surges ahead with large scale infrastructure projects and focusing on the development of industrial capacity [6]. The concept of green, although has been lauded by many countries, is quite new to the Sultanate of Oman. As a booming country, Oman has many new projects in store which provides huge opportunity for the green concept to be integrated alongside the growth of the country. As such, it is wise to push the green concept as fast as possible into the construction industry to optimize the opportunity of building green for many new projects.

The objectives of this paper are to present the current progress of green construction in Oman and the implementation challenges. A survey has been conducted to explore the perception of the construction practitioners in Oman regarding the level of knowledge and to understand the hindering factors of green development. The findings suggest that many more efforts are necessary to push green construction to the forefront and the government should play a major role in this development.

2.0 Progress of Sustainable Construction in Oman

In its Eighth Five-Year Plan (2011-2015) Oman has allocated US$ 6.5 billion a year to Investment projects, US$ 1.2 billion to housing, US$ 4.2 billion to airport expansion and construction, US$ 3.1 billion to road projects, US$ 1.3 billion to seaports, and US$ 1.2 billion to the utilities sector. It plans to spend US$ 110.9 billion during this period [7]. Intensive development without proper attention to environmental aspects shall spring many environment-related problems as experienced by many countries [8-11]. As Oman is still at early stages of construction booming, the impact related to environmental problem may not be apparent yet but if it continues to ignore the importance of being green, the country shall suffer with environmental related issues.

Currently, Oman faces water scarcity and periodic droughts. Irrigation-intensive agriculture in the most fertile region, the Batinah coast, often leads to the saline intrusion via aquifers [12]. Due to the geophysical and climatic setting of the country, 75% of the country total surface is covered by desert. Thus the remaining land has to be shared for different purposes of the Omani society. While land consumption by residential use, services and transport infrastructure is increasing steadily, suitable land for agriculture is decreasing. This supports rural-to urban migration or transition and thus recreates new demands on shrinking land resources. Urban settlement patterns are created through random additional of freestanding villas on walled plots, consuming high amount of energy for conditioning [13]. Buildings are usually designed by engineer, thus creating stereotype in design, poor design quality and lack of design ingenuity normally found in architecturally-designed buildings. Oman' economy is heavily dependent on oil and gas sectors. However, Oman's oil fields are generally smaller, more widely scattered and pose higher production costs than other Arabian Gulf countries. With the fast growing energy demand in the country, the Government of Oman has beginning to venture into alternative energies which mainly focus on solar and wind projects [14]. The residential sector is the largest consumer category with its consumption taking more than half of the total system energy (67%) in 2011 [15]. As Oman experienced less rain in the past, attention was not given to drainage system. But the change of climate in the last 3 years has seen a more frequent rain which leads to scattered flooding due to poor or no drainage system.

Energy saving has been identified as most important sustainable factor, which will also bring down the utility bills [16]. Reference [17] mentioned that various wastes and by-product materials are generated in the Sultanate of Oman including reclaimed asphalt pavement aggregate, demolition concrete, cement by-pass dust, copper slag, petroleum-contaminated soils, discarded tires, incinerator ash, and others. Recycling of such materials in construction is not practiced even though laboratory data generally indicate that it is feasible to partially reuse some of these materials in construction provided that economic incentives and environmental concerns are taken into consideration.

In principle, the Sultan has demonstrated a personal commitment to environmental protection, and the country has a national integrated environmental policy with high-level, cross-cutting bodies under the direct control of the Government. In 1984, Oman became the first country in the region to establish a specific Ministry for the Environment, which is now reconstituted as the Ministry for the Environment and Climate Affairs (MECA). The inclusion of "climate" within the title of the Ministry suggests a renewed focus upon climate and "green energy" projects. Oman's Public Authority for Electricity and Water has recently opened a tender for advisory services in relation to a large-scale concentrated solar power project. Non-governmental organisations with an environmental agenda are also flourishing in Oman, such as the Environmental Society of Oman, which was founded in 2000 under Sultanate Decree 14/2000 and has conducted a series of high-profile campaigns including raising awareness of environmental issues such as the generation and disposal of waste plastic bags. A legislative reform and regulatory framework is necessary to encourage investments in renewable technologies and carbon reduction projects in Oman [18].

Environmental concern tends to lose pre-eminence with increasing frequency in the face of private interests and the development of tourism, real estate and industrial projects. Research institutes are still nonexistent and R&D was evaluated at 0.1% of GDP in 2007. The government has started to address this deficiency by creating a Scientific Research Council (established in 2005 by royal decree) [12]. The Research Council aims to promote research and innovation to increase the awareness of Omani on the importance of environmental protection, green building and eco-design. One of the way to promote green construction is by introducing Oman Eco house Design competition where Higher Education Institutions competed in designing houses that are innovative in architecture, engineering and inspiration in 2013 [19].

Oman Green Building Council (OGBC) has been newly established in 2012. It is a non-profit non-governmental organisation established with the objective of promoting green building concept and its principles, save the environment and ensuring sustainable development. It has been actively promoting green concept through conferences and seminars. Builders' international gathering called as The BIG Show exhibition and The Oman Builders Forum is held annually. It brings together the local and international construction and interiors experts from the public and private sectors to discuss the new trends and technologies, address the challenges and issues facing the industry in Oman and provide possible solutions to these issues. In 2010, there are 15 projects registered with LEED but none of the projects has been certified yet [20].

Energy subsidies in Oman and other GCC countries are driving consumers to excessive consumption. Some GCC countries are also looking at reducing subsidies on fuel, electricity and water. The 2013 budget included RO740 million (1.6 million USD) relates to fuel subsidies. These subsidies distort energy markets, making competing alternative energy sources appear much more expensive in comparison to highly subsidized gas and liquid fuels [21]. Oman is relying for its natural gas for its electricity generation. Over-dependence on a single fuel type is not a prudent step, and thus, a review of alternatives to domestic gas is important. Oman is targeting alternative energy sources to meet its rapidly growing energy demand (8 – 10 percent annually). Through its Public Authority for Electricity & Water (PAEW), locations for four large-scale solar projects of Oman have been identified. Some parts in south Oman was also found to be suitable for wind energy projects [22].Recent building by Majan Electricity Company have used green techniques such as cavity walls, double glazing for windows and roof insulation to obtain 15 percent energy saving. They also used recycling of waste water for gardening purposed and solar power for energy generation [23].

3.0 Overviews on Green Construction Barriers

The construction industry as a whole has to rapidly come to terms with the broader environmental and social agenda as the built environment affects all human activity. Bringing change to the construction industry requires multiple efforts from various angles especially to overcome any resistances in the form of technology, human aspects, financial and resources. The transition from conventional to sustainable approach consumes time as it requires changes from different facets in the industry ranging from individual, organizational to industry level.

Green construction is commonly associated with the use of green technology especially to reduce the energy consumption, indoor cooling, water saving, green material etc. However, the use of new technology is perceived as expensive due to the need of technical capacity, lacks competition and commonly manufactured abroad [24]. The challenge is to produce this technology at the minimal cost possible as building which is too expensive will have marketability problem.

The lack of knowledge, information and understanding are another worrying barrier to the success of delivering sustainable buildings [25]. The factors that will jumpstart sustainable movement are awareness and knowledge [10]. It is important for construction practitioners to understand sustainable construction sufficiently to be able to ensure that their actions and decisions add as little as possible to the total burden on the environment.

Commitment of construction organisation will lead to actions that will expedite the move towards sustainable direction. Reference [26] stated that it is within the professional responsibility of the construction players to show leadership in creating a step for better sustainability in construction. In most cases, the fear of upfront cost is the reason for construction practitioners to ignore the implementation of sustainable design and technology in the construction industry [27].

Public awareness is one of the significant factors affecting the lack of responsibility towards nature [28]. Greater involvement and constructive interaction from the demand side which includes the clients, buyers and users, will inevitably improve good practice initiatives, drawing closer linkages to the supply side and consequently in the delivery of improved sustainable construction.

The government is responsible for enforcing better regulation, revising legislation and policy and introducing building codes, regulations, voluntary actions, incentives and fiscal instrument [29]. If the governments are not proactive, the progress of green construction will face many hurdles.

4.0 Field Study and Findings

4.1 Research methodology

The concept of green construction is new to the country but Oman has initiated several efforts to push the concept into the construction industry. To understand how far this concept has penetrated the industry, a survey was conducted among construction practitioners in Oman focusing on the current progress and barriers of implementation from the practitioners' point of view. The questionnaire was first distributed via post and email. However, due to very poor response, a snowball sampling style is later adapted to gain more response as it enables prior interactions with respondents. A total of 67 questionnaires were returned for analysis. The data gathered was analysed qualitatively (open-ended questions) and quantitatively (scale-typed questions). The results are discussed next.

A total of 67 construction practitioners have responded to this survey. Majority of the respondents are from contractors company (47.8%), followed by developers (17.9%), engineering consultants (16.4%), university and government sector (16.4%) and lastly architectural firms (1.5%). A total of 50.8% respondents have more than 10 years of experience, working in the construction industry and another 32.8% have work experience between 5 to 10 years. Their range of project profiles varies from individual villas to commercial buildings, governmental offices and infrastructures projects.

4.2 Overview of current progress

To overview the current progress based on the perception of the construction practitioners, the respondents were asked 4 straightforward questions: level of environmental consideration in Oman's construction industry; government actions to ensure construction projects protects the environment; level of knowledge of construction practitioners on environmental issues; and level of environmental practice perceived within the construction projects. The respondents were asked to provide their answer in the scale of 1 (very poor) – 5 (excellent).

From the survey, no respondent believes that the environmental consideration in Oman construction industry is 'excellent'. Majority of the respondents (40%) rated the consideration as 'moderate' followed by 'poor' level (25%). Around 19% respondents rated the performance as 'good' while 15% rated it as 'very poor'. The average mean is 2.66, placing the consideration between poor to moderate level. Majority of respondents (40%) also rated the level of actions by the Omani government to ensure the construction projects protects the environment as 'moderate' and followed by those who ranked it as 'good' (30%). A total of 9% respondents believed that the action by Omani government is 'excellent' and another 4% ranked is as 'very poor'. The remaining 16% chose 'poor' level. The average mean is 3.16, placing the consideration at moderate level. For the third question, majority of the respondents (43%) also perceived that the level of knowledge of construction practitioners in Oman on environmental practices as 'moderate'. Another 30% respondents viewed it as 'poor' while another 9% as 'very poor'. The remaining 18% perceived it as 'good'. The average mean is 2.72%, in line with the respondents' view on the level of environmental consideration in Oman construction industry. For the last question, majority of the respondents (43%) also perceived that the level of environmental practice within the projects in Oman as 'moderate'. Another 31% respondents viewed it as 'good' while another 4% as 'very good'. The remaining 15% perceived it as 'poor' and another 6% viewed it as 'very poor'.

From the survey, generally, the progress of Oman construction industry is perceived as between 'very poor' to 'moderate'. Several challenges that contributed to this situation are discussed next.

4.3 Barriers of green construction

A total of 12 barriers have been identified as potential barriers to effective implementation of green construction. Each impending factors can be rated in a scale between 'very low' (1) to 'very high' (5). Further support to this findings were available through open-ended question while enable the respondents to express their opinion on the matter. The result from the survey is summarized in Figure 1. Apart from barrier (a), all other barriers are within 3.0 to 4.0 in average mean ('moderate' level). Barriers (b) to (j) are above 3.5 while the last 2 barriers are lower than 3.5.

a) **Lack of demand for green construction**– This barrier is ranked as no. 1 in the list with the average mean of 4.10. A total of 46% respondents perceived this factor poses 'high' level of problem followed by 33% at 'very high' level. Some respondents stated that people are indirectly aware on this issue but there is very poor demand for green construction in the market either from public sector or private sector. A few respondents argued that the lack of awareness among the construction practitioners contributed to the poor demand for this concept. One respondent stated that the lack of demand is because it depends on individual effort and currently there is no interest to pursue it. A few respondents stated that the initiation from the authorities did not reach the greater population of construction practitioners.

b) **Lack of pressure by the government**– Many respondents believed that the government should play a bigger role in promoting green construction, through actions such as through strong enforcement of legislation, devising new policy, or demanding green concept in their new governmental buildings. As shown in Table 2, the average mean is 3.96, placing this barrier as no. 2. A total of 52% respondents perceived this factor poses 'high' level of problem followed by 24% at 'very high' level. Many respondents stated that there is no pressure from the authorities to do green buildings even for governmental projects. Aspects of environmental issues have been initiated for a few major projects. However, the initiations were mostly from the private sector.

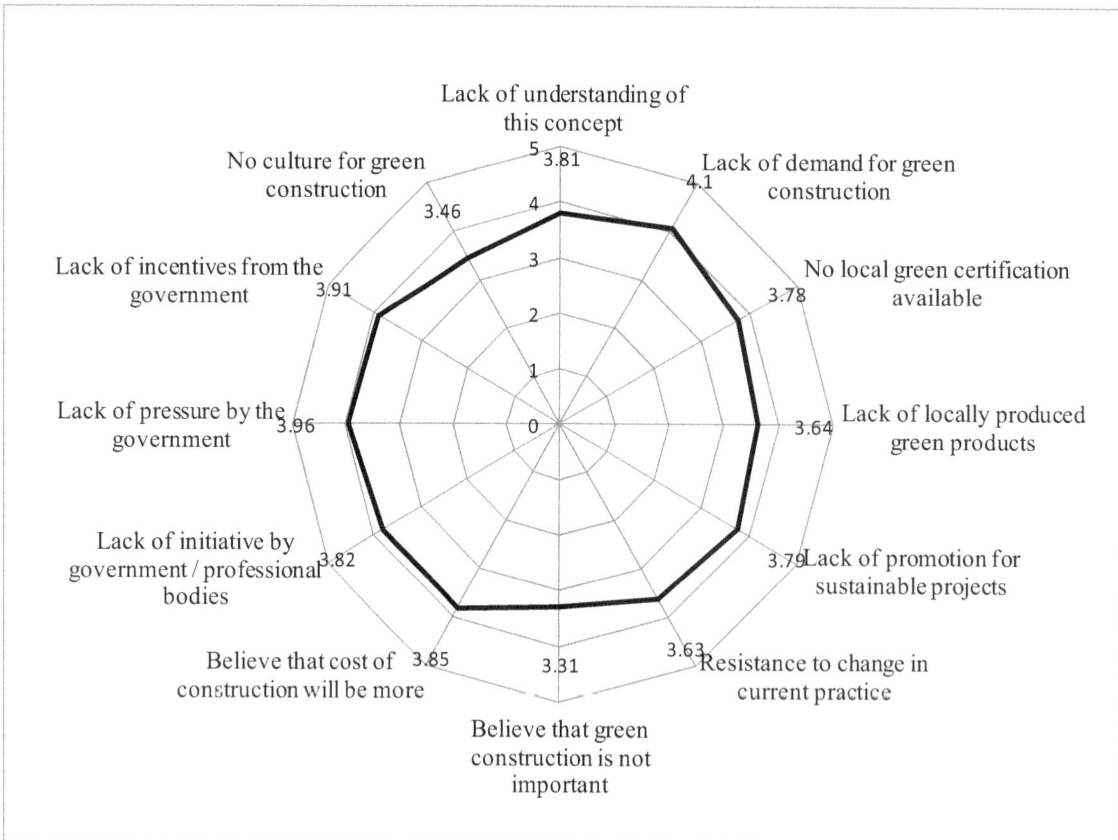

Figure. 1 Barriers for green construction in Oman

c) **Lack of incentive by the government**– Many respondents argued that the lack of pressure by the government is related to the poor incentives introduced by the government to the construction practitioners who want to pursue green in their projects. The average mean is 3.91. A total of 40% respondents perceived this problem as 'high', followed by 30% who perceived it as 'moderate'. One respondent stated that having government LEED building projects will help propagate green architecture in Oman. Active involvement from the government will enable them to see what sorts of incentive they can offer to the industry to boost up green construction.

d) **Believe that cost of construction will be more** – This fourth barrier has an average mean of 3.85. Majority respondents (48%) perceived this problem as high followed by 31% who rated it as moderate. Another 19% rated it as 'very high'. Many respondents believed that green construction is economically non-viable. It increases project cost, thus making it non-attractive. One respondent stated that incentive from the government can help to reduce the increment in construction cost. Another respondent argued why the cost will be more is because of the lack of basic infrastructure such as recycling centre, public transport network and easy access which will prohibit the achievement of good environmental standards in the country and increases the cost for the project if the basic infrastructure is to be added to the project cost.

e) **Lack of initiative by the government/ professional bodies**– respondents argued that the government has been supportive to the actions by the private sectors that are pursuing green certification and has been involved in collaborative efforts with professional bodies in organizing conferences and seminars on green construction. However, those efforts have limited impact as only those who have been involved or planning to involve with green projects are participating in it. Another respondent argued that it is crucial that government commitment for green construction is made clear to all construction practitioners. At present, the government desire for green construction is not clear. The average mean is 3.82. A total of 46% respondents perceived this problem as 'high', followed by 31% who perceived it as 'moderate'. Another 19% perceived the barrier as 'very high'.

f) **Lack of understanding of green concept** – This barrier is ranked as no. 6 in the list with the average mean of 3.81. A total of 54% respondents perceived this factor poses 'high' level of problem. This is followed by 22% respondents perceiving the problem as 'moderate' and another 16% as 'very high' level. One respondent stated that green architecture is making a push into the country, but as many do not understand what it entails, the acceptance has been affected. Another respondent argued that it is a common practice in Oman that civil engineers are taking on the work of an architect such as designing the building. Without good designing background, these engineers are unable to provide advice to the clients on eco-friendly architecture as they themselves are not knowledgeable on the subject. As such, not only the building is not 'green' in design, the quality of the finished product is also of lower standard as compared to other developed countries.

g) **Lack of promotion for sustainable projects** – This barrier is ranked as no. 7 in the list with the average mean of 3.79. A total of 39% respondents perceived this factor poses 'high' level of problem. This is followed by 34% respondents perceiving the problem as 'moderate'. It is difficult to promote sustainable projects when the number of such projects is very few. Currently, only 15 projects have registered with LEED but none of those projects has been certified yet. The progress of these projects, however, has been informed through newspapers and national conferences.

h) **No local green certification available** – The average mean is 3.78 (no. 8). Majority respondents (40%) perceived this factor poses 'moderate' level of problem. This is followed by 33% respondents perceiving the problem as 'high'. A few respondents argued that although there is no local green certification available, the construction practitioners can always opt for other countries green assessment system such as LEED. However, other respondents also argued that having local green certification will speed up the green progress in the country as it will cater for the need of the local such as Pearl Rating System in U.A.E. One respondent stated that LEED certification is based on American standard which has different climate than Oman, thus Oman should establish its own rating system.

i) **Lack of locally-produced green products** – The average mean is 3.64 (no. 9). Majority respondents (39%) perceived this factor poses 'moderate' level of problem. This is followed closely by 37% respondents perceiving the problem as 'high'. One respondent stated that research and development (R&D) culture is not yet present in Oman. As such, the growth of high-tech locally-made product is very slow. Another respondent stated that the poor demand for green construction also discourages the introduction of local green products.

j) **Resistance to change in current practice** – This is barrier no. 10 with average mean of 3.63. Majority respondents (40%) perceived this factor poses 'moderate' level of problem. This is followed closely by 28% respondents perceiving the problem as 'high'. One respondent stated that it is difficult to change the mindset of the construction practitioners as there is no clear support from the government and no apparent demand from the society. Another respondent stated that lack of knowledge for this subject is another hurdle that contributes to the resistance for change. Other respondent stated that green movement usually come from the rise in cost of energy, as this factor is not critical in Oman, they are not desperate to change their way of life or work pace.

k) **No culture for green construction**– This is the second lowest barrier with the average mean of 3.46. Majority respondents (36%) perceived this factor poses 'moderate' level of problem. This is followed by 33% respondents perceiving the problem as 'high'. According to a respondent, traditional way of building houses has incorporated many eco-friendly functions to reduce indoor heat by having small openings, courtyard, maximize natural ventilation. But with the fast growth of the residential industry and because many builders in Oman do not have qualified architects, the task of building design is taken over by civil engineers who can build structure to serve as building but lack of knowledge on ways to make a building eco-friendly.

l) **Believe that green construction is not important**– This is the lowest barrier with the average mean of 3.31. Majority respondents (40%) perceived this factor poses 'high' level of problem. This is followed by 34% respondents perceiving the problem as 'moderate'. Another 12% perceived the issue is 'low' and 6% perceived as 'very low'. The remaining 7% selected 'very high'. It seems that respondents are generally aware that green construction is important, placing this factor as the least aspect to worry about.

5.0 Conclusions

The push towards green construction is relatively new in Oman, but actions have been initiated by several parties to bring this concept to the forefront of the country agenda at par with other developing countries. In a nutshell, the progress of green construction is still at initiation stage. There are many aspects that need further improvement such as governmental participation and involvement, professional bodies' proactive movement, introduction of local rating system, changing the primary task of design from engineers to architects, raising more awareness and knowledge on the subject and many more.

Acknowledgement

The authors acknowledge the financial support provided by Universiti Sains Malaysia (*Science University of Malaysia*) (USM) and appreciate the cooperation provided by Muscat College, Sultanate of Oman throughout the duration of this research.

References

[1] Al Hathloul, S., "Planning in the Middle East: Moving towards the future", *Habitat International*, 28, pp. 641 – 643,2004.

[2] Carbon Dioxide Information Analysis Centre., "Ranking of the World's Countries by 2020: Total CO_2 emission", http://www.cdiac.ornl.gov (accessed 5 December 2012), 2007.

[3] Al Saleh Y.M., and Taleb H.M., "The integration of sustainability within value management practices: A study of experienced value managers in the GCC Countries", *Project Management Journal*, 41 (2), pp. 50 – 59, 2010.

[4] Alnaser, N.W., and Flanagan, R., "The need of sustainable building construction in the Kingdom of Bahrain", *Building and Environment*, 42 (1), pp. 495 – 506, 2007.

[5] Samarai, M., and Qudah, L.M., "Planning sustainable mega projects in UAE", World Housing Congress 2007: Affordable Quality Housing, Terengganu, Malaysia, 2007.

[6] Thebigshow-oman, "Builders international gathering",www.Thebigshow-oman.com (accessed 04 January 2014), 2014.

[7] Oxford Business Group, "The Report: Oman 2012",http://www.oxfordbusinessgroup.com/country/oman (accessed 1 August 2013), 2012.

[8] Ling F.Y.Y, and Gunawansa A., "Strategies for potential owners in Singapore to own environmentally sustainable homes", *Engineering, Construction and Architectural Management,* 18 (6), pp. 579-594, 2011.

[9] Qi G.Y., Shen L.Y., Zeng S.X., and Jorje O.J., "The drivers for contractors' green innovation: An industry perspective", *Journal of Cleaner Production*, 18, pp. 1358 – 1365, 2010.

[10] Zainul Abidin, N., "Investigating the awareness and application of sustainable construction concept by Malaysian developers", *Habitat International*, 34(4), pp. 421 – 426, 2010.

[11] Majdalani, Z., Ajam, M. and Mezher, T., "Sustainability in the construction industry: A Lebanese case study", *Construction Innovation*, 6, pp. 33 – 46, 2006.

[12] BTI., "Oman country report", http://www.bti-project.de/fileadmin/Inhalte/reports/2012/pdf/BT%202012%20Oman.pdf (accessed 7 November 2012), 2012.

[13] Nebel, S., "Towards sustainable patterns of urbanisation in Oman", http://www.gutech.edu.om (accessed 6 May 2013), 2013.

[14] Al-Badi A.H., Malik, A., and Gastli, A., "Sustainable energy usage in Oman – opportunities and barriers", *Renewable and Sustainable Energy Reviews*, 15, pp. 3780 – 3788, 2011.

[15] Authority for Electricity Regulation, "Annual Report 2011-Oman",http://www.aer-oman.org (accessed 5 August 2013), 2011.

[16] HasanAlfarra, Vicki Stevenson and Phil Jones., "The Architectural Perception of Incorporating Innovative Solar Energy technologies in the Built Environment", Paper presented at the Sustainable Buildings: International Conference- SB 13, 8-10 December, Dubai, paper 004, 2013.

[17] Taha R, A Al-Rawas, K Al-Jabri, A Al-Harthy, H Hassan, S Al-Oraimi., "An overview of waste materials recycling in the Sultanate of Oman", *Resources, Conservation and Recycling*, 41(4), pp. 293–306, 2004.

[18] Eldridge, Tom., and Rudd, Michael., "The Environmental and renewables regime in Oman", http://www.lexology.com/library/detail.aspx?g=fb2e1c42-65ee-4195-9e4c-9a42f6925c19 (accessed 04 January 2014), 2009.

[19] The Research Council, "Oman eco house design competition". http://home.trc.gov.om/tabid/402/language/en-US/Default.aspx (accessed 23 April 2013), 2013.

[20] Senerivatne, M., "Green buildings in the GCC countries", Paper Presented in Development of Economic and Innovation, Montreal, Canada, 2010.

[21] Albawaba Business, "Should the GCC's new year's resolution be throwing energy subsidies out the window?",http://www.albawaba.com/business/gcc-consumer-spending-oman--544435 (accessed 04 January 2014), 2014.

[22] Kevin T Connor and Ziad G El-Khoury, "The future for renewable energy in the MENA region", http://www.lexology.com/library/detail.aspx?g=759c143f-41e1-46fb-8194-542928e59a6a (accessed 04 January 2014), 2013.

[23] Muscatdaily.com, "Oman's first green building by 2013", http://www.muscatdaily.com/Archive/Features/Oman-s-first-green-building-by-2013 (accessed 04 January 2014), 2011.

[24] Robichand, L.B. and Anantatmula, V.S., "Greening project management practices for sustainable construction", *Journal of Management in Engineering*, pp. 48 – 57, 2011.

[25] William, K. and Dair, C., "What is stopping sustainable building in England? Barriers experienced by stakeholders in delivering sustainable development", *Sustainable Development*, 15, pp. 135 – 147, 2007.

[26] Manly, G., "Creating a Step Change for Sustainable Construction", CIBSE National Conference on Delivering Sustainable Construction.London, 29 – 30 September, 2004.

[27] Sonagar, B. and Fieldson, R., "Towards a sustainable construction practices", *Construction Information Quarterly*. 10, pp. 101 – 108, 2008.

[28] Zainul Abidin, N., Yusof, N., and Awang, H., "A foresight into green housing industry in Malaysia", *In Proceeding of World of Science, Engineering and Technology: International Conference of Housing and Urban Environments,* Stockholm, Issue 66, July 2012, pp. 465-473, 2012.

[29] Meacham, B., Bowen, R., Traw, J., and Moore, A., "Performance-based building regulation: current situation and future needs", *Building Research & Information*, 33, pp. 91-106, 2005.

22

EFFECT OF SELECTED PROCUREMENT SYSTEMS ON BUILDING PROJECT PERFORMANCE IN NIGERIA

Olugbenga Timo Oladinrin, Samuel Olusola Olatunji and Basir Tomiwa Hamza

Department of Quantity Surveying, Federal University of Technology, PMB 704, Akure, Ondo State Nigeria

*Corresponding E-mail: gbengaoladirin@gmail.com

Abstract

Every client at the beginning of any project aims at having value for the money spent via a quality structure delivered on time and within budget by the contractor. However, researches have shown that in most cases this aim is not met. Therefore, this study examines the effect of procurement systems on building project performance in Nigeria, with a view to assess their effect on cost and quality. The data for this study were collected with the aid of structured questionnaires which were administered to actors in the construction industry in Lagos state being the major hub of construction activities in Nigeria. The questionnaires were related to the variants of procurement systems common to the Nigerian construction industry. Data analysis was done using descriptive statistics. The result revealed that the traditional system of procurement is the most adopted option in project execution in Nigeria. Meanwhile, design and build system performs better in cost, but lag construction management system in quality achievement. The study concluded that no procurement system is a do it all in that a procurement system may perform better than the other in an instant and fail in others as revealed in the findings. The study finally recommends that consultants and other stakeholders in the construction industry and particularly those in the building sector should be up to date as to be able to suitably advise clients when it comes to building procurement, and that before choosing a procurement system the main objective and even supporting objectives should be established.

Keywords: *Building project, Construction management, Design and build, Nigeria, Procurement*

1.0 Introduction

Project procurement has been described as an organized methods or process and procedure for clients to obtain or acquire construction products [1]. The procurement of construction project is vast in scope because it involves the gathering and organizing of myriads of separate individuals, firms and companies to design manage and build construction products such as houses, office buildings, shopping complex, roads, bridges etc. for specific clients or "customers". Masterman [2], described project procurement as the organizational structure needed to design and build construction projects for a specific client. It is in a sense very true because the process of "obtaining" a building by a client involves a group of people who are brought together and organized systematically in term of their roles, duties, responsibilities and interrelationship between them.

Apart from the traditional approach, there are now other "fast-tracking" or innovative procurement systems used by the construction industry worldwide. The different procurement systems differ from each other in term of allocation of responsibilities, activities sequencing, process and procedure and organizational approach in project delivery. These differences have invariably affected the project performance. Project performance has been defined as "the degree of achievement of certain effort or undertaking" [3]. It relates to the prescribed goals and objectives which form the project parameters [4]. From project management perspective, it is all about meeting or exceeding stake holders' needs and expectations from a project. It invariably involves placing consideration on three major project elements i.e. time, cost and quality [3].

There are many other factors that determine project success, but the focus of this research is on the two critical parameters of project performance i.e. cost and quality. The aim of this research is to assess the effect of the different procurement systems on project performance. Since there are many different project procurement systems, it is appropriate for the purpose of this research to limit it to the common ones practiced here in Nigeria i.e. traditional system, design and build, management contracting, and construction management.

2.0 Literature Review

The Nigeria construction industry is modeled after the British system being our colonial master, although, since independence in 1960, it has incorporated the styles of other European countries, such as Italy, Germany and France [5]. This industry is of paramount importance for employment and economic growth [6]. The Nigerian construction industry forms nearly 70% of the nation's fixed capital formation [7], yet its performance within the economy has been, and continues to be, very poor. For example, the Nigerian construction industry's contribution to employment has remained consistently at 1.0% over the last decade against the World Bank's average observation of about 3.2% in developing countries [8]. The traditional design-bid-build system of procurement is still dominant in the Nigerian construction sector and this may likely continue to be the trend. In addition, the Nigerian construction sector comprises the clients, contractors, subcontractors, suppliers, and key professional actors responsible for design and supervision of projects. The professionals includes architects, engineers (structural and services), and Quantity Surveyors. There are professional bodies that regulate the activities of these professionals.

Delay in project execution is a major problem in the Nigerian construction industry. This occurs both in small and large projects. Virtually, all the projects executed over the years in Nigeria were faced with problem of delay in delivery. Odeyinka and Yusuf [9] observed that seven out of every ten projects suffer delay in Nigeria. Nigerian construction industry is faced with problem of cost overrun. Ogunsemi and Jagboro [6] noted that one of the most serious problems the Nigerian construction industry is faced with is the project cost overrun, with attendant consequence of completing projects at sums higher than the initial sum. Therefore, working with realistic project estimate is necessary at the outset of a project work, which would eliminate uncertainty and as well provide a platform for project success. Idrus and Sodangi [8] also observed that the last decade has however exposed the declining level of clients' satisfaction from the built facilities as a result of poor quality performance in addition to the perennial problems of time and cost overruns in the Nigerian construction industry.

The Nigerian construction industry continues to occupy an important position in the nation's economy even though it contributes less than the manufacturing or other service industries, [10]. This industry plays an important role in the economy, and the products of its activities are so vital to the achievement of national socio-economic development goals of creating job opportunities and social amenities and infrastructures [11].

2.1 Construction Project Performance

Performance has been described as "the degree of achievement of certain effort or undertaking" [3]. It relates to the prescribed goals or objectives which form the project parameters [4]. From project management perspective, it is all about meeting or exceeding stake holders' needs and expectations from a project. It invariably involves placing consideration on three major project elements i.e. time, cost and quality [3]. Yates and Eskander [12] defined a successful project as a project that has been completed on schedule, within budget, within scope and satisfied the required quality.

Project performance remains a prominent issue in project delivery because projects involve defined objectives which must be achieved and numerous resources which need to be efficiently utilized. Robinson, Anumba, Carillo & Al-Ghassani, [13] emphasized the need to develop and use tools for performance measurement and Ling & Chan [14], Thomas, Macken, Chung, & Kim, [15], Naoum, Fong, & Walker, [16] and Josephson & Lindstrom [17] developed numerous parameters for measuring project performance. Josephson and Lindstrom [17] identified 250 parameters while Ling [18] evaluated 70 potential factors for measuring project performance. These parameters can be classified into two broad categories namely: subjective and objective parameters. Ling [18] stated that the performance of a project is multifaceted and may include unit cost, construction and delivery speeds and the level of clients' satisfaction. Pinto and Slevin (1998) classified project performance parameters into (1) internal factors which are project variables namely: schedule, cost and quality and (2) external factors which are concerned with stakeholders' satisfaction with the performance of a project and the perceived impact on organisation's effectiveness. Ling, Chan, Chong, & Ee, [19] identified two categories of indicators of project success namely: product success which consists of measures of achievement of quality standards and process success which is made up of variables that measure the achievement of time and cost.

On stakeholders' satisfaction, clients remain the most important stakeholder when considering project performance. Neto, Mourao, Ferreira de Freitas & Aves, [20] stated that matching or exceeding the client's expectations result in a satisfied client. They opined further that this can reflect on how loyal a client becomes to a provider or a brand and result in higher sales volumes, lower levels of sensitivity to price and generates positive comments about the provider and the brand. Clients' satisfaction can be measured from several perspectives [21], however time, cost and quality have remained the most prominent in research studies. Josephson and Lindstrom [17] maintained that project goal which considers clients' goals, is measured from several perspectives but the main aim is to stimulate clients to identify and clearly present their goals and to stimulate all managers involved to inform and remind all individuals of the goals. Hatush and Skitmore [22] maintained that success in a project is generally operationalized into time, cost and quality.

Michell, Bowen, Cattell, Edward & Pearl, [23] remarked that the primary concern of construction clients is that their projects are completed within budget, on time and at the required level of quality. On objective measurement of project outcome, two parameters namely: schedule and cost: are common with research studies. Michell, Bowen, Cattell, Edward & Pearl, [23] identified time and cost as the principal factors. The third parameter (quality) is not a common objective parameter in research studies because as Vincent and Joel [24] put it: stakeholders see the goal of quality management as customer satisfaction. From the perspective of previous studies, two parameters namely: time-overrun and cost-overrun remain the prominent indicators of objective measurement of project outcome. However, these two parameters have their limitation because their values rely on the initial contract period or cost of a project.

Construction project performance is influenced by many factors due to the interactions and interrelationships of the stakeholders. A construction project is commonly acknowledge as successful, when it is completed on time, within budget, profitability to contractors, absence of claims and court proceeding and "fitness for purpose" for occupiers have also been used as measures of project [25]. In developing countries, poor project performance has been attributed to inadequate mechanisms and systems for: Land allocation, funding, mortgage institutions, infrastructure, procurement systems etc. Project delivery is also affected by the poor performance of the construction industry as reported in literature [26-29, 5].

The poor performance associated with small to medium sized indigenous contractors include protracted delay in payment for work done, lack of capital, high fluctuations in work load, inadequate resources, technical expertise, managerial skills and other items. According to Yng, Lean, Wai, Ping and Min [30], project owners', contractors', and consultants' characteristics, procurement systems and other factors affect project performance. Cost and quality are among the major consideration throughout the project management life cycle and can be regarded as one of the most important parameters of a project and the driving force of project success. Morris and

Hough [31] examined the records of more than 4000 construction projects and found that projects rarely finished within the allocated budget and to specification. Despite their proven importance, it is not uncommon to see a construction project failing to achieve its objective within the specified cost and quality. Cost overrun can be simply defined as when the final cost of the project exceeds the original estimates [32]. Most of the significant factors affecting project costs are qualitative such as client priority on construction time; contractor's planning capability, procurement methods and market conditions including the level of construction activity [33].

2.2 Factors Influencing Choice of Procurement System

The following criteria were used to examine client requirements and experts preferences for the performance of each procurement method as cited in [34-37] suggest employing the following criteria to establish a profile of the clients requirements: speed (during both design and construction); certainty (price and the stipulated time and knowledge of how much the client has to pay at each period during the construction phase); flexibility in accommodating design changes; quality (contractors reputation, aesthetics and confidence in design); complexity (client may specify particular subcontractor, or buildability analysis); risk allocation/avoidance; responsibility (completion of program, price, product quality, design and construction); price competition (covering such issues as value for money, maintenance costs and competitive tendering); and disputes and arbitration.

Similarly, Arazi I., Mahmoud S. and Mohamad H.H. [38] analyzed project performance criteria using severity index method as follows; Construction cost; Construction time; Quality of finish project; Occupational health and safety; Level of technology; Environment friendliness; Contractor's flexibility; Labour dependency; Quality of coordination by construction team; Contractor's project management and Contractor's capacity of manpower.

2.3 Project Procurement Systems

Procurement systems have become an important issue in the construction industry because of two reasons. Firstly, the procurement of construction projects involves a series of processes that are interrelated and sequential. The effectiveness and efficiency of the processes have considerable impact on the success or failure of projects. Secondly, there are several procurement methods that are available for a developer to adopt in procuring a project. For this reason, one major challenge that the project developer faces is the method to adopt among the available procurement options [1]. Ogunsanmi and Bamisele [39] and Ashworth and Hogg [1] defined procurement method as the management of the total process involved in construction project delivery. It is also ways in which a client or a sub-client may procure a building or other construction work varied and complex. According to [1], different variants of procurement are available for meeting different clients' needs and projects specifics. A number of factors have to be taken into account in determining the best method for a specific project. The variants of procurement methods available today metamorphosized from the need to improve construction project delivery, that is, project completion within budget and time.

Daniel [40] emphasized that procurement methods is on optimizing all parameters involved in project delivery namely, time, cost and quality. Procurement of projects within these constraints has continued to be a challenge to the design team, the contractors, and managers of investments [41]. Traditionally, construction projects starts with the client's brief on which designs are based. The Architect and engineers prepare designs, in collaboration with quantity surveyor who advises on the cost implications of design variables. Tender process afterwards produces the contractor for the execution of the work. On the award, the successful contractor executes the work as designed under the supervision of the consultants. Thus, the approach separates the design, tendering process and construction as separate tasks. This separation of activities also led to sequencing of activities in which design is completed before construction commences.

This became the traditional' sequence and it is now referred to as Design-Bid-Build [40]. Other variants of procurement method not following this format became the 'non-conventional' procurement method. Concepts of project delivery have been developed to compress the time required to realize a constructed facility which focuses on simplifying the project delivery process, with emphasis on optimizing the parameters (e.g. quality, cost, time of completion, meeting market needs, and safety among others). In non-conventional procurement methods, the grounds are gradually shifting from just meeting clients' needs into apportionment of risk, as the contractors are gradually taking their stance as business organizations with the aim of making optimum profits at the minimum risk, and this has led to the development of integrated methods of procurement which are hybrids of both traditional and non-conventional procurement methods [1, 42-44], classified construction procurement methods into two broad categories as: traditional procurement method, and non-conventional procurement method.

2.4 Traditional Procurement System

According to Seeley [43]; and Kadiri and Odusami [45] The main variants of traditional procurement method are: bills of firm quantities; bills of approximate quantities; drawings and specification; schedule of rates; cost reimbursement; and labour only. The traditional method as the name implies, is a project procurement method where the three sequential phases of design, bid and build are identified as separate tasks. It is traditionally referred to as the competitively bid contract. This method allows for all contractors that fill competent to bid for projects in a free and competitive atmosphere similar to competitive market environment. In a typical traditional approach, the client initiates the project and produces a written scope statement, identifying the project's objectives and verifying the scope definition by the architect. The architect is responsible for defining the project scope in order to facilitate a clear assignment of responsibilities and to monitor the scope change control with the project team. The design team produces complete design documents before engaging the contractor, often affecting the quality by not taking into consideration build- ability, constructability and life-cycle costing.

Certain conditions warrant the use of Traditional procurement as opined by Turner, [46] these include when:

- A programme allows sufficient time;
- Consultant design is warranted;
- A client wishes to appoint designers and contractors separately;
- Price certainty is wanted before the start of construction;
- Product quality is required; and
- A balance of risk is to be placed between the client and constructor.

2.5 Design and Build

This approach gives the client a single point of contact. However, the client commits to the cost of construction, as well as the cost of design, much earlier than with the traditional approach. In this method, the contracting organization is responsible for design and construction [44]. In this system of procurement, all phases of a project, from conception through design and construction are handled by the same organisation. This form of procurement has been used for the majority of process-oriented heavy industrial project. Projects using a design-build approach are designed and constructed by a single company or a partnership of companies. Several varieties of Design-Build have evolved including Design-Build-Maintain, Design-Build-Operate-Maintain, and Design-Build-Operate-Maintain-Warrant. Each version of Design-Build provides the government or owner with one source of responsibility for the project. Design-build can be specified in many different ways based on the magnitude of the project.

2.6 Management Contracting

According to Seeley [19], Management contracting is a system whereby a main contractor is appointed, either by negotiation or in competition, and works closely with the team of professionals. Also, Oyegoke [47] opined that "in a management contract, the permanent works are constructed under a series of construction contracts placed by the management contractor after approval by the client." All physical construction is undertaken by sub-contractors selected in competitive bidding. This system usually has the main contractor called the management contractor who provides the management expertise in the construction of the project for a fee. This Manager is appointed at the inception or better still feasibility stage to join the client's team of consultants, to help work out the design programme and site operations. He manages and co-ordinates the work packages to individual sub-contractors and equally provides on the site service, plant and equipment, amenities etc for the work. The fee paid to the management contractor depends on the nature and extent of the work done and not on the cost of the work.

However, management contracting system is most appropriate for large and complex projects which exhibit particular problems that militate against the employment of fixed price contract procedures. Typical examples of which are: Projects for which complicated machinery and / heavy equipment are to be installed concurrently with the building works; Projects for which the design process will of necessity continue throughout most of the construction periods; Projects on which construction problems are such that it is necessary or desirable that the design and management team includes a suitably experienced building contractor appointed on such a basis that his interests are largely synonymous with those of the employer's professional consultants.

Though, there is a wide range of views as to the best procedures to be adapted in management contracting [43], but they usually incorporates the following activities and requirement: The management contractor is precluded from carrying out any of the physical works using directly employed labour; His role is primarily that of a planner, manager and organizer; The works are divided into packages agreed by the professional team and the management contractor as being most appropriate for the particular projects; The management contractor provides from his own resources the following: Site supervisor, technical and administrative staff to run the contract.

2.7 Construction Management

Construction management is that group of services over and above the normal Architectural and Engineering services related to the construction programme executed during the pre – design, design and construction phases, that contribute to the control time, cost and quality of new facility. Professional construction management treats the project planning, design, and construction phases as integrated tasks. This approach unites a three –party team consisting of owner, designer, and construction manager in a non-adversary relationship, and it provides the owner with an opportunity to participate fully in the construction process [41]. Construction Management is a fee-based arrangement in which the construction manager is responsible exclusively to the owner and acts in the owner's interest at every stage of the project. He offers advice on: the optimum use of the available funds, control of the scope of work, project scheduling, avoidance of delays, changes and disputes, enhancing project design and construction qualities, and optimum flexibilities in contracting and procurement [48]. A prime construction contractor or funding agency may also be part of the team. The team works together from the beginning of design to project completion, with the common objective of best serving the owner's interest.

Construction Management Under a construction management approach, professional expertise in the specialized areas of systems analysis, value engineering, "construct-ability" review, activities scheduling, procurement systems, and construction coordination and supervision is added to the capabilities of the traditional project team of client and architect. The involvement of a Construction Manager during the entire design process as a collaborative yet independent member of the professional team helps ensure that every major design decision is balanced by proper analysis of its cost consequences, and impact on project schedule [48].

The role of the construction manager on a building project may vary substantially, and can be performed under a variety of contractual terms. The most traditional or purest form of construction management is that where the Construction Manager acts as the client's agent as a professional consultant, providing estimating, cost control and scheduling services and undertaking administrative responsibilities during construction. Under this arrangement, all construction contracts are executed directly between the owner and contractors. The approach permits the construction work to be broken down into a number of trade contracts thereby eliminating the need for one or more "general" contractors. The elimination of the general contractor avoids a duplication of fees, cost mark-ups and general conditions costs otherwise incurred by the client.

2.8 Project Cost and the Effect of Certain Procurement System

Cost has been defined as the degree to which the general conditions promote the completion of a project within the estimated budget [49]. It covers overall costs incurred from project inception to completion. This highlights the importance that has to be attached to every project management activity carried out through every stage of the project development up to completion. [50] also argues that cost is not only confined to the tender sum and that it is the overall cost that a project incurs from inception to completion, which includes any cost arising from variations, modifications during construction period. These cost variables give indication of certain additional practices that when engaged in during the project management process would have both direct and indirect implications for the project cost performance.

The number and manner in which variation orders are issued by consultants during construction is an important practice to look at. Clients who often engage in the habit of agitating for numerous design changes before practical completion also play great role in the influences on project cost. The way contractors respond to variation orders may also have implications for the project performance. In predicting the performance of design-build and design-bid-build projects, [19] identified certain variables that affect cost performance. These include: the number of repetitive elements contained in a project, the extent of design completion when bids are invited, and the level of paid up capital of contractors engaged. These variables bring to bear certain related practices that may affect the performance of project cost. For instance the kind of procurement method usually adopted by clients; traditional procurement or design and build will determine the extent of completion of designs to be used for bidding. Moreover the kind of project consultants selected by clients for design of a particular kind of project will also have influence on the way the design will be made (i. e. whether repetitive elements will be brought into the design or not).

The attitude of client towards the project cost will also determine whether he or she will adhere to the advice given by designers concerning the cost advantage of having repetitive elements in designs. How contractors are usually selected (i. e. always selecting through competitive tendering or negotiated tendering) will also determine the kind of contractors that are employed to execute the projects. The presence of certain features within a particular contract also goes a long way to determine the kind of contractors that would tender for the job and eventually win. For instance the availability of certain facilities (such as payment of advance mobilization by client) within a given building contract may attract contractors who have low level of paid up capital or low level of ability to pre-finance a project. The level of financial capability of the winning contractor would have bearing on project performance.

2.9 Project Quality and the Effect of Certain Procurement System

Construction quality is defined as "the totality of the features required to satisfy a given need; fitness for purpose. The extent to which projects are monitored, the experience of project consultants, quality and past performance record of contractors and the number of variation orders issued all have effect on quality. How all these factors can be competently coordinated would be relevant to achieving satisfactory quality performance. The project management team leader has the responsibility to ensure that these factors combine well to yield good quality performance.

Quality performance has been considered as a function of the procedures adopted during the construction process [51]. Those procedures comprise the concept of procurement form and the method of tendering. The fragmented nature of the construction industry and the fact that every building project is unique places great responsibility on the project management team in setting up the building process that will bring the project to a successful conclusion. The emphasis here is on process and procedures having influence on quality of a building project. The subsequent issue that arises is how often project managers, having a sense of the uniqueness of every project, tailor certain procurement practices to correspond with the uniqueness of a project in order to yield good quality performance. Some of the procedures to be given recognition may therefore include the selection procedure of organizations required to perform the design and supervision and those responsible for the construction of the particular project too. Usually, the construction team would be appointed under competition through competitive tendering process. Sometimes, a contractor may be appointed by negotiation on the basis of a fee. In cases where the design and construction is done as a complete package, both may be let by competition.

The selection procedures applied to contractors are therefore by no means always the same. Different methods have different levels of impact on project success. For instance it was noted from previous research that "competitive tendering can adversely affect the outcome of major projects and the number of separate contracts is related to the chances of success; different selection methods will pose different levels of risk to the project team members" [52]. The selection procedures adopted by clients for project consultants should also not be overlooked since less attention has been given to this aspect of project management by several research works.

Quality is considered as a very crucial issue in project development. [53] opined that the importance of quality performance and safety at times surpass cost and time of delivery in civil engineering projects although, these factors are interrelated and interdependent. [54] maintained that the issue of quality performance of construction projects in Nigeria has resulted in colossal waste of human and material resources and in most of the cases, the indigenous contractors are found culpable. The degree to which a project's quality objective is attained which is subjectively measured on a ranking scale. In view of the numerous parties and factors whose requirements should be met, the several activities, actions, processes and techniques involved in meeting these requirements and the many individuals and bodies concerned with planning and implementing them, [55] maintained that quality management is far more difficult to achieve in construction than in other industries.

3.0 Methodology

The data used for this study were collected via well structured questionnaires which were self administered to construction professionals involved in the usage of various procurement systems in Lagos state, Nigeria. Lagos state was chosen for the study because of its characteristic as the major hub of construction activities in Nigeria. A total of 100 (one hundred) questionnaires were distributed and administered to the targeted professionals using decisive convenient sampling technique out of which 76 numbers were returned and suitable for analysis.

The questionnaires were of two parts: the first part identified the demographic features of the respondents, and the second part related to the performance of selected procurement methods available in the Nigerian construction industry and peculiar factors that affect cost and quality achievement. Eight major factors relating to assessment of procurement methods, fourteen and eleven factors relating to cost and quality performance respectively were identified from literatures and some were added having established them through pilot survey. The respondents were asked to rank these factors accordingly. The respondents were asked to rate on a 5-point Likert scale rating with 5 being the highest of the rating, for example 1= very low, 2= low, 3= neutral, 4= high 5= very high. The mean rank for a group of respondents is obtained as follows:

$$Ms = \frac{\sum (F * S)}{N}$$

Where, MS = Mean score

F = Frequency of responses

S = Score given to the criteria which ranges from 1-5

N = Total number of respondents.

4.0 Results and Discussions

The results obtained from the analysis of gathered data are presented in this section where table 1 shows the back ground of the respondents who participated in survey.

Table 1: Professional Background of Respondents

Respondents	Distribution	Responses	% of Response
Quantity Surveyors	35	30	39.50
Architects	25	19	25.00
Builders	30	22	28.90
Civil Engineers	10	5	6.60
Total	100	76	100

From the table above, the majority of the respondents i.e. 39.5% have Quantity Surveying as their professional background, 25% of the respondents are of Architectural background, while Building technology is 28.9% and Civil engineering has the least with 6.6%. The result expressed the generation of adequate opinion of the construction industry in the study area as majority of construction professionals are represented

Table 2: Professional Qualification of Respondents

Respondent Professional Body	Frequency	Percentage %
MNIQS	21	27.60
MNIA	10	13.20
MNIOB	17	22.40
MNSE	3	3.90
Others	7	9.20
None	18	23.70
Total	76	100

Among the professionals in the table above who responded to the questionnaire, 27.6% are members of the Nigeria institute of Quantity Surveyors (NIQS), 13.2% belong to Nigeria institute of Architects (NIA), and 22.4% belong to the Nigeria institute of Builders (NIOB), while 3.9% are members Nigeria society of Engineers (NSE), those who belong to other professional body other than those above are 9.2% of the respondents and 23.7% of the respondents are not registered with the professional bodies thus expressing their position as able to supply reliable data for the study. Table 3 shows the experience of the respondents where about 37% of the respondents have 1 to 5 years of experience, 33% had between 6 to 10 years of experience, and 29% had 11 to 16 years of experience and 1% of the respondents have above 20 years of experience in the construction industry indicating that these professionals are of current experiences and training in the examined systems. The assessment of procurement system is presented in Table 4.

Table 3: Years of Experience of Respondents

Years of Experience	% of Response
1 – 5	37.00
6 – 10	33.00
11 – 15	29.00
16 – 20	0.00
Above 20	1.00
Total	100

Table 4: Assessment of the Procurement Systems

Assessment Criteria	Mean score	Ranking
Client Satisfaction		
Construction Management	4.51	1
Management Contracting	4.28	2
Traditional System	4.12	3
Design and Build	3.31	4
Cost Effectiveness		
Design and Build	4.29	1
Construction Management	4.14	2
Traditional System	3.98	3
Management Contracting	3.71	4`
Extent of Usage		
Traditional System	4.67	1
Design and Build	3.78	2
Management Contracting	3.47	3
Construction Management	3.01	4
Level of Awareness		
Traditional System	4.59	1
Design and Build	4.11	2
Management Contracting	3.87	3
Construction Management	3.74	4
Level of Competition		
Traditional System	4.51	1
Management Contracting	3.56	2
Construction Management	3.36	3
Design and Build	2.98	4
Price certainty		
Design and Build	4.45	1
Traditional System	3.76	2
Construction Management	3.71	3
Management Contracting	3.37	4
Quality Assurance		
Construction Management	4.44	1
Design and Build	4.33	2
Traditional System	4.13	3
Management Contracting	4.04	4
Quality Sacrifice		
Design and Build	4.34	1
Traditional System	4.19	2
Construction Management	3.86	3
Management Contracting	3.81	4

From the table above, assessing the selected procurement systems on the listed criteria, it was discovered that the traditional procurement system rank highest with mean score of 4.59 and 4.67 in their level of awareness and usage respectively. This means that Traditional system is the most patronized system of procurement in Nigeria due to to its highest level of awareness. Construction management system with mean score 3.74 and 3.01 rank least in the level of awareness and extent of usage respectively. Design and build top the ranking for price certainty with a mean score of 4.45, followed by the traditional system with a mean score of 3.78 while management contracting rank least. Also considering the quality assurance and client satisfaction, construction management rank highest with mean score of 4.44 and 4.51 respectively, while management contracting ranked least in quality assurance; design and build ranked least in the client satisfaction criteria.

Table 5 presents means of the various procurement systems as they are individually investigated under various cost criteria. The analysis shows that design and build has the least mean of 3.218 while traditional system has the highest mean of 3.898 which implies that Traditional system is the most flexible procurement system in terms of cost.

Table 5: Selected procurement systems as they affect project cost

Assessment Criteria	**C.M Mean**	**D & B Mean**	**M.C Mean**	**TS Mean**
Accident / Delay	3.40	2.82	2.76	3.04
Claims	3.42	3.01	3.03	3.78
Contingencies	3.46	3.11	3.26	3.55
Cost related to Environmental issues	2.91	4.34	2.85	4.01
Cost related to Insurance	3.17	2.98	3.43	3.67
Dispute	3.87	3.18	3.55	3.84
Legal Costs	4.26	2.99	3.16	4.19
Managerial Cost (Consultancies)	3.36	2.98	3.86	4.44
Variation between contract sum and final account	3.10	3.52	3.87	4.00
Variation in Design / Change orders	3.80	3.49	3.71	4.14
Retention	3.12	2.91	2.71	4.09
Rework	3.75	3.28	3.49	4.03
Total mean score of each on the Cost criteria	41.62	38.61	39.68	46.78
Average Mean of the Procurement systems	**3.47**	**3.22**	**3.31**	**3.90**

* C.M = Construction Management, D & B = Design and Build, M.C = Management Contracting and TS = Traditional System.

Table 6 presents mean of various procurement systems as they are individually investigated under various quality criteria. The analysis shows that construction management has the highest rank of (3.722), follow closely by management contracting with mean 3.697, while traditional system came third with a mean of (3.627) and lastly the design and build has the lowest mean score (3.482). Thus Construction Management is the most suitable procurement system whenever quality becomes the most priotized objective of the construction project in Nigeria.

Table 6: Effect of procurement systems on quality of work

Assessment Criteria	**C.M Mean**	**D & B Mean**	**M.C Mean**	**TS Mean**
Adherence/Compliance to specification	3.97	3.66	4.10	3.64
Competence of contractor and his team	3.50	3.76	3.61	3.29
Inconsistency of variation / change orders	3.62	2.91	3.42	3.88

Insistence on specification	4.04	3.20	3.63	3.33
Major Variation between Original Design and the actual completed work	3.75	3.80	3.87	4.08
Material test	3.64	3.43	3.84	3.45
Number of Rework	3.99	3.89	3.93	4.07
Number of Variation / Change Orders	3.27	3.35	3.19	3.41
Supervision of works	3.72	3.34	3.68	3.49
Total mean score of each on Quality criteria	33.50	31.34	33.27	32.64
Average Mean of each Procurement systems	3.72	3.48	3.70	3.63

* C.M = Construction Management, D & B = Design and Build, M.C = Management Contracting and TS = Traditional System.

4.1 Discussion on Findings

The aim of this research work was to investigate the effect of the selected procurement systems on the cost and quality performance of building projects in the Nigeria construction industry. The result of the data retrieved from the questionnaires indicates in table 1 that majority of the respondents are Quantity Surveyors, followed closely by the Builders, and then the Architects and the Civil Engineers were the least. Table 2 reveals that majority of the respondents are Qualified with NIQS, followed by those with none, also those with Qualified Certificate of NIOB. Table 3 indicates that among the selected procurement systems as used in Nigeria, the Traditional system is and still remain the most popular and frequently use system, despite its shortcomings, which is in agreement with Adedokun [56] and Babatunde, Opawole and Ujaddughe [57] that approximately half of construction projects are executed using traditional procurement system which is presumably due to long existence of the traditional procurement systems, less awareness of other procurement systems available, little or less information available on the systems and Nigerians and the construction industry adamant to changes.

It was discovered from this research that none of the selected procurement systems can provide all the cost and quality performance criteria always required by the clients, supported by the submissions of Adedokun [56] who posited that there is nothing like 'the best' procurement system and that the best we could say is that this procurement is better than this if this or that is the objective of the project. Also it was discovered from the general assessment of the procurement systems that the traditional system gives and provides the clients a good competition from contractors for the work which is also one of the rules in the new procurement acts as enacted by the federal government of Nigeria while the design and build system does not give room for that. Again the analysis shows that design and build system is highly cost effective as it ranked first while the cost effectiveness of the management contracting is low in the face of quality sacrifice. The design and build system top the ranking with the traditional system follow closely and the construction management ranked third and lastly is the management contracting in this regard.

Finally, the result obtained from the average mean rankings of the selected procurement systems against cost criteria as shown in table 5 shows that all the cost criteria listed are cost burden to the clients and or contractors and since the economic purpose of all investment and project is to maximize profit and avoid loss and or expenses as much as possible then the system with the least mean ranking is considered the most cost effective, while that with the greatest or highest mean ranking will be less cost effective so going by the result of the analysis (Traditional system = 3.898, Design and Build = 3.218, Management Contracting = 3.307, Construction Management = 3.468) and comparing it with that obtained in table 3 under cost effectiveness, it is discovered that both results proved that the design and build system performs better in terms of cost. Also comparing the level of satisfaction enjoyed by the client, quality assurance guaranteed the client and less quality sacrifice suffered by client as presented in table 4 under headings; client satisfaction, quality assurance and quality sacrifice it shows that Construction Management with the mean score of 4.51, 4.44 and 3.81 respectively and this also agree with [30] that the choice of variants of the non-conventional procurement system is made in order of consideration of quality assurance; and a consideration of project been completed at the estimated budget, Management Contracting is next in client satisfaction with 4.28 mean score, traditional system is next to the least is client satisfaction with mean score of 4.12, and Design and Build system satisfy client the least with its lowest mean score of 3.61. it is also discovered and thereby established that the construction management system of procurement has the highest mean of 3.722 to rank first and thereby pass to be said that it performs better in quality among the selected procurement systems.

5.0 Conclusions

Based on the findings of this study, the following conclusions are drawn:

a. That despite its problems and shortcomings which includes long and bureaucratic processes, lots of variation and change orders and the resultant disputes, the traditional system of procurement still remain most popular, prevalent and frequently used system.

b. That none of the selected procurement systems could be called 'the best' but one can be better than the other the other in term of specific performance.

c. Finally that among the selected procurement systems which are Construction management, Design and Build, Management contracting and Traditional system, the Design and Build is most cost effective as it gives no room to claims, extra and external managerial cost in form of consultancies on the part of both the client and contractor, legal cost is minimal because the procurement is single source. Also from the findings on quality, if quality slippage must be curtail the Construction Management system is the most appropriate as it ranked highest from the field survey.

References

[1] A. Ashworth, and K. Hogg, "Willis's Practice and Procedure for Quantity Surveyor", Blackwell Publishing Ltd, Oxford, UK. 2007.
[2] J. W. E. Masterman, "An Introduction to Building Procurement Systems, 2nd ed.", Spon Press, London, 2002.
[3] Project Management Institute, "A Guide to the Project Management Body of Knowledge", 3rd Ed, Global Standard, Four Campus Boulevard, Newtown Square, PA 19073-3299, USA, 2004.
[4] W. Chitkara, "On the Development of Project Management Research: Schools of Thought and Critique. International Project Management Journal, vol. 8, pp. 20-31, 2005.
[5] N. R. Mansfield, O. O Ugwu. and T. Doran, "Causes of Delay and Cost Overruns In Nigerian Construction Projects." International Journal of Project Management, Vol. 12, pp. 254–260, 1994.
[6] D. R. Ogunsemi, and G. O. Jagboro, "Time-Cost Model For Building Projects In Nigeria". Journal of Construction Management and Economics, vol. 2, pp. 253-258, 2006.
[7] Federal Office of Statistics (FOS) Quarterly Report, "Federal Ministry of Works and Housing, Abuja, Nigeria", 2004.
[8] A. B. Idrus, and M. Sodangi, "Framework for Evaluating Quality Performance of Contractors in Nigeria". International Journal of Civil & Environmental Engineering IJCEE-IJENS, vol. 10, pp. 34 – 39, 2008.
[9] H.A. Odeyinka, and A. Yusuf, "The causes and effects of construction Delays on cost of housing project in Nigeria," Journal of Financial Management and Property and Construction. vol. 2, pp. 31-41, 1997.
[10] A. A. Aibinu, and G. O. Jagboro, "The effect of construction delays on project delivery in Nigeria construction industry". International journal of project management, vol. 112, pp. 45-55, 2002.

[11] Anaman and Osei-AmponsahAn, "Evaluation of the Delivery Performance of Construction Projects Funded by the District Assemblies' Common Fund" (A case Study of 4 Districts in Ashanti Region), Unpublished M Sc. Thesis submitted to the Department of Building Technology, KNUST, 2007.
[12] J. Yates and A. Eskander, "Construction Total Project Management Planning Issues". International Journal of Project Management, vol. 33, pp. 37-48, 2002.
[13] Robinson, Anumba, Carillo and Al-Ghassani, "Measuring and classifying construction field rework: a pilot study". Department of Civil and Environmental Engineering, University of Alberta. Presented to the Construction Field Rework Committee, Construction Owners Association of Alberta, 2005.
[14] F. Y. Y. Ling and S.L. Chan, "Performance evaluation of alternative project procurement methods". Research Brief. Singapore: National University of Singapore. 55 p. http://dx.doi.org/10.1080/096132103200048497, 2002.
[15] H. Thomas, D.G. Macken, L.Y. Chung, and Y.C. Kim, "Factors influencing contractor performance": an international investigation. Engineering, Construction and Architectural Management, vol. 10, pp. 322 – 332, 2002.
[16] S. Naoum, D. Fong and G. Walker, "Critical success factors in project management, in Proc. of the International Symposium on Globalization and Construction". CIB 2004. W107, TG23, 17-19 September, 2004, School of Civil Engineering, Asian Institute of Technology, Thailand, 2004.
[17] P. E Josephson and J. Lindstrom J, "Measuring performance in construction projects, in Proc. of the CIB World Building Conference on Construction for Developmen"t, 14-18 May, 2007, Cape Town, South Africa, 383-394, 2007.
[18] F. Y. Y. Ling, "Key determinants of performance of DBB projects in Singapore", Building Research & Information 32(2): 128-139. http://dx.doi.org/10.1061/(ASCE)0733-9364 (2004)130:1(75), 2004.
[19] F. Y. Y. Ling, S. L. Chan, E. Chong and L.P. Ee, "Predicting performance of Design-Build and Design-Bid-Build Projects". Journal of Construction Engineering and Management, vol. 130, pp. 75-83, 2004.
[20] J. B. Neto, Y. R. Mourao, A. A. Ferreira de Freitas and T. L. Aves, "A method to evaluate and manage client requirements in housing projects, in Proc. of the CIB World Building Conference on Construction for Development", 14-18 May, 2007, Cape Town, South Africa, 310-321, 2007.
[21] G. I. Idoro, "A comparative study of direct labour and design-tender-construct procurement systems in Nigeria". PhD Thesis. Lagos: University of Lagos, Nigeria, 2007.
[22] Z. Hatush and M. Skitmore, "Evaluating contractor prequalification data: selection criteria and project success factors", Construction Management and Economics vol. 15, pp. 129-147. http://dx.doi.org/10.1080/01446199700000002, 1997.
[23] K. Michell, P. Bowen, K. Cattell, P. Edward and R. Pearl, "Stakeholder perceptions of contractor time, cost and quality management on building project, in Proc. of the CIB World Building Conference on Construction for Development", 14-18 May, 2007, Cape Town, South Africa. 231-240, 2007.
[24] K.O. Vincent and E.R. Joel, "Principles of Total Quality". Kogan Page, London, 1995.
[25] P A Koushki, K. Al-Rashid, and N. Kartam, "Delays and cost increases in the construction of private residential projects in Kuwait", Construction Management and Economics, vol. 2, pp. 285–94, 2005.
[26] O. A. Adams, "Perceived Relative Importance of Training in Contractor Development in Nigeria, The Organization and Management of Construction: Shaping Theory and Practice", Volume Two, Managing the Construction Project and Managing Risk; London; E&FN Spon; pp 640 – 649, 1996.
[27] A.N. Aniekwu, and D.C. Okpala, "The Effect of Systematic Factors on Contract Services in Nigeria"; Construction Management and Economics 6; London, E&FN Spon; 171-182, 1988.
[28] S. Ogunlana, K. Promkuntong and V. Jearkjirm, "Construction delays in a fast-growing economy: comparing Thailand with other economics", International Journal of Project Management, vol. 14, pp. 37-45, 1996.
[29] J. Wells, "The Construction Industry in Developing Countries: Alternative Strategies for Development", London; Croom Helm, 1986.
[30] F.L.Y. Yng, C.S. Lean, E.C.C. Wai, E.L. Ping and L. Min, "Performance evaluation of alternative project procurement methods", Project Report, National University of Singapore, 2002.
[31] P.W.G. Morris and G.H. Hugh, "reconditions of success and failure in major projects. Templeton college", The Qxford Centre for Management Studies, Kinnington Oxford, UK, 1983; Technical Paper No. 3, 1983.
[32] I. Avots, Cost-Relevance Analysis for Overrun Control. International Journal of Project Management, vol. 1, pp. 142-148, 1983.
[33] J. Elfving, I. Tommelein I. And G. Ballard, "Consequences of Competitive Bidding in Project-Based Production", Journal of Purchasing & Supply Management, vol. 11, pp. 173-181, 2005.

[34] P. E.D. Love, M.R. Skitmore, and G. Earl "Selecting an appropriate procurement method for the construction process: an empirical study" Construction Management and Economics vol. 16, pp. 221-233, 1998.
[35] National Economic Development Office (NEDO), "Think About Building", NEDC, London, 1985.
[36] R.M. Skitmore and D.E. Marsden "Which procurement system? Towards a universal procurement selection technique", Construction Management and Economics, 6, 71Ð89, 1988.
[37] S. Singh, "Selection of appropriate project delivery system for building construction projects", Proc. CIB W-90 International Symposium on Building Economics and Construction Management, Sydney, Australia, pp. 469Ð480, 1990.
[38] I. Arazi, S. Mahmoud and H.H. Mohamad, "Prioritizing Project Performance Criteria within Client Perspective", Research Journal of Applied Sciences, Engineering and Technology vol. 3, pp. 1142-1151, 2011.
[39] O.E. Ogunsanmi and A. Bamisele, "Factors Affecting the Selection of Project Procurement Methods", Builder Magazine, vol. 12, pp. 11-16, 1997.
[40] W.H. Daniel, "Construction Management" USA, John Wiley & Sons Inc, 2006.
[41] O. Adesanya, "Project Procurement Paths", The Journal of the Federation of Construction Industry, vol. 2, pp. 6-21, 2008.
[42] J. Egan, "Rethinking construction, Construction Task Force report", Department of the Environment, Transport and the Regions. HMSO, London, 1998.
[43] I.H. Seeley, "Quantity Surveying Practice", Macmillan Publishers Ltd, London, 1997.
[44] A. Turner, "Building Procurement" Palgrave Macmillan Ltd., UK, 1997.
[45] D.S. Kadiri and K.T. Odusami, "Comparative Study of Time and Cost Performance of Direct Labour and Labour Only Procurement System", Journal of the Nigerian Institute of Quantity Surveyors, vol. 44, 2003.
[46] A. Turner, "Building Procurement", Macmillan, UK, 1990.
[47] A.S. Oyegoke, "Construction Management Contracting Systems in the UK and US Practices: Consulting and Contracting Perspectives", Journal of the Nigerian Institute of Quantity Surveyors, vol. 34, pp. 22-34, 2001.
[48] T. Alhazmi and R. McCaffer, "Project Procurement System Selection Model", Journal of Construction, Engineering and Management. Vol. May/June, 2000.
[49] A. A. Bubashit and S. A. Almohawis, Evaluating the general conditions of a construction contract. International Journal of Project Management, vol. 12, pp. 133-35, 1994.
[50] A. P. C. Chan and A. P. L. Chan, "Key Performance Indicators for measuring construction success Benchmarking", International Journal of Project Management, vol. 11, pp. 203-221, 2004.
[51] L.F. Alarcon and D. A. Sarpel, "Improving the design-construction interface", In: Proceedings of the sixth Annual Conference of the International Group for Lean Construction, Guaruja- Brazil, 1998.
[52] A.P.C. Chan, "Towards an expert system on project procurement", Journal of Construction Procurement, vol. 1, pp. 111-23, 1995.
[53] F. Harris, and R. McCaffer, "Modern Construction Management", 5th edition, London, Blackwell Science Limited, 2002.
[54] Financial Standard, "Construction Industry Operators call for Review of Road Maintenance Act. Financial Standard", Dec. 20, Lagos, Financial Standard Newspapers, pp 2, 2004.
[55] C.M. Tam, Z.M. Deng, S.X. Zeng and C.S. Ho, "Quest for Continuous Quality Improvement for Public Housing Construction in Hong Kong", Construction Management and Economics,vol. 18, pp. 437-446, 2000.
[56] G. A. Adedokun, "Assessment of Variation on different Contract Procurement systems", An unpublished B. Tech. Thesis submitted to the department of Quantity Surveying, Federal University of Technology, Akure, 2009.
[57] S.O. Babatunde, A. Opawole and I.C. Ujaddughe, "An appraisal of project procurement methods in the Nigerian construction industry", Civil Engineering Dimension, vol. 12, PP. 1-7, 2010.

FRP SHEETS CONTRIBUTION IN COMMON REPAIR TECHNIQUES OF CONCRETE STRUCTURES WITH EMPHASIS ON CONCRETE COLUMNS

Mohammad Panjehpour[1*], Nima Farzadnia[1], Mohammed Parvez Anwar[2], and Abang Abdullah Abang Ali[1]

[1]Housing Research Centre, University Putra Malaysia, 43400, Serdang, Selangor.
[2]Department of Civil Engineering, University of Nottingham Malaysia Campus, Malaysia

*Corresponding Email: mdpanjehpour2006@yahoo.com

Abstract

The history of composites dates back to few thousand years ago. Actually, natural fibrous composites were used by ancient Egyptians to build small houses. Numerous studies have revealed that Fibre Reinforced Polymer (FRP) is a convenient material for repair and strengthening of concrete structures compared to the traditional materials. Since the presentation of fibre reinforced polymer (FRP) in the concrete structures, the need for practice codes containing FRP in field of strengthening and repair has emerged. Many parts of structures are usually replaced simply, instead of repair due to the lack of knowledge about the techniques of repair. Hence, in this review, advantages and disadvantages of FRP repair, different types of FRP repair systems, repair stages and principles of repair theory for concrete structures with FRP are reviewed. In addition, modern repair techniques are reviewed in detail for different damaged levels of concrete structures. Recent developments in the field of repair with FRP have highlighted the need for assessment of repaired concrete columns. Thus, in one part of this review, the authors emphasise different damaged concrete column repaired with CFRP. Finally, the recent needs for further researches in field of repair with FRP are discussed.

Keywords: *Column, Concrete, FRP, Repair*

1.0 INTRODUCTION

The history of composites dates back to few thousand years ago. For example natural fibrous composites were used by ancient Egyptians for building small houses in form of straw-reinforced-clay bricks. Besides, fabricating natural compression was used as a cross-ply papyrus papers in improving the linen wrapping method for their mummies to increase the flexibility of the dried brittle dead bodies as shown in Figure-1 [1].

(a) **(b)** **(c)**

Figure 1: Ancient Egyptians utilization of natural composites: a) Fibrous clay blocks, b) Compression moulded cross-ply papyrus sheets, c) Mummies linen wrapping systems [1].

Recent developments have concentrated on the mixing of continued fibre-base textiles with mortars instead of resins which is leading to the textile-reinforced mortars (TRM). Both FRP and TRM material are called, continuous fibre composites. They are also known as advanced composites or simply composites. The applications of FRP as structural reinforcement have increasingly become common in construction in many countries [2]. The development of practice codes containing FRPs has its limitations in the wider range of application. Nowadays, the market for externally bonded FRP has a dynamic situation. Several types of manufactures are working in this market part, while researchers and institutes are trying to obtain design guidelines for FRP. The ranges of FRP applications in civil structures have been become very wide. Hence, various difficult projects have been done by using FRP, such as tripling the bearing capacity of a floor slab in Belgium, strengthening of silos in Sweden, seismic strengthening in Greece and Italy [3].

Many researches about the application of FRP have been carried out in Europe since 25 years ago. Most of the European activities have focused on externally bonded FRP reinforcement (FRP EBR). In Switzerland alone, the amount of FRP material used for strengthening in structures reached 30 to 50 km per year until 2000 [4]. From 1980s FRP has been utilized in Japan for different application of repair and strengthening. Its application has increasingly developed to the level that nowadays it is using in many modern construction projects and researches in the world due to their low labours costs, light weight, high tensile strength to weight, and easy installation [5]. Moreover, new types of products and methods have been introduced in the market to develop the feasibility of FRP repair and strengthening methods such as L shaped carbon FRP strips for shear strengthening, methods for improving anchorage capacity by mechanical devices, methods to use prestressed FRP for external bonded reinforce [4]. In this review, advantages and disadvantages of FRP repair, different types of FRP repair systems, repair stages and principles of repair theory for concrete structures with FRP are reviewed. In addition, modern repair techniques are reviewed in detail for different damaged levels of concrete structures.

2.0 APPLICATION OF FRP

FRPs have many advantages in outdoor and indoor for structure applications. Firstly they are very rigid, strong with outstanding strength compared to their weight. Secondly, they present a high creep resistance in the long term. They are highly resistant to temperature change without softening or brittleness. They are also good resistance to UV radiation, humidity, atmospheric pollution. Apart from having a low flammability depending on the choice of resin, the material has also a good thermal resistance and a good dimensional stability. Users use a wide range of possibilities in terms of colour and different form of good design. Furthermore, FRPs can be used in construction as lightweight materials. Hence, they are easy to carry. FRPs can also be used as prefabricated elements, which are easily and rapidly joined without the need of special handling equipment [6, 7].

3.0 REPAIR CONCEPT

Repair of defective parts of concrete structures is still an issue among the researchers due to the different levels of damage. The method of repair depends on the type and size of damaged area. For example a small damaged area can be simply repaired with epoxy and resin [14]. For large area of damaged in terms of depth, usually the repair mortar are utilised to fill the defective part of concrete member. For this type of defect, two parts of material from different source are joined together. Hence, this type of damages is weaker than small size damage after repair. The most common types of FRP repair systems are wet lay-up, pre-cured, and near surface method [14, 15].

The quality of repair depends on the bonding between surface and FRP. Hence, the quality of the epoxy and resin affects the bonding stress. For original parts of structure, the resin which is used between FRP and surface cures chemically as a uniform unit. This is primary and strongest bonding between concrete and FRP. Since usually some parts of concrete structures are damaged, the repaired part becomes secondary bonding connected to the original primary structure. Thus, repairs parts are only as strong as the adhesive utilised to connect the damaged parts to the original parts. As a result, for repair, the stronger adhesive resins should be used [6, 16-18].

Repair influences by quality of adhesion to the primary structure. Hence, durability and strength of repair depends on the amount of surface area where is utilised for bonding. Nowadays different style of FRP structures are used for strengthening and repair of concrete structures such as fully wrap, spiral strap, district strap. These differences are mostly due to FRP price which is expensive. In the strap style, the amount of utilised FRP is less than fully wrap. For repair, the amount of surface area where is repaired by FRP affect the bonding. Thus, among these styles, fully wrap style is convenient for repair. As a repaired part of structure becomes thicker, it becomes stiffer. Hence, due to the lack of flexibility, the FRP are peeled. As a result increase of number of FRP sheets layer should be avoided for repair [17, 19, 20].

4.0 REPAIRS ON CONCRETE STRUCTURES

Based on the demand, in different situations, repair of concrete structures is needed due to the lack of ductility, durability, stiffness and strength. The most common situations that a concrete member needs repair are listed below [8-11].

- Seismic repair
- Damages due to the accidents and environmental conditions
- Change of application
- Initial design flaws

FRP repair method is more convenient and economic compared to the traditional repair methods. It can be utilised either to repair some damaged parts of concrete structures or to strengthen concrete structural members such as beam, column, wall, and slab to resist higher load especially for seismic upgrade [5]. FRPs are convenient material for corrosion problems of concrete structures. Khaled et al points out FRP sheets can be used to repair of corroded reinforced concrete beams to maintain structural integrity. However, these experimental results are achieved based on variable chloride levels from 0 to 3% [12]. Carbon and glass FRP are convenient material to repair of unreinforced masonry walls. By utilising carbon or glass FRP on the exterior surface of the wall, both flexural and shear capacity of masonry walls can be increased. The advantages of FRP using for masonry walls are increasing out-of-plane flexural strength, increasing stiffness at service loads, adding very little weight to the wall, and limited access requirements.

However, evaluation of long-term properties of FRP is difficult. Additionally fatigue, potential of FRP to be damaged against the ultraviolet radiation, high price of FRP and low modulus of elasticity for aramid and glass fibres are still potential disadvantages of FRP repair methods. Hence, application of FRP repair method is limited due to these disadvantages [13].

5.0 METHODOLOGY OF REPAIR ON CONCRETE STRUCTURES

5.1 TO FIND THE LEVELS OF DAMAGE

Damage usually devotes into four groups. The most common types of structure damaged levels are tears, crushed cores, punctures, and delaminations. For realize the type of damages, coin tap test is commonly used. Detection of crushed material and solid laminate is easy with audible differences [7].

5.2 DETERMINATION OF REPLACEMENT OR REPAIR OF DAMAGED PART

To inspect the level of damaged area, the size of damaged area is important. Estimating of the type of material and labour time needed for repair is the best way to find which option of replacement or repair is more convenient according to the estimated price [6, 7, 17].

5.3 TO PREPARE THE AREA SURFACE FOR BONDING WITH REMOVING THE DAMAGED MATERIAL

Coin tapping should be used until all damaged concrete layers are removed from surface. As far as possible, little material must be removed to avoid repair grows up too large. Nevertheless, concrete surface must be exposed for a strong repair. In this stage, concrete surface must be prepared with sandpaper or cup brush depending on the surface area before repair with FRP. Any remaining dust must be vacuumed and the concrete surface should be wiped with solvent rag. Normally, air-compressor is used to remove the small pieces of materials inside the cracks. Acetone is convenient to remove dusts, grease, oil, waxes, or other contaminants which would affect repair adhesion. Open time, pot life, and curing duration must be considered for mix of epoxy and resin before sticking the FRP sheets. For over coating with FRP, when a part of FRP is damaged, it is difficult to realign the fibres together, because after damage, FRP composites fibres tend to hang up on each other [7, 21-23].

5.4 LOOKING OVER FOR REPAIRING

Coin tap method should be utilised to check the cured repair. The entire parts of concrete structure should resonate the same sound. Non-destructive load test should also be used for more confidence [23-25].

6.0 CONSIDERATION ON COLUMN REPAIRWORK

Recently, the technique of repair and strengthening of RC column with steel jackets has been widely replaced by FRP sheets. The most common form of FRP for repair of column involves the external wrapping of FRP sheets. According to the previous researches, FRP-confined concrete behaviour is different from steel confined concrete. Hence, design manual and standards for steel-confined concrete column cannot be used for FRP-confined concrete column [26-32].

FRP repair method is based on providing confinement for column to increase its strength and ductility. The significant of FRP confinement action is enhancement of the ductility and strength of concrete column. It should be emphasised that enhancement of ductility is more important than strength in terms of seismic upgrading. It also prevents slippage and buckling of the longitudinal reinforcement for concrete columns. In seismic problems, repair techniques for upgrading are usually based on confinement pressure for the entire column or potential plastic hinge regions. FRP fully wrap can only increase tri-axial compression for concrete column which is not enough for seismic upgrade [13, 33].

Numerous studies have attempted to model compressive behaviour of FRP-confined concrete. In reinforced concrete columns confined with FRP jackets, concrete is loaded under tri-axial compression and FRP is loaded in hoop tension. These FRP jackets are usually oriented only in the hoop direction for circular section columns. Hence, the best properties of both of concrete and FRP can be used. Confinement of FRP increases greatly concrete column ductility due to thigh tensile strength of FRP. Thus, both strength and ultimate strain of concrete can be enhanced with FRP confinement [34].

Nowadays, FRP repair is very useful due to the loss of security in order to preserve the structural serviceability. Hence, due to the lack of guidelines for repair with FRP, the need for numerical model to assess different damaged concrete members is feeling. However, analytical and numerical models to assess repaired concrete members are still limited [15]. According to this need, the authors are doing some experiments to evaluate the effects of different pre-damaged levels of cylindrical concrete specimen repaired with CFRP on axial compressive strength of concrete specimen; Figures-2,3. With simulation of behaviour of cylindrical concrete specimens to the concrete columns, the results of this research can be developed from cylindrical specimens to the concrete columns. Hence, the authors are going to investigate stress- strain curves and compare them for different pre-damaged levels of cylindrical concrete specimens.

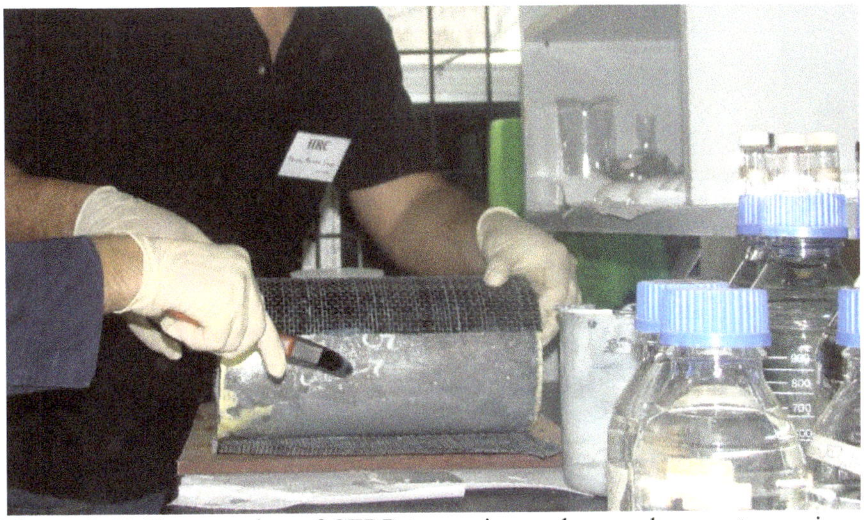

Figure 2: Fully wrapping of CFRP to repair pre-damaged concrete specimen

Figure 3: CFRP rupture after compressive test

Basically, two parts of damage are considered in repair based on small and large amount of damage. The large amount of damage is considered as defect which is shown in Figure-4.

Figure 4. Pre-defected concrete specimens

For repair of defected concrete members, repair mortar must be used before wrapping with FRP. To repair defected parts of concrete members, ASTM C 928-05 has some specifications about rapid repairs for structure materials that include organic compounds [35].

9.0 CONCLUSION

According to this review, lack of knowledge in some parts of this topic has emerged which is described in below.

- For repaired concrete structures, experimental results are not available in terms of different types of damaged.
- There is a need for numerical models for different types of damaged concrete structures to assess strength of repaired concrete members.
- In terms of fire resistance for repaired concrete structures, there are still some gaps about experimental results.
- There is no numerical model to assess repaired concrete members in terms of fire resistance.

ACKNOWLEDGEMENTS

The authors appreciate Hosing Research Centre (HRC) for financial support and providing the requirements for this research.

REFERENCES

[1] Hulatt, et al. (2004), "A novel advanced polymer composite/concrete structural element", Advanced Polymer Composites for Structural Applications in Construction, pages: 9-17.

[2] Mosallam, A.S. and K.M. Mosalam (2003), "Strengthening of two-way concrete slabs with FRP composite laminates", Construction and Building Materials, Vol 17(1), pages: 43-54.

[3] Cromwell, J.R., K.A. Harries, and B.M. Shahrooz (2011), "Environmental durability of externally bonded FRP materials intended for repair of concrete structures", Construction and Building Materials, Vol 25(5), pages: 2528-2539.
[4] John, B. (2001), "fibre reinforced plastics for reinforced concrete structures", proceedings of the international conference on fibre-reinforced plastics for reinforced concrete structures, FRPRCS-5, Cambridge, UK.
[5] MERTZ., D.R., et al. (2003), "Application of Fiber Reinforced Polymer Composites to the Highway Infrastructure", in NCHRP Report 503, TRANSPORTATION RESEARCH BOARD: WASHINGTON.
[6] Attard, T.L., C.M. Abela, and K. Dhiradhamvit (2011), "Seismic FRP retrofit of circular single-column bents using a ductility wrap envelope to alter failure modes", Engineering Structures, Vol 33(5), pages: 1553-1564.
[7] Benzarti, K., F. Freddi, and M. Frémond (2011), "A damage model to predict the durability of bonded assemblies. Part I: Debonding behavior of FRP strengthened concrete structures", Construction and Building Materials, Vol 25(2), pages: 547-555.
[8] American Concrete Institute (2007), "State-of-the Art Report on Fiber Reinforced Plastic Reinforcement for Concrete Structures", Farmington Hills, Michigan.
[9] Bakis, C.E. and L.C. Bank (2002), "Fiber-Reinforced Polymer Composites for Construction", Journal of Composites for Construction, pages: 73-87.
[10] CEB-FIP (2006), "Retrofitting of Concrete Structures by Externally Bonded FRPs, with Emphasis on Seismic Applications", International Federation for Structural Concrete: Geneva, Switzerland.
[11] Czaderski, C. and M. Motavalli (2007), "40-Year-Old Full- Scale Concrete Bridge Girder Strengthened with Prestressed CFRP Plates Anchored Using Gradient Method", Journal of Engineering, Vol 38, pages: 878-886.
[12] Khaled, et al. (2008), "FRP Repair of Corrosion-Damaged Reinforced Concrete Beams", Department of Civil Engineering, University of Waterloo: Waterloo, Canada.
[13] CEB-FIP (2001), "Externally Bonded FRP Reinforcement for RC Structures", International Federation for Structural Concrete Geneva, Switzerland.
[14] Mirmiran, A. (2004), "Bonded repair and retrofit of concrete structures using FRP composites, in recommended construction specifications and process control manual", Technology & Engineering: washington.
[15] Rougier, V.C. and B.M. Luccioni (2007), "Numerical assessment of FRP retrofitting systems for reinforced concrete elements", Engineering Structures, Vol 29(8), pages: 1664-1675.
[16] Au, C. and O. Büyüköztürk (2006), "Debonding of FRP plated concrete: A tri-layer fracture treatment", Engineering Fracture Mechanics, Vol 73(3), pages: 348-365.
[17] Boyd, A.J., et al. (2008), "Sprayed FRP repair of simulated impact in prestressed concrete girders", Construction and Building Materials, Vol 22(3), pages: 411-416.
[18] Christopher K.Y, L. (2006), "FRP debonding from a concrete substrate: Some recent findings against conventional belief", Cement and Concrete Composites, 28(8), pages: 742-748.
[19] Constantin E, C. (2007), "Analytical model for the torsional behaviour of reinforced concrete beams retrofitted with FRP materials", Engineering Structures, Vol 29(12), pages: 3263-3276.
[20] Gadve, S., A. Mukherjee, and S.N. Malhotra (2009), "Corrosion of steel reinforcements embedded in FRP wrapped concrete", Construction and Building Materials, Vol (1), pages: 153-161.
[21] Colalillo, M.A. and S.A. Sheikh, "Seismic retrofit of shear-critical reinforced concrete beams using CFRP", Construction and Building Materials.
[22] Wei, Y.-Y. and Y.-F. Wu (2012), "Unified stress–strain model of concrete for FRP-confined columns", Construction and Building Materials, Vol (1), pages: 381-392.
[23] Yaqub, M. and C.G. Bailey (2011), "Repair of fire damaged circular reinforced concrete columns with FRP composites", Construction and Building Materials, Vol 25(1), pages: 359-370.
[24] Walker, R.A. and V.M. Karbhari (2007), "Durability based design of FRP jackets for seismic retrofit", Composite Structures, Vol 80(4), pages: 553-568.

[25] Wu, G., Z.S. Wu, and Z.T. Lü (2007), "Design-oriented stress–strain model for concrete prisms confined with FRP composites", Construction and Building Materials, Vol 21(5), pages: 1107-1121.

[26] Ibell, T., A. Darby, and S. Denton (2009), "Research issues related to the appropriate use of FRP in concrete structures", Construction and Building Materials, Vol 23(4), pages: 1521-1528.

[27] Karimi, K., M.J. Tait, and W.W. El-Dakhakhni (2011), "Testing and modeling of a novel FRP-encased steel–concrete composite column", Composite Structures, Vol 93(5), pages: 1463-1473.

[28] Lam, L., et al. (2006), "FRP-confined concrete under axial cyclic compression", Cement and Concrete Composites, Vol 28(10), pages: 949-958.

[29] Laoubi, K., E. El-Salakawy, and B. Benmokrane (2006), "Creep and durability of sand-coated glass FRP bars in concrete elements under freeze/thaw cycling and sustained loads", Cement and Concrete Composites, Vol 28(10), pages: 869-878.

[30] Lu, X.Z., et al. (2006), "Finite element simulation of debonding in FRP-to-concrete bonded joints", Construction and Building Materials, Vol 20(6), pages: 412-424.

[31] Mazzotti, C., M. Savoia, and B. Ferracuti (2008), "An experimental study on delamination of FRP plates bonded to concrete. Construction and Building Materials", Vol 22(7), pages: 1409-1421.

[32] Mohamed H, H. (2009), "Bond strengthening of lap spliced reinforcement using external FRP jackets: An effective technique for seismic retrofit of rectangular or circular RC columns", Construction and Building Materials, Vol 23(3), pages: 1265-1278.

[33] Chai, Y.H. (1991), "Steel Jacketing of Circular Reinforced Concrete Bridge. Columns for Enhanced Flexural Performance, in Civil Engineering", University of California: San Diego.

[34] Teng, J.G., (2002), "FRP-strengthened RC structures", John Wiley and Sons,Technology & Engineering.

[35] ASTM (2005), "Standard specification for packaged, dry, rapid-hardening cementitious materials for concrete repairs"

Factors affecting Safety Performance in Repair, Maintenance, Alteration, and Addition (RMAA) Projects

Adnan Enshassi[1,*], Ehsan Risqa[1], Faisal Arain[2]

[1]Department of Civil Engineering, IUG, Gaza, Palestine
[2]School of Sustainable Building and Environmental Management, NAIT, Edmonton, Canada

*Corresponding E-mail: aenshassi@gmail.com

Abstract

Repair, Maintenance, Alteration and Addition (RMAA) works are playing an increasingly important role in developing countries. The accidents and fatalities records of RMAA sector in Gaza Strip have been alarmingly high; however, research in the RMAA sector remains limited. Safety of RMAA works has long been neglected because the project sizes of RMAA are small and only last for a short period of time, which make the working environment of RMAA works more difficult to control than new building works. The aim of this paper is to identify, valuate and rank the most important factors that affect safety performance and the most important causes of fatal accidents in RMAA projects. A questionnaire survey was used in this study. The results revealed that poor safety awareness of managers in maintenance firms and lack of training of RMAA workers for handling multi-tasks were the most important factors that affecting safety performance of RMAA works. The results showed that ineffectiveness of lack of training and certification of competence; immature corporate systems of firms which does not care with safety and health through RMAA works, and lack of leadership from government as a key client are the most significant causes of construction fatal accidents of RMAA projects. The results also indicated that the macro level factor is the most important category that causes fatal accidents in RMAA works. It is recommended to enhance the awareness of construction firms, project managers and workers regarding the importance of safety performance in repair and maintenance works and strengthen site monitoring and supervision system in construction firms. Safety training courses should be organized for workers and project managers in order to improve their safety culture and competence regarding safety performance through repair and maintenance works. Furthermore the RMAA subcontractors should be selected according to their good records of safety performance.

Keywords: *Repair, maintenance, safety performance, construction*

1.0 Introduction

With shrinkage of the construction market and reduced volume of new works, repair, maintenance, minor alteration and addition (RMAA) works are playing an increasingly important role in the local construction industry (Chan et al., 2009). RMAA works refers to those minor works such as construction projects for village type houses in the new territories, minor alterations, repairs, maintenance and interior decoration of existing buildings (Labour Department, 2008). RMAA work involves different types of workers as compared with new building works. Most RMAA contracting companies found in the construction market are subsidiaries of general building contractors or small specialty contractors of RMAA works (Hon et al., 2010).

Most of RMAA contracting companies are small and medium-sized enterprises (SMEs). Poor safety practices are more prevalent in SMEs than the large size ones (Brace et al., 2009).Small companies tend to work in greater isolation. Thus, owing to limited resources, these companies may experience more difficulty in meeting health and safety requirements (Lam, 1997; cited by Hung, 2012).RMAA works involve more small contractors than new buildings works do. Many of these contractors have less knowledge and experience in health and safety. The small project size and short duration of RMAA works tend to make the working environment more difficult to control than new building works (Hung, 2012).The risk of falls is high in RMAA works because of temporary, short-term, and precarious nature of access to the task. When the task duration is short, workers are tempted to take higher risks (Cameron et al., 2007).Safety of RMAA works has long been neglected because RMAA works have a fragmented nature and only last for a short period of time (Hon and Chan, 2013b).The accident record of the RMAA sector has been alarmingly high; however, research in the RMAA sector remains limited (Hon et al., 2013). With rising importance of the RMAA sector in many developed societies, safety of RMAA works has begun to draw attention (Hon and Chan, 2013a).

Importance of RMAA works to the whole construction industry is expected to increase continuously in the foreseeable future (Chan et al., 2009).Safety performance of the construction industry still needs improvement. Hon et al (2012) conducted an empirical study to investigate the difficulties of implementing safety practices in the repair and maintenance sector in Hong Kong. They concluded that the most common difficulties are: difficult to change the mindset of RMAA workers, difficult to conduct safety supervision due to scattered locations, limited safety resources for RMAA projects undertaken by small/medium-sized contractors, difficult to standardize the operational procedures of RMAA works due to ad hoc site problems, shortage of time to deal with safety issues, high turnover rate of RMAA workers, small scale and short duration of RMAA projects, influx of illegal workers, and difficult to control self-employed workers. These difficulties of implementing good safety practices in the construction industry largely hinder further safety performance improvement and undermine the effectiveness of safety measures. Many RMAA contracting companies are small and medium-sized enterprises (SMEs) which do not have comprehensive safety management systems. Hon and Chan, (2013b) concluded that existing safety legislation and regulations for new construction sites are not fully applicable to RMAA works. RMAA subcontractors with good track records of safety performance should be selected for bidding.

Enshassi (2003) illustrated the factors affecting the construction safety on Gaza Strip such as, the severe competitive tendering methods, the age of the workers, experience, and the concern of management on productivity with the ignorance of safety issues, and the lack of training. But unfortunately, there was a shortage in safety applications in the Gaza Strip construction industry. Statistics showed that more than one third of fatalities among workers were dead during the working in construction site. Falls and excavations were the main factors causes for the death of constriction workers. The main causes of injuries in the RMAA sector in Gaza Strip are classified into five categories, the categories are, falls, struck by falling object, struck by moving or stable object, caught in/between, machines, and others (Hassona, 2005). The absence of a unified set of safety regulations adversely affects the enforcement of safety on the job site (Alaqqad, 2009).For these reasons, more research is needed to be done in the RMAA sector, where small companies dominate the industrial sector(Hung, 2012).

Little studies have been done in this field in Gaza Strip. Health and safety in Gaza Strip is not widely recognized as inherent characteristic of RMAA projects. Contractors consider health and safety a legal requirement that means spending money without any profit. This situation resulted in the increased number of accidents in RMAA sector (Abu Alqumboz, 2007).The objectives of this paper are:

- To identify, evaluate and rank the most important factors that affecting safety performance in RMAA sector in Gaza strip,
- To identify most important causes of construction fatal accidents in RMAA projects,
- To identify the most common personal protective equipment provided by contractors in RMAA works,
- To highlight the adverse impacts of the occurrence of incidents in maintenance projects, identify most types of serious accidents of repair and maintenance projects and propose preventive measures to improve safety practice in RMAA sector in Gaza Strip.

2.0 Repair, Maintenance, Alteration and Addition (RMAA)

As the RMAA sector expands, safety of the RMAA sector deserves greater attention (Hon and Chan, 2013a). Despite research studies on improving safety performance of the construction industry are readily available; most of them are based on new construction projects. There is a very limited research on the RMAA sector, not to mention studies focusing on investigating strategies to curb the worsening safety problems of the rising RMAA sector (Hon et al., 2010). Nature of works, tasks undertaken and workers involved in RMAA projects are entirely different from those in new construction projects. Hence, strategies for resolving safety problems of RMAA projects are also incongruent to new construction projects.

While RMAA sector is expanding in many developed cities; occurrences of fatalities and injuries in this sector are also soaring (Hon et al., 2011). Fatalities in the construction industry are often the highest among all industries in a number of developed countries (Hung, 2012). Statistics indicated that the percentage of fatal industrial accidents arising from RMAA works in Hong Kong was disturbingly high and was over 56% in 2006 (Chan et al., 2009). Fatal accident case analysis in RMAA sector is important in safety research because it can reveal the root causes of accidents, providing valuable information for designing future preventive measures. Hon and Chan (2013a)found that the behavior of contractors who work in RMAA sector on safety management are of grave concern, including the lack of provision of personal protection equipment, regular safety meetings, and safety training. The main factors affecting safety performance in RMAA projects include poor safety awareness of top management, lack of training, poor safety awareness of project managers, reluctance to input resources to safety, and reckless operations (Tam et al., 2004).

Hon and Chan (2013b) concluded that construction safety performances of many developed countries with regard to RMAA sector have been improving. The prevailing challenge of this industry is to find out ways for further safety improvement. By the same token, proper safety knowledge management may be one of the ways to improve construction companies' safety performance in RMAA projects (Hon and Chan, 2013b). Hallowell (2012) asserted that better safety knowledge management can improve the organization's capacity of adapting to changes, resulting in better safety performance. Term contract nature of some RMAA projects makes safety supervision difficult because RMAA work tasks are widely scattered (Hon et al., 2012). Different subcontracting works of new construction projects are likely to be responsible by individual specialty subcontractors of a single trade. However, nature of RMAA works generally requires workers to have multi-skills to perform multi-tasks. RMAA workers may be endangered when performing tasks that they are not as familiar with (Hon et al., 2010 as cited by Hon and Chan, 2013b). Safety knowledge is important to successful safety management and good safety performance (Hallowell, 2012). Safety legislation, regulations, and guidelines provide the minimum requirements of site safety. However, meeting these stipulated standards are not enough to maintain a safe working environment (Hon and Chan, 2013b).

Brace et al.,(2009) have unveiled the causes of fatal construction accidents in the RMAA sector in United Kingdom. They identified three levels of factors: macro, mezzo, and micro. Accidents occur in RMAA projects when these three levels of factors are concurrently aligned. Six out of the nine micro-level factors are related to human behavior: lack of individual competency and understanding of workers and supervisors; ineffectiveness of lack of training and competence; lack of individual competency and understanding of workers and supervisors; ineffectiveness of lack of training and certification of competence; lack of ownership, engagement and empowerment of communication with, and responsibility for workers and supervisors; poor behavior; misuse of equipment (including personal protective equipment); and itinerant workforce. Most macro and mezzo level factors are related to the safety attitudes and safety practices of RMAA organizations.

The exploratory study of Cheng et al. (2004) revealed six root causes of construction accidents in RMAA sector in China. Amongst them, three are related to workers, namely: lack of attention to personal safety protection by workers, insufficient safety training, and tiredness of workers. Another two are related to organizational management, namely: lack of attention to safety management by main contractors/ project managers, and inadequate setting of safety level. The sixth cause was poor quality of construction materials and equipment. Mullen (2004) has provided an insightful discussion of underlying factors leading to unsafe behavior in a workplace of RMAA projects. He has synthesized the factors into three broad categories: organizational factors, image, and avoiding negative consequences. Organizational factors substantially contribute to unsafe behavior. The organizational factors include role overload production performance over safety, socialization influence, safety attitudes, and perceived risks. Image negatively influences the safety behavior of workers because workers want to be seen as "macho" or tough and competent. To avoid negative consequences, workers want to avoid teasing and harassment from co-workers, and do not want to lose their jobs.

Haslam et al.(2005) have identified a hierarchy of factors leading to accidents in the RMAA sector in United Kingdom. These factors included problems arising from workers or the work team, workplace issues, shortcomings with equipment (including personnel protective equipment (PPE)), problems with suitability and condition of materials, and deficiencies with risk management. They attributed over two-thirds of the accidents to the behavior and capabilities of workers. As in the construction industry, a number of safety researchers have identified human and organizational factors to be the key causes leading to accidents in RMAA projects. Kartam et al. (2000) summarized nine causes of safety problems in RMAA projects in Kuwait construction industry. Amongst them, disorganized labor, extensive use of foreign labor, extensive use of subcontractor, lack of safety regulation and legislation, low priority given to safety, small size of most construction firms and competitive tendering are human and organizational factors contributing to unsafe behaviors.

3.0 Research Methodology

A questionnaire was designed based on previous studies (Abdel hamid and Everett, 2000; Brace et al., 2009; Chan et al., 2008; Cheng et al., 2004; Tam et al., 2004; Haslamet. al., 2005; Brace et al., 2009; Hon et al., 2010; Hon et al., 2011; Hon et al., 2012; Hung, 2012, Hon and Chan, 2013a; Hon and Chan, 2013b; Hon et al., 2013). A set of 23 factors that affected safety performance through RMAA works were identified. A total of 19 causes of fatal accidents through repair and maintenance projects were also identified. These 19 factors were further grouped under three major categories: macro level factors, mezzo level factors, and micro level factors. Five-point Likert scale was used in this questionnaire (1 = very low effect, 2 = low effect, 3 = neutral effect, 4 = strong effect, and 5 = very strong effect). Likert scale was chosen in order to expand the way the respondents would reply.

The target population of the study comprised of: governmental institutions, private institutions and international institutions. A total of 38 questionnaires were distributed to selected respondents and 30 were retrieved which were used for the analysis giving a response rate of 78.94%. Respondents were selected based on their level of education and experience in the construction field. The following formula was used to determine the sample size (Ayyub and Mccuen, 2003).

$$SS = \frac{Z^2 \times P \times (1 - P)}{C^2}$$

Where: SS = Sample size

Z = Z value (e.g. 1.96 for 95% confidence level)

P = Degree of variance between the elements of population percentage

C = Confidence interval (margin of error), expressed as decimal (e.g., .05 = ±5)

Correction for Finite Population

$$SS_{correcte} = \frac{SS}{1 + \frac{(SS - 1)}{N}}$$

Content validity test was conducted by sending the questionnaire to two experts in construction to evaluate the questionnaire validity, clarity, comprehensive, readability and reliability and to add more information or delete unacceptable wording if needed. They presented their opinions and comments about content of questionnaire, the degree to which questionnaire paragraphs are relevant to their groups, and lingual clarity of words. In this regard, few paragraphs have been dropped while others have been altered. Two statistical methods were used to analyze the data which obtained from the questionnaires. The first was to obtain percentage values through frequencies of the answers received. The other was to calculate a Relative Importance Index (RII). According to Tam et al. (2000) the RII was evaluated using the following expression.

$$RII = \sum w/A * N$$

where w is the weighting given to each factor by the respondent, ranging from 1 to 5; '1' is the least strong effect and '5'' is the extremely strong effect, A is the highest weight; in this study it is 5; and N is the total number of samples. The RII shall be a variable ranging from 0 to 1.

4.0 Results and Discussion

Table 1 illustrates the institution profile of the study sample.

Table 1: Institution profile

Item	Category	Frequency	Percentage
Types of institution	Governmental	12	40%
	Private	16	53.33%
	International	2	6.66%
Is the institution specialized in maintenance field	yes	12	40%
	no	18	60%
Experience of the institution	Less than 5 years	6	20%
	From 5 years – 15 years	10	33.33%
	From 16 years – 25 years	6	20%
	Over 25 years	8	26.66%
Number of executed projects in the last 5 years	Less than10 projects	10	33.33%
	From 11 – 20 project	5	16.66%
	From 21 – 30 projects	3	10%
	More than30 project	12	40%
Work monetary volume in the last 5 years (USD)	Less than $500000	11	36.66%
	$500000 – less than $1 million	3	10%
	$1 million- less than $5 million	5	16.66%
	$5 million – less than $10 million	10	33.33%
	More than $10 million	1	3.33%
The position of the respondent	Director	4	13.33%
	Project Manager	6	20%
	Site/Office Engineer	20	66.66%
The experience of the respondent	less 5	14	46.66%
	from 5-10	9	30%
	from10-15	2	6.66%
	over than 15	5	16.66%

4.1 Cracking and Failure Patterns – Numerical Observation

Table 2 illustrates the result regarding the safety practices in repair and maintenance projects. As shown in Table 2, the majority of people (83.33%) who worked in repair and maintenance projects are not committed to health and safety. Only 43.33% of companies encourage suggestions on how to improve health and safety in repair and maintenance projects. This means that the awareness of institution and workers with regards to safety in maintenance projects need to be enhanced. The majority of people (63.33%) who worked in maintenance projects haven't received safety trainings courses related to repair and maintenance works. This means that they are exposing every day to injuries and accidents because of the lack of training and experience in the maintenance field. It is unfortunate to notice that only (26.66%) from companies has a good preparedness for emergencies that may happened through execution of repair and maintenance projects. This means that every worker is exposed to death, deteriorating of his health, or aggravation of his injury because of the lack of first-aid in the maintenance projects in Gaza Strip.

However 83.33% of people who are working in maintenance field are fully understand the health and safety risks associated with repair and maintenance projects. Results showed that 46.66% from respondents stated that sometimes it is necessary to take risks to get the job done. This result may appear because of the special nature of maintenance works which differ from new construction works and may have a high level of risk. The findings also showed that 60% of respondent agreed that some health and safety rules or procedures are difficult to follow through repair and maintenance projects. This result can be traced to the special nature maintenance works, which characterized with unclear scope, short duration, and fragmented nature. This results sets alarm of bells that peoples who are working in maintenance field have a more chance to have injuries or accidents.

Table 2: Safety practices in repair and maintenance projects

Question	Response	Frequency	Percentage
I have noticed that all the people who work in repair and maintenance projects are fully committed to health and safety	Agree	4	13.33%
	Disagree	25	83.33%
	Don't know	1	3.33%
I am sure that my company encourages suggestions on how to improve health and safety in repair and maintenance projects.	Agree	13	43.33%
	Disagree	15	50%
	Don't know	2	6.66%
I have received safety trainings courses related to repair and maintenance projects.	Agree	7	23.33%
	Disagree	19	63.33%
	Don't know	4	13.33%
I believe that there is good preparedness for emergencies that may happened in execution of repair and maintenance projects in my company.	Agree	8	26.66%
	Disagree	14	46.66%
	Don't know	8	26.66%
I fully understand the health and safety risks associated with repair and maintenance projects	Agree	25	83.33%
	Disagree	2	6.66%
	Don't know	3	10%
I believe that sometimes it is necessary to take risks to get the job done.	Agree	14	46.66%
	Disagree	15	50%
	Don't know	1	3.33%
I believe that some health and safety rules or procedures are difficult to follow through repair and maintenance projects.	Agree	18	60%
	Disagree	11	36.66%
	Don't know	1	3.33%

4.2 Protective equipment provided by contractors in RMAA works

Figure 1 illustrates the results of personal protective equipment (PPE) provided by contractors through RMAA works in Gaza Strip. The results show that "hard hats, gloves, eye goggles" is the most PPE provided by contractors through RMAA works. Results also showed that "ear plugs" have never provided by any contractor through RMAA works. Tam et al., (2004) concluded that "Eye goggles, Gloves, and Hard hats" are provided at a percentage 100% by contractors through RMAA works in China. This result reflects the high importance of this equipment's.

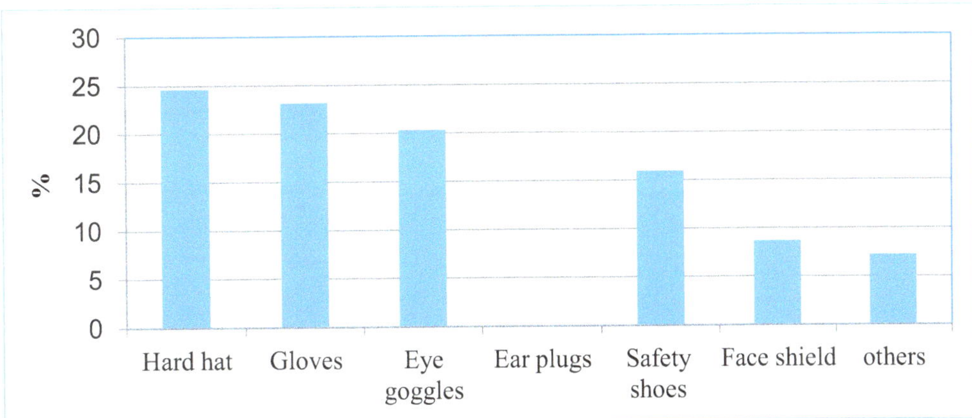

Figure 1: Personal protective equipment provided by contractors in RMAA works

4.4 Most types of serious accidents of repair and maintenance projects

Figure 2 illustrate the results regarding the adverse impacts of the occurrence of incidents in maintenance projects. Results showed that "increase in cost" and "imposing psychological burden on workers" are the most common and important impacts of accidents in RMAA works. It is axiomatic that if any worker has injured or suffers from an accident, the moral of workers will be affected, and they will feel so bad. This means that their productivity will be reduced. Tam et. al., (2004) showed that the majority of respondent (68%) believed that site accidents caused impairing reputation of firms. However this impact is in the least position in the current study. This result may appear because of the poor safety behavior and culture of firms and top leaders in Gaza Strip who don't care if the selected contractor is from those who have good record of safety performance or not.

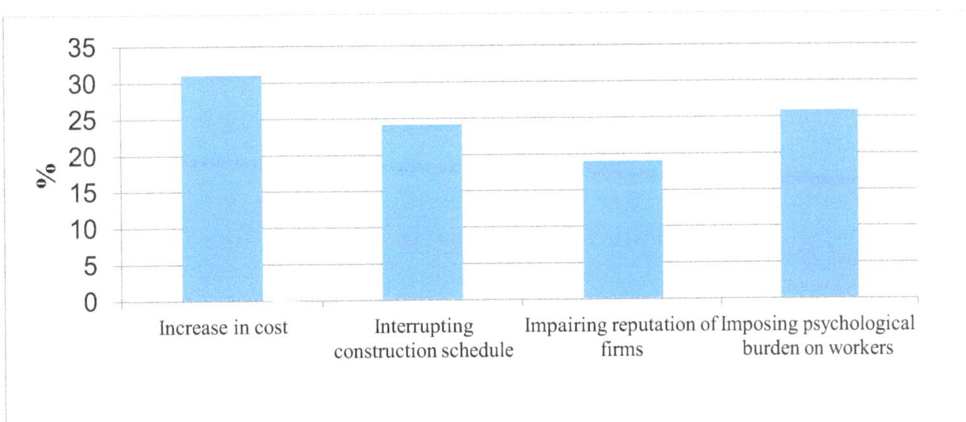

Figure 2: adverse impacts of the occurrence of incidents in maintenance projects

4.5 Most types of serious accidents of repair and maintenance projects

Figure 3 illustrate the results regarding the most types of serious accidents through repair and maintenance projects in Gaza strip. The results revealed that "fall from height" and "slip, trip or fall on same level" are the most types of serious accidents through RMAA works. This result may appear because of the lack of experience, and safety course training of workers. Additionally, most of construction sites are suffering from lack of cleanliness. Nails and toolboxes are sparse around the sites of most projects in Gaza Strip. Hung (2012) concluded that falling from height is in the first position with parentage (63.1 %) and contact with electric shock is in the second position with percentage (15.5%).

The results of Hung (2012) and this study regard to type of accident in RMAA works are very similar and consistent. This means that these types of accidents is serious, widespread, and have extended roots all over many countries. It is clear that there is a massive need to enhance the awareness of RMAA workers regard to safety performance through RMAA works. Every worker should receive a safety course training to avoid injuries and accidents through RMAA works.

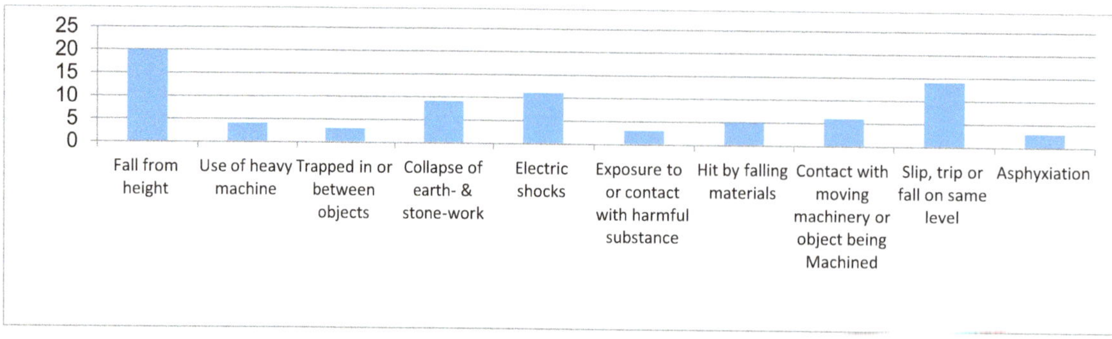

Figure 3: most types of serious accidents of repair and maintenance projects in Gaza strip

4.6 Causes of fatal accidents in RMAA projects

Table 3 illustrates the results regarding the causes of fatal accidents in repair and maintenance projects.

4.6.1 Macro level factors group

As shown in Table 3 the respondents ranked that "Immature corporate systems of firms which doesn't care with safety and health in repair and maintenance projects" in the first position with RII=0.8133. This result may interpret as most of firms in Gaza Strip are suffering from lack of experience, training, and awareness regards to the importance of safety through RMAA projects, which leads workers often to exposing to injuries and accidents.

4.6.2 Mezzo level factors group

The respondents ranked "Lack of organizational learning to avoid injures" in the first position in the mezzo group with RII=0.786.This can be due to the lack of education and concern of workers awareness regard to safety performance through RMAA works. The firms should have an organized learning plan to educate workers how to deal with emergencies and avoid accidents through RMAA works. Every worker should fully understand all the risk associated with repair and maintenance works.

4.6.3 About Micro level factor group

The respondents ranked "Ineffectiveness of lack of training and certification of competence" in the first position in the micro group. This factor is also ranked in the first position in all groups and have the highest relative important index (RII=0.820). This result is due to the lack of training, concerning, educating, and improving abilities of workers through RMAA works. Workers in RMAA field should receive course training to improve their competence regard to how to deal with and avoid injuries through RMAA works

Table 3: Causes of fatal accidents in repair and maintenance projects

No	Causes of construction fatal accidents in repair and maintenance projects	RII	Rank	Rank in total
Macro level factors				
1	Immature corporate systems of firms which doesn't care with safety and health through repair and maintenance projects	0.813	1	**2**
2	Lack of leadership from 'Government's a key client	0.793	2	3
3	Lack of influence of trades unions	0.773	3	6
4	Inappropriate enforcement	0.726	4	11
5	Lack of proper accident data	0.713	5	12
Mezzo level factors				
6	Lack of organizational learning to avoid injures	0.786	1	5
7	Immature project system and process which doesn't care with workers safety	0.746	2	9
8	Lack of proper accident investigation/data	0.700	3	13
9	Lack of understanding and engagement by some of the design community	0.626	4	17
10	Inappropriate procurement and supply chain arrangements	0.593	5	18
Micro level factors				
11	Ineffectiveness of lack of training and certification of competence	0.820	1	**1**
12	Lack of individual competency and understanding of workers and supervisors	0.793	2	4
13	Poor behavior	0.773	3	7
14	Site hazards	0.760	4	8
15	Poor equipment or misuse of equipment (including PPE)	0.740	5	10
16	Poor employment practices	0.700	6	14
17	High Cost of safety requirements	0.666	7	15
18	Lack of ownership, engagement and empowerment of communication with, and responsibility for workers and supervisors	0.660	8	16
19	Itinerant workforce (Wandering the workers at the site)	0.586	9	19

4.7 Factors affecting site safety in repair and maintenance projects

Table 4 illustrates the results regarding factors affecting site safety in repair and maintenance projects. As shown in Table 4, the respondents ranked" Poor safety awareness of firms top leaders " and "Insufficient safety training of RMAA workers for handling multi-tasks" in the first position with same RII=0.806. This result may be explained because most of construction firms in Gaza Strip are suffering from lack of awareness, experience, culture, concern and efficiency regard to safety procedures through repair and maintenance projects. The majority of workers haven't received any safety training course. This means that most of construction workers are exposing every day to injuries and accidents.

Nature of RMAA work generally requires workers to have multi skills to perform multi-tasks. Wherefore, the awareness of firm top leaders regarding safety in maintenance projects should be enhanced. Every worker should receive training courses on how to work safely in RMAA projects, and how to deal with emergencies in RMAA works. Results also showed that "Low educational level of workers" is in the last position with RII=0.56.This result may appear because of the literalism nature of maintenance works. Tam et al., (2012) conducted a survey in china and found that poor safety awareness of firms top leaders ranked in the first position with RII=0.93, and lack of training ranked in the second position with RII=0.90. This result is consistent with the current study. Hung (2012) concluded that poor safety conscientiousness of workers ranked in the first position with mean (4.64). However this factor was ranked in the eighth position in the current study.

Table 4: Factors affecting site safety in repair and maintenance projects

No	Cause	St.Dev.	w	RII	Rank
1	Poor safety awareness of firms top leaders	6.2849	121	0.806	1
2	Insufficient safety training of RMAA workers for handling multi-tasks	6.0415	121	0.806	1
3	Lowest bid tendering method without pricing for safety items	6.7823	120	0.800	3
4	Lack of rigorous enforcement of safety regulations	6.0000	118	0.786	4
5	RMAA workers underestimate potential risks when performing small tasks for a short period of time	5.0497	117	0.780	5
6	Poor safety awareness of RMAA managers	7.2456	115	0.766	6
7	Lack of personal protective equipment	4.6904	114	0.760	7
8	Poor safety conscientiousness of workers	7.3824	113	0.753	8
9	Lack of certified skill labor	6.5954	113	0.753	8
10	Low safety awareness of small/medium sized contractors on RMAA works	4.7958	112	0.746	10
11	Reckless operation	5.0497	111	0.740	11
12	Poor housekeeping and congested working environment	7.3484	110	0.733	12
13	Inadequate safety supervision	7.1063	109	0.726	13
14	Inadequate site safety planning and hazard assessment	5.6568	108	0.720	14
15	Shortage of safety management manual	4.0620	107	0.713	15
16	Lack of protection in material transportation	5.3385	107	0.713	15
17	Poor equipment	6.5192	107	0.713	15
18	Lack of first aid measures	5.2440	106	0.706	18
19	Lack of protection in material storage	5.3385	105	0.700	19
20	Poor information flow	3.0822	103	0.686	20
21	Excessive overtime work for labor	3.6742	102	0.680	21
22	Lack of teamwork spirits	3.7416	100	0.666	22
23	Low education level of workers	3.8729	84	0.560	23

4.8 Problems of running building maintenance projects

Figure 4 illustrate the results regarding to the problems of running building maintenance projects through repair and maintenance projects in Gaza strip. Results showed that "fragmented nature" and "unclear scope" are the most common problems that facing maintenance projects. This result may be explained because of the difference nature between maintenance works and new construction works. New construction projects have a clear and limited scope (tasks and construction activities are precisely defined in time, place and nature of the work), and characterized with a sequent nature compared with the fragmented nature of maintenance projects. This nature of maintenance works forms a lot of problems for workers, and caused confusing and distractibility for them on which element need to be done, and which procedures or methods should be used. Hon et al., (2012) conducted a survey in Hong Kong, and adopted mixed methods approach for the data collection, semi structured interviews and a two-round Delphi survey. Major problems were identified, including limited safety resources for small and medium enterprises (SMEs), difficulty in changing the mindset of RMAA workers, and difficulty in performing safety supervision.

Figure 4 Problems of running building maintenance projects

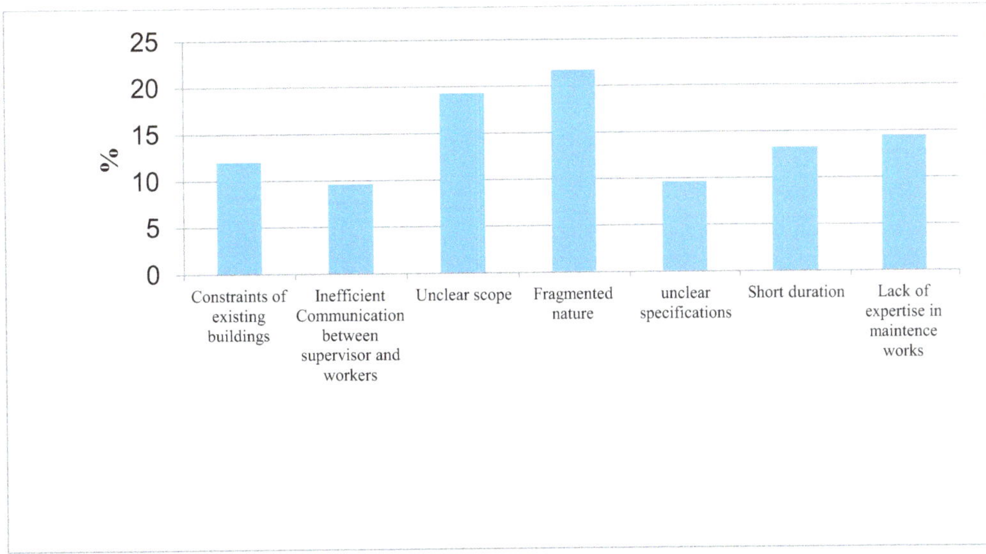

4.9 Recommended preventive measures to improve safety practice in RMAA works

Table 5 illustrates the results regarding the recommended preventive measures to improve safety practice through repair and maintenance projects. The respondents ranked "Strengthen site monitoring and supervision "in the first position with RII=0.88. This result may appear because most of projects in Gaza Strip are suffering from poor monitoring and supervision system. Supervisors don't care about the safety methods and procedures used by workers through RMAA works. A lot of supervisors also doesn't care if the workers wear protective equipment or not through execution of RMAA works. There is an urgent need to strengthen the site monitoring and supervision system. The results also showed that "Safety promotion and education towards RMAA sector" was ranked in the second position with RII=0.873. This result appeared because of the lack of experience, awareness, and training of workers regard to safety procedures through RMAA works.

Mahalingam and Levitt (2007) pointed out in their study that education and training can change one's mindset and attitude towards safety and thus enhancing safety awareness but it takes time. Raising safety awareness of RMAA workers and selecting contractors with a good record of safety performance were the two most important strategies to improve the safety performance in RMAA works. However selecting contractors with a good record of safety performance was in the tenth position in this study with RII=0.806; which reflected the poor safety culture in Gaza strip regard selecting the appropriate contractor to perform RMAA works. A technology innovation was ranked in the least important strategy in Hon et al., (2011) study. Their result is consistent with the result of this study which ranked technology innovation factor in the 13th position with RII=0.76. That means that RMAA work are traditional works and innovation region in this area is very limited.

Table 5: Recommended preventive measures to improve safety practice in RMAA works

Recommended preventive measures	Degree of Approval					w	R.I.I	Rank
	5	4	3	2	1			
Strengthen site monitoring and supervision	15	13	1	1	0	132	0.880	1
Safety promotion and education towards RMAA sector	14	13	3	0	0	131	0.873	2
Legislative control.	12	15	3	0	0	129	0.860	3
Improvement of site tidiness and housekeeping.	11	16	3	0	0	128	0.853	4
Implement pay for safety scheme for RMAA works.	15	9	4	2	0	127	0.846	5
Raise safety awareness of RMAA workers.	10	16	4	0	0	126	0.840	6
Relevant safety training for specific trades of RMAA works	9	19		2	0	125	0.833	7
Build up good company safety culture	10	12	8	0	0	122	0.813	8
Clear working procedures and guidance for RMAA workers	8	16	6	0	0	122	0.813	8
Select RMAA subcontractors with good record of safety performance.	15	5	8	0	2	121	0.806	10
Design for safety of RMAA works.	7	14	9	0	0	118	0.786	11
Award and penalty scheme.	7	16	5	1	1	117	0.780	12
Technology innovations for better safety.	8	13	6	2	1	115	0.766	13
A mandatory licensing system for RMAA workers	5	12	10	3	0	109	0.726	14

5.0 Conclusion

The results illustrated the most common and important factors that affected safety performance in repair and maintenance works. The results also illustrated the most important causes of fatal accidents through repair and maintenance projects. Gaza Strip is suffering from a poor safety culture, performance, and awareness regard to safety and health through RMAA works. The site safety monitoring and supervision should be improved, and the awareness of firms, project managers, and workers regard to safety issue should be enhanced. The results indicated that both of "Poor safety awareness of firms top leaders "and "Insufficient safety training of RMAA workers for handling multi-tasks" have been ranked in the 1st position regarding to its high effect on the performance of safety in repair and maintenance works. This result reflect the urgent need to enhance the awareness of firms regarding safety performance through RMAA works, and the importance of improving the workers and project managers safety culture and performance.

The results revealed that "lack of training and certification of competence" has been ranked in the 1st position in causing injuries and accidents. This factor belongs to micro level factor group. "Strengthen site monitoring and supervision" is considered the most important Preventive measures for improving safety practice through RMAA works. The RMAA subcontractors should be selected from those who have good record of safety performance. Results indicated that "hard hats, gloves, eye goggles" are the most equipment's provided by contractors through RMAA works. Results also showed that "ear plugs" have never provided by any contractor through RMAA works. The results revealed that "increase in cost" and "imposing psychological burden on workers" are the most common impacts of accidents in RMAA works.

The results indicated that "fall from height" and "slip, trip or fall on same level" are the most serious accidents in RMAA works. Nails and toolboxes are sparse around the sites of most projects in Gaza Strip. This result reflects the massive need to enhance the awareness of RMAA workers regard to safety performance through RMAA works. Every worker should receive a safety course training to avoid injuries and accidents through RMAA works."Strengthen site monitoring and supervision "is considered the most important preventive measures in improving safety practice. The awareness of workers regard to safety performance through RMAA project need to be enhanced. Safety awareness is intrinsic and incubated by one’s mindset and attitude towards safety. It is recommended that RMAA subcontractors should be selected from those who have good record of safety performance. Technology innovations should be encouraged to obtain better safety performance through RMAA works.

References

[1] Abdelhamid T. S., and Everett J. G. (2000). Identifying root causes of construction accidents. Journal of Construction Engineering and Management, 126 (1):52–60.

[2] Abu Alqumboz M.(2007). Developing a model for integrating safety, quality and productivity in building projects in Gaza Strip. Master thesis. Islamic university of Gaza. Palestine.

[3] Alaqqad M. (2009). Assessment of the Factors Affecting Safety Performance on Construction Projects in Gaza Strip. Master thesis. Islamic university of Gaza. Palestine.

[4] Ayyub B., Mccuen R. (2003).Probability, statistics and reliability for engineers and scientists. Chapman & Hall/CRC, London, NY.

[5] Brace C., Gibb A., Pendlebury M., and Bust P. (2009). Inquiry into the underlying causes of construction fatal accidents. Phase 2 Report: Health and Safety in the Construction Industry: Underlying Causes of Construction Fatal Accidents-External Research. Secretary of State for Work and Pensions Inquiry into the underlying causes of construction fatal accidents.

[6] Cameron I., Gillan G., and Duff A. R. (2007). Issues in the selection of fall prevention and arrest equipment. Engineering, Construction and Architectural Management, 14(4):363-374.

[7] Chan, A. P. C., Wong, F. K. W., Chan, D. W. M., Yam, M. C. H., Kwok, A. W. K., Lam, E. W. M. and Cheung, E. (2008). Work at height fatalities in the repair, maintenance, alteration and addition (RMAA) Works. Journal of Construction Engineering and Management, ASCE, 134(7):527-535.

[8] Chan A. P. C., Wong F. K. W., Yam M. C. H., Chan D. W. M., and Hon C. K. H., Dingsdag,D., and Biggs, H. C. (2009). Construction safety in the repair and maintenance Sector. In: Proceedings of ISEC-5, 21-27 September, 2009, Las Vegas

[9] Cheng, E. W. L., Li, H., Fang, D. P., and Xie, F., (2004). Construction safety management: an exploratory study from China. Construction Innovation 4(4):229–241.

[10] Dept. for Business, Enterprise and Regulatory Reform. (2007). "Construction statistics annual 2007." The stationary office, ⟨http://www. berr.gov.uk⟩(Mar. 4, 2010).

[11] Enshassi, A., (2003). Factors Affecting Safety on Construction Projects, Islamic University of Gaza, Gaza Strip, Palestine.

[12] Hallowell, M. R. (2012). Safety-knowledge management in American construction organizations. Journal of Management in Engineering, 28(2):203-211.

[13] Haslam, R.A.; Hide, S.A.; Gibb, A.G.F.; Gyi, D.E.; Pavitt, T.; Atkinson, S.; Duff, A.R.(2005). Contributing factors in construction accidents. Applied Ergonomics, 36(4):401–415.

[14] Hassona A. (2005). Improving Safety Performance in Gaza Strip. Master Thesis. Islamic University of Gaza. Palestine.

[15] Hon C. K. H, Chan A. P. C., and Wong F. K. W. (2010). An analysis for the causes of accidents of repair, maintenance, alteration and addition works in Hong Kong. Safety Science, 48(7):894–901.

[16] Hon C. K. H. and Chan A. P. C. (2013a).Fatalities of repair, maintenance, minor alteration, and addition works in Hong Kong. Safety Science, 51(1):85–93.

[17] Hon, C. K. H., Chan, A. P. C. and Chan, D. W. M. (2011).Strategies for Improving Safety Performance of Repair, Maintenance, Minor Alteration and Addition (RMAA) Works.Facilities - Special Issue on Infrastructure Management, UK, 29(13/14):591-610.

[18] Hon C. K. H., Chan A. P. C., and Yam M. C. H. (2013). Determining Safety Climate Factors in the Repair, Maintenance, Minor Alteration, and Addition Sector of Hong Kong. Journal of Construction Engineering and Management, 139(5):519-528.

[19] Hon, C. K. H. and Chan A. P. C. (2013b). Safety management in repair, maintenance, minor alteration and addition works: A knowledge management perspective, Journal of Management in Engineering. doi:10.1061/(ASCE)ME.1943-5479.0000233.

[20] Hon, C. K. H., Chan, A. P. C., and Yam, M. C. H. (2012). An empirical study to investigate the difficulties of implementing safety practices in the repair and maintenance sector: a case of Hong Kong. ASCE Journal of Construction Engineering and Management, 138(7):871-884.

[21] Hung, K. H. (2012). Relationships between safety climate and safety performance of repair, maintenance, minor alteration and addition (RMAA) works; Ph.D Thesis, The Hong Kong Polytechnic University.

[22] Kartam N. A., Flood I., and Koushki P. (2000). Construction safety in Kuwait: issues, procedures, problems, and recommendations. Safety Science 36(3):163-184.

[23] Labour Department, HKSAR. (2008). Accidents in the Construction Industry of Hong Kong (1998-2007). Occupational Safety and Health Branch, Labour Department, HKSAR Government.http://www.labour.gov.hk/eng/major/300805.htm

[24] Mahalingam, A. and Levitt, R. (2007). Safety issues on global projects, Journal of Construction Engineering and Management, 133(7):506-516.

[25] Mullen J. (2004). Investigating factors that influence individual safety behavior at work; Journal of Safety Research, 35:275– 285.

[26] Tam C. M., Deng Z. M., Zeng S.X., and Ho C. S. (2000).Quest for continuous quality improvement for public housing construction in Hong Kong. Journal of Construction Management and Economics, 18(4):437–46.

[27] Tam C. M., Zeng S. X., and Deng Z. M. (2004). Identifying elements of poor construction safety management in China. Safety Science, 42 (7):569–586.

Permissions

All chapters in this book were first published in IJSCET, by Penerbit UTHM; hereby published with permission under the Creative Commons Attribution License or equivalent. Every chapter published in this book has been scrutinized by our experts. Their significance has been extensively debated. The topics covered herein carry significant findings which will fuel the growth of the discipline. They may even be implemented as practical applications or may be referred to as a beginning point for another development.

The contributors of this book come from diverse backgrounds, making this book a truly international effort. This book will bring forth new frontiers with its revolutionizing research information and detailed analysis of the nascent developments around the world.

We would like to thank all the contributing authors for lending their expertise to make the book truly unique. They have played a crucial role in the development of this book. Without their invaluable contributions this book wouldn't have been possible. They have made vital efforts to compile up to date information on the varied aspects of this subject to make this book a valuable addition to the collection of many professionals and students.

This book was conceptualized with the vision of imparting up-to-date information and advanced data in this field. To ensure the same, a matchless editorial board was set up. Every individual on the board went through rigorous rounds of assessment to prove their worth. After which they invested a large part of their time researching and compiling the most relevant data for our readers.

The editorial board has been involved in producing this book since its inception. They have spent rigorous hours researching and exploring the diverse topics which have resulted in the successful publishing of this book. They have passed on their knowledge of decades through this book. To expedite this challenging task, the publisher supported the team at every step. A small team of assistant editors was also appointed to further simplify the editing procedure and attain best results for the readers.

Apart from the editorial board, the designing team has also invested a significant amount of their time in understanding the subject and creating the most relevant covers. They scrutinized every image to scout for the most suitable representation of the subject and create an appropriate cover for the book.

The publishing team has been an ardent support to the editorial, designing and production team. Their endless efforts to recruit the best for this project, has resulted in the accomplishment of this book. They are a veteran in the field of academics and their pool of knowledge is as vast as their experience in printing. Their expertise and guidance has proved useful at every step. Their uncompromising quality standards have made this book an exceptional effort. Their encouragement from time to time has been an inspiration for everyone.

The publisher and the editorial board hope that this book will prove to be a valuable piece of knowledge for researchers, students, practitioners and scholars across the globe.

List of Contributors

Abdul Halim Mohd Yusof
Advanced Membrane Technology Research Centre (AMTEC), UniversitiTeknologi Malaysia, 81310 Skudai, Johor DarulTa'zim, Malaysia
Department of Biology, Building 68-370, Massachusetts Institute of Technology, Cambridge, Massachussetts 02139, United States of America

Mohamed Afizal Mohamed Amin
Advanced Membrane Technology Research Centre (AMTEC), UniversitiTeknologi Malaysia, 81310 Skudai, Johor DarulTa'zim, Malaysia
Faculty of Chemical Engineering (FKK), UniversitiTeknologi Malaysia, 81310 Skudai, Johor DarulTa'zim, Malaysia
Department of Chemical Engineering and Sustainability Energy, Faculty of Engineering, Universiti Malaysia Sarawak, 94300 Kota Samarahan, Sarawak, Malaysia

Ahmad Fauzi Ismail
Advanced Membrane Technology Research Centre (AMTEC), UniversitiTeknologi Malaysia, 81310 Skudai, Johor DarulTázim, Malaysia

Mohd Noorul Anam
Advanced Membrane Technology Research Centre (AMTEC), UniversitiTeknologi Malaysia, 81310 Skudai, Johor DarulTázim, Malaysia

Nik Azmi N
Faculty of Chemical Engineering (FKK), UniversitiTeknologi Malaysia, 81310 Skudai, Johor DarulTázim, Malaysia

ik Mahmud
Faculty of Chemical Engineering (FKK), UniversitiTeknologi Malaysia, 81310 Skudai, Johor DarulTázim, Malaysia

S. N. Eluozo
Department of Civil Engineering, Faculty of Engineering, University of Nigeria Nsukka

Chetan S. Dhanjode
M Tech. Construction Technology and Management, VNIT, Nagpur, India

R. V. Ralegaonkar
Associate Professor, Civil Engineering Department, VNIT, Nagpur, India

Vaidehi A. Dakwale
Research Scholar, Civil Engineering Department, VNIT, Nagpur and Assistant Professor Civil Engineering Department, RCOEM, Nagpur, India

Olufisayo Adewumi Adedokun
Department of Quantity Surveying, Federal University of Technology, Akure, Nigeria

Olajide Timothy Ibironke
Department of Quantity Surveying, Federal University of Technology, Akure, Nigeria

Ayokunle Olubunmi Olanipekun
Department of Quantity Surveying, Federal University of Technology, Akure, Nigeria

Solomon Olusola Babatunde
Department of Quantity Surveying, Obafemi Awolowo University, Ile-Ife, Nigeria

Rajesh s. Gujar
Research Scholar, SVNIT, Surat, Gujarat, India

Krupesh A. Chauhan
Associate Professor, Civil Engineering Department, SVNIT, Surat, India

Gautam Dadhich
UG Student, Pandit Deendayal Petroleum University, Gandhinagar, Gujarat, India

Maroof Opeyemi Anifowose
Department of Quantity Surveying, Federal University of Technology Minna, Nigeria

Wasiu Adeniran Ola-awo
Department of Quantity Surveying, Federal University of Technology Minna, Nigeria

Tanimu U Mohammed
Department of Quantity Surveying, Federal University of Technology Minna, Nigeria

Akeem Ayinde Raheem
Civil Engineering Department, Ladoke Akintola University of Technology, Ogbomoso, Nigeria

Olumide A. Adenuga
Building Department, University of Lagos, Akoka-Lagos, Nigeria

Baba Shehu Waziri
Department of Civil and Water Resources Engineering, University of Maiduguri, Maiduguri, Nigeria

Zanna Alhaji Lawan
Works Department University of Maiduguri, Maiduguri, Nigeria

Ma'aji Mala Mustapha
Central Bank of Nigeria, Corporate office, Abuja, Nigeria

Anwar Khitab
Department of Civil Engineering, Mirpur University of Science & Technology, Mirpur, Azad Kashmir, Pakistan

Muhammad Tausif Arshad
Department of Civil Engineering, Mirpur University of Science & Technology, Mirpur, Azad Kashmir, Pakistan

Faisal Mushtaq Awan
Department of Civil Engineering, Mirpur University of Science & Technology, Mirpur, Azad Kashmir, Pakistan

Imran Khan
Department of Civil Engineering, Mirpur University of Science & Technology, Mirpur, Azad Kashmir, Pakistan

Miguel Pires Amado
Department of Civil Engineering, Universidade Nova de Lisboa, Caparica, Portugal

Luis Marques Barroso
GEOTPU, Faculdade de Ciencias e Tecnologia, Universidade Nova de Lisboa, Caparica, Portugal

Adnan Enshassi
Department of Civil Engineering, Islamic University IUG, Gaza, Palestine

Ahmed S. Kullab
Department of Civil Engineering, Islamic University IUG, Gaza, Palestine

Benviolent Chigara
Department of Landscape Architecture and Urban Design, National University of Science and Technology, Bulawayo, Zimbabwe

Tirivavi Moyo
Department of Quantity Surveying, National University of Science and Technology, Bulawayo, Zimbabwe

Fungai Hamilton Mudzengerere
Faculty of The Built Environment, National University of Science and Technology, Bulawayo, Zimbabwe

Jack C.P. Cheng
Department of Civil and Environmental Engineering, The Hong Kong University of Science and Technology, Hong Kong

Vignesh Venkataraman
Department of Civil and Environmental Engineering, The Hong Kong University of Science and Technology, Hong Kong

Oyedolapo Ekundayo Ogunbiyi
University of Central Lancashire, United Kingdom

Adebayo Akanbi oladapo
University of Central Lancashire, United Kingdom

Jack Steven Goulding
University of Central Lancashire, United Kingdom

Elvis M. Mbadike
Department of Civil Engineering, Michael Okpara University of Agriculture, Umudike, Umuahia, Nigeria

N. N. Osadebe
Department of Civil Engineering, University of Nigeria, Nsukka

J. Jayaprakash
Department of Civil Engineering, The University of Nottingham, Malaysia

Abdul Aziz Abdul Samad
Faculty of Civil and Environmental Engineering, Universiti Tun Hussein Onn, Malaysia

M. Y Mustafasanie
School of Civil Engineering, Universiti Sains Malaysia, Malaysia

M. P Anwar
Department of Civil Engineering, The University of Nottingham, Malaysia

A. A Mohamed
Department of Civil Engineering, The University of Nottingham, Malaysia

Reza Taheriattar
School of Civil and Environmental Engineering, University of New South Wales, Sydney, Australia

Morvarid Farzanehrafat
School of Civil Engineering, Iran University of Science and Technology, Tehran, Iran

Tirivavi Moyo
Department of Quantity Surveying

Ellen Mangore
Department of Civil and Water Engineering

Benviolent Chigara
Department of Landscape Architecture and Urban Design National University of Science and Technology, P.O. Box AC 939, Ascot, Bulawayo, Zimbabwe

Renard Yung Jhien Siew
Centre for Energy and Environmental Markets, University of New South Wales, Australia

Ayisha Powmya
Built Environment Engineering Department, Muscat College, Oman

Nazirah Zainul Abidin
School of Housing, Building and Planning, Science University of Malaysia (USM), Penang, Malaysia

Olugbenga Timo Oladinrin
Department of Quantity Surveying, Federal University of Technology, PMB 704, Akure, Ondo State Nigeria

Samuel Olusola Olatunji
Department of Quantity Surveying, Federal University of Technology, PMB 704, Akure, Ondo State Nigeria

Basir Tomiwa Hamza
Department of Quantity Surveying, Federal University of Technology, PMB 704, Akure, Ondo State Nigeria

Richard O. Asiedu
Chair of Construction Economics, Bauhaus University, Weimar, Germany

Hans W. Alfen
Chair of Construction Economics, Bauhaus University, Weimar, Germany

Adnan Enshassi
Department of Civil Engineering, IUG, Gaza, Palestine

Ehsan Risqa
Department of Civil Engineering, IUG, Gaza, Palestine

Faisal Arain
School of Sustainable Building and Environmental Management, NAIT, Edmonton, Canada